A Dictionary of
BOTANICAL
TERMS

A Dictionary of

BOTANICAL
TERMS

Compiled by
A.V.S.S. SAMMBAMURTY
Reader in Botany, Sri Venkateswara College
Dhaula Kuan, New Delhi—110 021

CBS

CBS PUBLISHERS & DISTRIBUTORS

4596/1-A, 11-DARYAGANJ, NEW DELHI-110002

ISBN : 81-239-0631-5

First Edition : 1999
Reprint : 2004
Reprint : 2005
Reprint : 2007

Publishing Director : Vinod K. Jain

Published by :
Satish Kumar Jain for CBS Publishers & Distributors,
4596/1-A, 11 Darya Ganj, New Delhi - 110 002 (India)
E-mail : cbspubs@del3.vsnl.net.in
Website : http://www.cbspd.com

Branch Office :
Seema House, 2975, 17th Cross, K.R. Road,
Bansankari 2nd Stage, Bangalore - 560070
Fax : 080-6771680 • E-mail : cbsbng@vsnl.net

Printed at :
Asia Printograph, Shahdara, Delhi - 110 032 (India)

PREFACE

Botany is a vast field and embodies many branches, each of which are specialized in their own way, like Algae, Fungi, Taxonomy, Genetics, Cytogenetics, Plant Physiology, Ecology, Biotechnology etc. It is for this reason, that the present book on the *'Glossary of Botany'* is arranged branch-wise and not as a combined one like in any Dictionary.

The present work is an exhaustive treatment of the entire subject of Botany, giving the meanings of several technical terms used in different disciplines of Botany. However, elaborate explanations and details are avoided since it will swell into an encyclopaedia which is not in the scope of this book.

The reason for dissecting out into several branches is that there are many overlapping terms in different disciplines and their meanings are often misleading if they are not specified in which branch they occur. It is better these terms are spelled at their respective branches more specifically than leaving it to the reader. It will also be helpful for the reader, that the term that is used will directly mean the specified topic, without searching here and there, if it were to be in a dictionary form. Undoubtedly such an arrangement into different sections will have some repititions, due to overlapping subjects like Genetics, Cytogenetics, Cell Biology and Biotechnology; similarly Morphology, Taxonomy, Biosystematics and Evolution have certain overlapping terms.

The present book on Glossary of Botany will cater to the needs of a general reader in Botany and also to those interested in the border line sciences like Agriculture, Medicine, Pharmacology and Pharmacognosy, Biophysics, Biochemistry, Environmental sciences and Life sciences. This book can be used as a reference text for students, teachers in schools and colleges where Botany is taught. Students appearing for Entrance tests especially in Medicine, Agriculture, Horticulture, Biotechnology and Forestry will be greatly benefitted by this book.

There is always room for improvement for a work of this type which is slightly encyclopaedic in nature and efforts will be made in future editions to improve the subject matter. Suggestions form all quarters of readers are welcome.

—A.V.S.S. Sammbamurty

CONTENTS

ALGAE

A

Acronematic – Flagella with slender and smooth surface and ending in thin hair.

Acyle lipid – Lipids joined to acids through anhyride bonds.

Akinete – Vegetative cell that becomes converted into thick-walled non motile resting spore; wall of cell becomes wall of spore.

Algicide – Substance highly toxic to algae.

Algicolous – Living on or attached to algae.

α-granules – Submicroscopic granules rich in glycogen and found in cells of blue-green algae.

Anastomosing – Plants or plant parts intermittently joining and sepa-rating to form network.

Androsporangium – Sporangium producing androspore.

Androspore – Antherozoid-like zoospore formed singly in androsporangium; produces dwarf male filament, as in Oedogoniaceae.

Anisogamy – Union between two mor phologically dissimilar gametes.

Annular – Ring-like.

Antapical – A way from apex; basal.

Anterior – In or towards front.

Antheridium – Uni – or multicellular male gametangium.

Antherozids – Sperms; male gametes.

Anticlinal – Pertaining to pattern of cell division which occurs in direction per-pendicular to circumference of surface. Anticlinal cell division leads to increase in circumference of plant part.

Apanogametes – Non-flagellate gametes showing amoeboid movement.

Aplanospore – Non-motile spore in which spore wall is not derived from wall of its perent cell.

Apogamy – Development of diploid sporophyte from haploid gametophyte without fertilization.

Archegonium – Multicellular female gametangium producing egg and consisting of neck and venter; includes sterile cells in addition to fertile egg.

Areolae – Cavity-like depressions, each covered with perforated membrane, the sieve membrane.

Asexual – Lack of apparent sexual organs.

Assimilator – Upright leaf-like part of some siphonaceous green algae.

Autospores – Non-motile spores resembling parent cell in shape and structure.

Autotrophic – Capable of producing required food substances from inorganic raw materials.

Auxiliary cell – Cell that receives nucleus from zygote and then produces gonimoblast filaments in red algae.

Auxospore – Spores of diatoms.

Auxotrophic – Nutrition in which externally produced organic compounds such as amino acids and vitamins are required.

B

Benthos – Algae growing at bottom of sea or lakes.

β-granules – Granules of proteinaceous nature confined to

Cyanophyceae.

Bifurcate – Divided into two parts; forked.

Bipinnate – Twice pinnate; bearing on axes pinnate leaflets that grow opposite each other on main stem.

Biseriate – Arranged in two rows.

Blepharoplast – Granule lying at base of flagellum; gives rise to one flagellum.

Bloom – profuse development of one of a few species of algae in water body.

Boreal – Usually pertainaing to colder waters of Northern Hemisphere.

Bulbil – Small group of cells cut off from thallus; group functions as vegetative reproductive unit.

Byssoid – Very slender and resembling a cobweb.

C

Caducous – Seasonally shed.

Caespitose – Matted or tangled; tufted.

Calcareous – Impregnated with lime.

Calciphillc – Pertaining to organisms well adapted to calcium-rich water.

Canaliculi – Tubular canals present on valve surface of diatoms.

Carpogonium – Female gametangium in red algae; consists of swollen base and elongated neck or trichogyne.

Carposporangia – single-celled structures borne on minute, parasitic carposporophytic generation in Rhodophyta; these sporangia produce non-motile carpospores.

Carpospore – Spore produced within carpoporangium of red algae.

Carposporophyte – Carposporeprod ucing second generation plant or red algae; parasite on female gametophyte, arising directly or indirectly from zygote.

Cellular – Organisms made of conventional units of structure and function known as cells, each containing several proteins and both DNA and RNA; differ form acel-lular entities which have very few proteins and either DNA or RNA but not both together.

Central nodule – Wall thickening in centre of vale in diatoms.

Centric – Pertaining to diatioms that are circular in valve view and have radial symmetry.

Centrosome – Polar body at each pole of dividing cell; spindle is seen to diverge from these bodies.

Chromatic adaptation – Capacity of some algae to synthesize pigments that are complementary to quality of available light. Also known as Gaidukov phenomenon.

Chromatophore – Plastid containing chlorophyll-a and other pigments but not chlorophyll-b.

Chromocentres – Feulgen positive bodies of unknown function found on chromosomes of Phaeophyta.

Chrysolaminarin – Leucosin; food reserve of Chrysophyta.

Cistron – Functional unit of gene which codes for single polypeptide.

Clathrate – Having net-like appearance.

Coccoid – Pertaining to habit; non-motile unicells.

Coenobial – See coenobium.

Coenobium – colony consisting of definite number of cells arranged in specific manner.

Coenocytic – Multinucleate cell.

Colligate – Septum showing H-shaped structure.

Colonial – Habit showing number of cells held together within envelope.

Colony – Loose organization of similar·cells which have developed toge-ther from single, original parent plant or cell. Each cell is potentially capable of like activities independent of the others in the colony.

Conceptacles – Cavity-like depressions on receptacle of Fucales that contain gametangia.

Conjugation – Sexual union in Conjugales involving fusion between amoeboid gametes.

Contractile vacuoles – Organelles of osmo-regulation;also thought to play role in excretion of waste material.

Cortex – That tissue of a more or less solid alga which lies beneath epidermis, i.e., between epidermis and medulla.

Corticated – Convex or outer layer(s) of small cells covering that produced directly by apical cell.

Cosmopolitan – found in most or all parts or the world.

Costae – Thickened regions of valve that separate rows of areolae one from the other.

Cryptoblasts, cryptostomats–Sterile conceptacles found in Fucales.

Cuticle – Structureless, amorphous layer bounding outer surface of some algae.

Cyanophage – Virus that infects certain blue-green algae.

Cyanophycin, granules – Proteinaceous food reserve occurring in granular form in cells of blue-green algae.

Cyst – Resting cells with thick envelope.

Cystocarp – Fruiting body of red algae; aggregate structure consisting of carpo-sporangia and sterile covering cells.

Cytokinesis – Process of cell division.

Cytopharyngeal – Pertaining to cytopharynx, i.e., gullet, reservoir and vacuoles in *Euglena*.

D

Dendroid – Tree-like habit with much branching. In certain colonial algae tree-like habit is achieved by repeated branching of mucilaginous stalks.

Derepression – Release of inhibition of enzyme synthesis.

Developmental – Changes that organism undergoes during its vegetative or body organization.

Diaminopinopimelic acid – Amino acid with two amino groups; has definite chemical structure.

Dichotomy – Repeatedly bifurcating pattern of branching.

Dictyosomes – Vesicular structures of in cyto-plasm of eucaryotic cells. Also known as Golgi bodies.

Diffused centromeres – Chromosome containing many scattered centro-meres along its length.

Diffuse – Pertaining to growth in which every vegetative cell is capable of growth and division.

Dimorphic – Where gametophytic and sporo-phytic generations are morphologically distinct.

Dioecious – Location of male and female sex organs on separate plants.

Diplobiontic – Type of life history having two free living morphological phases.

Diplohaplont – Where diploid and haploid generations alternate succesively.

Diplont – Diploid plant in which the only haploid stage consists of gametes.

Distal – Opposite of basal.

Distichous – Arranged in two vertical rows, along opposite sides or axis.

Doliform – Cask-or barrel-shaped.

Dormant – Not growing or germinating; resting.

Dorsiventral – Having recognizable dorsal and ventral surfaces.

E

Ecad (ecotype; ecophene) – Growth form determined by ecological or habitat conditions; purely somatic (non-inheritable) modification.

Encapsulated – Pertaining to habit; sac-like covering enclosing organism.

Endemic – Native to certain region.

Endophyte – Plant living within another plant.

Endoplasmic reticulum – Fine tubular or vesicular structures traversing cytoplasm of eucaryotic cells.

Endospore – Internally formed thin-walled spores of Cyanophyta, analogous to aplanospores.

Endozoic (endozootic) – Living with-in tissues of animals but not necessarily parasitic.

Epitheca – Larger half diatom frustule which covers smaller half (hypotheca)

Epizoic – Growing attached to outer surface of animals.

Eucaryota – Cellular organisms, in which genetic, respiratory and photosynthetic apparatuses are organized into nucleus, mitochondrion and chromatophore, respectively.

Euryhaline – Tolerating broad range of salinity.

Eurythermal – Tolerating broad range of temperature.

Eutrophic – Pertaining to habitats rich in nutrients and organic matter for growth of algae and other plants and micro-organisms.

Evection – Pushing one branch or part above, or into greater prominence than, another.

Exogenous – Formed at surface; arising from without.

Exospores – Spores produced externally or outwardly as in Cyanophyta; analogous to aplanospores.

Eye spot – Red-coloured spot (stigma) generally believed to have visual function.

F

Facultative – Occasional; incidental; can live under different conditions of life.

False branching – Branching resulting from degeneration of cell in loop or form growth of free ends of trichome through filament sheath, as in some blue-green algae.

Fenestrate – Having several openings.

Fibrillar – Bearing hairs or fibres which are more or less parallel and not matted.

Filamentous – Thread-like photosynthetic plants.

Flagella – Fine, thread-like structures by activity of which cells move.

Fucosan vesicles – Sac-like structures in cells of brown algae; contain phenolics and tannins.

Fusiform – Spindle-shaped.

G

Gamete – Sex cell; two gametes of opposite sex unite to form zygote.

Gas vacuoles – Gas-filled cavities in cells of certain planktonic blue-green algae which disappear when subjected to pressure. Also known as pseudovacuoles.

Girdle view – Side view of diatom that reveals junction of epitheca and hypotheca.

Globule – Male reproductive organs of Charales having jacket of sterile cells around fertile cells; analogous to antheridium.

Glomerule – Cluster of branches at node, as in *Batrachospermum*.

Glucosamine – Glucose derivative containing amino group of second carbon of six-carbon molecule.

Glycolipids – Compounds made of lipids and galactose.

Gonidium – Thick-walled asexual reproductive cell.

Gonimoblasts – Filaments formed from zygote in red algae; bear carposporangia.

Gonospores – Spores produced following meiosis; also called meiospores.

Gregarious – Grouped closely together; clustered.

Gynandrosporous – Species bearing both o ogonia and androsporangia on the same filament.

H

Habit – General form or aspect or mode of growth of plant.

Habitat – Natural place of growth of alga or plant.

Haematochrome – Orange or red oily pigment related to carotene and present in resting cells and eye spots of certain algae.

Haplobiontic - With only one separate multicellular generation in life history.

Haplont - Haploid plant in which the only diploid stage is confined to zygote.

Haustorium - Cellular process serving parasite as organelle or attachment, and as means for food absorption form host tissues.

Heterocyst - Specialized cell found in certain blue-green algae.

Heterokaryosis - Association of nuclei of different genetical constitution in vegetative cell.

Heteromorphic - Life cycle involving alternation between morphologically dissimilar generations.

Heterothallic - Self-incompatibility; sexual fusion occurs only between gametes of different parentage or plants.

Heterotrichous - Thallus differentiated into prostrate and erect system of branching filaments.

Heterotrophic - Organisms dependent on exogenous organic sources for their metabolism and growth.

Holdfast - Single cell or group of cells that acts as organ of attachment.

Holocarpic - Phenomenon in which entire alga is transformed into gametangium.

Hologamy - Fusion of mature individuals, i.e., mature individuals directly act as gametes.

Holophytic - Plant-like mode of nutrition involving photosynthesis.

Holozoic - Feeding like animals by ingesting solid food.

Homothallic - Self-compatible; fusion can occur between gametes derived form the same plant.

Hormocyst - Thick-walled hormogonium or multicellular akinete found in a few blue-green algae.

Hormogone - Short piece of trichome consisting of undifferentiated vegetative cells which are moniliform; hormogonium is generally motile and meant for propagation.

Host - Plant that gives nourishment to parasite or just mechanical support or shelter to epiphyte.

Hypnospore - Thick-walled spore; meant for perennation.

Hypotheca - Inner and smaller valve of diatom frustule.

I

Idioandrosporous - Species bearing andro-sporangia and oogonia on separate filaments.

Intercalary - Growth pattern in which newly formed cells are produced between two existing cells, e.g., of filament.

Internode - Space between two joints or points of attachment in a filament; it is the part of a cell between two end walls.

Intertidal - Between low and high tide levels.

Iridescent - Glowing; shining; refelcting different hues when seen from various angles.

Isogamy - Fusion between morphologically and physiologically similar gametes.

Isomorphic - Life cycle involving alternation between two morphologically similar generations.

J

Juvenile - Small, not yet well-developed or fully mature; not having adult traits.

Juxtaposed - Side by side.

K

Karyokinesis - Process of nuclear division.

L

Lamellate (lamellose) - Made up of layers joined together.

Laminarin - Polysaccharide food reserve in brown algae.

Leucosin - Highly refractile polysaccharide with -1,3 linkages which forms food reserve in Chrysophyta and Xanthophyta; chemically similar to laminarin; sometimes also called chrysolaminarin.

Lithophilous – Preferring rocky substrat.

Lithopytic – Growing attached to rock.

Littoral – Pertaining to organisms growing between tide levels near shore.

Locule – Compartment; chamber.

Loop-formation – Arch of dome formation; achieved by rejuvenation of growth and cell division in certain cells or trichomes of Scytonemataceae.

Lunate – Crescent-shpaed.

M

Macrandrous – Filaments producing antheridia similar in size and morphology to those producing oogonia; sexually mono-morphic plants, as in *Oedogonium.*

Marl – Calcareous deposit formed by such algae as *Chara* and some cyanophytes.

Matrix – Enveloping or surrounding material, generally mucilaginous.

Meiospores – Spores formed after meiosis in zygote.

Meristoderm – Meristematic surface layer of tissue.

Mitochondria – Cytoplasmic double-membraned organelles concerned with energy release in respiration.

Mitospores – Spores fromed after mitosis; may be haploid or diploid.

Moniliform – Arranged like string or beads.

Monoecious – Plant bearing both male and female sex organs on the same individual.

Monomorphic – Formation of only one kind of plant in life cycle.

Motile – Capable of independent movement by means of flagella or some other device.

Mucopeptides – Compounds made of carbohydrates and amino acids; carbohydrates are N-acetylglucosamine and N-acetyl muramic acid; alanine, glycine, aspartic acid, lysine, and diaminopimelic acid.

Multiaxial – Formation of main axis or thallus by group of branched filaments.

Multiplicative – Process leading to increase in number of an organism or its cells.

Muramic acid – Glucosamine derivative containing carboxyethle group at 3-0 position, e.g, 3-0 carboxyethle-d-gluco-samine.

N

Nannandrous – Sexually dimorphic plants as in Oedogoniaceae with dwarf male.

Necridium – Dead or degenerated cell in Cyanophytes which is commonly site of branch formation.

Nitrogenase – Enzyme concerned with conversion of molecular nitrogen into ammonia.

Node – Point or area of axis where branching or leafing occures. In filament, it is location of septum.

Nucleolar organizing chromosome – Chromosome concerned with formation of nucleolus.

Nucule – Female reproductive organ of Charales.

Nurse cell – Cell laden with reserve food and border near carpogonium in red algae.

O

Obligate – Essential or necessary; not facultative.

Obovate – Inversely egg-shaped, ie. broader at distal end.

Obpyriform – Inversely pear-shaped.

Ocellus – Eye spot or photoreceptor with associated lens-like organelle.

Oligotrophic – Habits relatively poor in nutrients.

Ontogenetic – Pertaining to development of individual organism throught its various stages.

Ontogeny – Dvelopment of organism.

Ooblast – Connecting filament through which zygote nucleus migrates from carpogonium to auxiliary cell in Rhodophyta.

Oogamy – Fusion of motile sperm with large passive non-motile egg.

Ostiole – Opening or pore of conceptacle of fucales.

Ovoid – Like hen's egg.

Ovum – Female, non-motile gamete or egg cell.

Oxidation pond – Enclosure for sewage designed to promote digestion of sewage with help of oxygen released by algae during photosynthesis.

P

Palisade – Erect cylindrical cells laterally united into compact tissue.

Palmella stage – Temporarily non-motile sedentary stage in life history or certain motile algae;cells remian passive and embedded in gelatinous matrix.

Palmelloid – Palmella-like habit.

Pantonematic – Flagellum in which surface is covered with hair-like appendages.

Parophysis – Sterile filament borne near reproductive structure; paraphysis is hair-like and may be branched or unbranched.

Parasexual – Organisms showing genetic recombination not involving regular alternation between karyogamy and meiosis.

Parasporangia – Sporangia whose proto-plast divides into more than four spores but at least first division is mitotic; commonly found in Rhodophyta.

Parenchymatous – Pertaining to tissue composed of thin-walled and living cells.

Parietal – Positioned against inner wall of cell.

Parthenogenetically – Pertaining to mode of development in which female gamete germinates without undergoing fertilization.

Pectinate – Resembling a comb.

Pelagic – Living unattached in the open sea, e.g., *Sargassum natans*.

Pellicle – Outer bounding membrane found in cells lacking cell wall; resembles but is not equivalent to firm cell wall.

Pericarp – Urn-shaped sterile envelope surrounding cystocarp or carposporophyte.

Pericentral cell – One of a ring of cells out off form and surrounding cystocarp or carposporophyte in red algae.

Periclinal – Parallel with circumference.

Periphyses – Sterile, branched or unbranched, hair-like cells growing from conceptacle but not closely associated with reproductive organ.

Periphyton – Microalgae associated with or growing attached to stems, twigs and culms of larger aquatic plants.

Periplast – Delicate protective covering or flagellates that lack cell wall.

Phototaxis – Movement oriented in response to light.

Phragmoplast – Parallel arrangement of microfibrils to spindle axis at telophase; cell plate is deposited across these microfibrils.

Phycobilisomes – Particles containing phy-cobiling, phycocyanin and phycoerythrin pigments.

Phycobiont – Algal component of lichen.

Phycoplast – Arrangement of microfibrils perpendicular to spindle axis and at equator of cell at telophase.

Phylogeny – Racial history or evolutionary developement of species.

Pit-connection – Cytoplasmic strand connecting two adjoining cells throught pit in their respective cell walls.

Placenta – Composite, often conspicuous, multinucleate cell produced when carpogonium fuses with auxiliary cell and other sterile cells in Rhodophyta; gonimoblasts develop from this placenta.

Plakea stage – Curved plate-like, eight-celled stage in development of coenobium.

Plane – pertaining to septa with smooth flat surface.

Planktonic – Floating in water.

Plasmalemma – Living membrane separating protoplasm of cell form cell wall.

Plasmodesmata – Cytoplasmic strands connecting adjoing cells and linking cytoplasm.

Plasmogamy – Fusion of cytoplasmic materials (usually also associated with fusion of nuclear materials) in formation of zygote.

Plethysmothallus – Microscopic, ectocarpoid filament arising from zoospore and serving as accessory method of reproducing sporophyte generation vegetatively or by formation of reproductive bodies in Phaeophyta.

Polar nodule – Wall thickening at the two poles in diatom frustule.

Polyherdral bodies – Polygonal particles found in cells of blue-green algae; their function and chemistry are not known.

Polyphosphate granules – Polymers of in-organic phosphate stored in algal cells; occur in granular form.

Polyploidy – Cells containing three, four or more times the haploid number of chromosomes.

Polysiphonous – Consisting of definite number of pericentral cells cut off from and surrounding each cell in axial cell row; consisting of several coherent longitudinal rows of cells.

Polysporangium – Sporangium containing more than four spores; at least first division is meiotic.

Polystichous – mode of branching in which branches are inserted on axis in various planes or ranks.

Procarp – Carpogonium and auxiliary cell or cells when they are members of common branch system.

Procaryota – Cellular organisms lacking membrane-bound genetic, photosynthetic and respiratory organelles.

Propagule – Few-celled branch serving as means of vegetative propagation, as in Sphacelariales.

Prostrate – Lying flat on substratum

Protandrous – Hermaphrodite organisms in which male reproductive structure matures earlier than female.

Protonema – Juvenile, often filamentatious or vesicular, stage in life history of plant.

Proximal – Opposite of distal; nearest to point of origin or attachment.

Pseudoparenchymatous – Collecti on of cells; filaments or hyphae forming tissue that resembles paren-chyma.

Pseudoraphe – False raphe; longitudinal space on valve of diatom bounded on both sides by striae.

Punctae – Perforations in wall of diatoms.

Pyriform – pear-shapped,with broad end towards base.

Q

Quadrate – squarish; four-sided or rectangular.

R

Raphe – Longitudinal cleft seen on valve surface of diatom.

Receptalce – tip of branch of thallus bearing conceptacles as in Fucales.

Recombinants – Offspring with new combination of genes different form either parent.

Regeneration – Vegetative development from dormant or old part of thallus, often terminated by breakage, senescence or wounding.

Replicate – Septum showing two orbicular in-foldings, one on eight aide.

Repression – Inhibition of enzyme synthesis.

Reproductive – processes leading to continuation of species or races.

Revolute – Rolled back form apex or margin.

Rhizopodia – Unicellular organisms

capable of forming pseudopodia or false feet for locomotion or anchorage.

Rhizopodial – Type of habit in which uni-cellular organisms form pseudopodia as locomotory organs.

S

Saprobic – Pertaining to organisms that feed on decaying organic matter.

Satellite chromosomes – Short segments or chromosomes constricted from rest of chromosomes.

Scalariform – Ladder-like arrangement.

Sedentary – Stationary and attached, whether stalled or sessile.

Semireplicate – Only one side of septum wall showing circular infolding.

Sessile – Attached directly to substratum;not stalked.

Silicalemma – Triple membrane system concerned with deposition of silica walls in diatoms.

Simultaneous division – Nuclear and cytoplasmic divisions occurring concurrently.

Siphoneous – Tubular thallus in algae lacking sept or cross walls during vegetative phase of growth.

Spermatium – Non-flagellated naked male gamete of red alga.

Spermocarp – Fruiting body consisting of sterile jacket of cells around fertilized oogonium, as in *Coleochaete*.

Sporophyte – Usually diploid, spore-producing phase of life history.

Stenohaline – Tolerant of narrow range of salinity.

Stenothermal – Tolerant or narrow range of temperature.

Stephanokont – Condition in which swarmers have grown of flagella, e.g, in *Oedogonium*.

Sterile – Vegetative; lacking reproductive organs.

Stigma – Eye spot or red spot of flagellates found within cell near or inside chromato-phore. Believed to have regulatory function in cell motility.

Stipe – Stalk; portion of kelp between hold -fast and blade.

Stratified – Layered.

Subaerial – Referring to habitats well raised above soil surface, e.g., tree barks, rocks and stones.

Sublittoral – Bottom or benthic region of ocean; usually refers to deeper or inshore layers of seawater below lowest tide.

Substratum – Surface or object upon or within which organism is growing.

Successive division – Nuclear and cytoplasmic divisions occurring one after the other.

Supporting cell – Cell that gives rise to carpogonial branch.

Supralittoral – Beyond reach of highest tide on seashore; spray zone.

Swarmer – Flagellated reproductive cell whether gamete or spore.

Synaptonemal complex – Characteristic structures connecting bivalents in meiotic nuclei and seen under electron microscope.

Syntagmic germination – Formation of single thallus by germination of several tetra-spores within tetrasporangium in Rhodo-phyta.

Synzoospore – Compound, multi-flagellate and multinucleate asexual spore found in *Vaucheria*.

T

Tetraspore – One of the four non-motile haploid spores formed in tetra sporangium in red algae.

Tetrasporophyte – Diploid phase producing tetraspores by meiotic division in tetra-sporangium.

Thalloid – Like plant or alga that is not differentiasted into roots, stem and leaves.

Thallus – Relatively undiffrentiated plant body that is not divided into roots, stem and leaves.

Thylakoid – Structural unit of lamellar system in chloroplast forming double membrane-disc; photosynthetic lamella.

Torulose – Knobby or twisted.

Trabeculae – Anastomosing projections of wall material which traverse central vacuole and cytoplasm, e.g., in *Caulerpa*.

Transcription – Process in which DNA is copied into messenger RNA.

Transduction – Process of genetic recombination in which DNA or donor bacterium is carried to recipient by virus.

Transformation – Process of genetic recombination in bacteria affected by incorporation of naked DNA from donor bacterium.

Trichoblast– Filamentous lateral outgrowth, consisting or single row of unpigmented cells, usually branched, borne from surface thallus layer.

Trichogyne – Hair-like, receptive, prolongation of carpogonium with which spermatium fuses.

Trichome – Row of cells without investing sheath (as in Cyanophyta); any sterile filamentous branch arising from thallus(as in algal divisions other than Cyano-phyta).

Trichothallic – Intercalary growth in which meristem is located at base of a hair.

True branching– Branched by lateral division of cell in main filament.

Trumpet hyphae – Long tubular achlorophyllous cells of medulla found in brown algae; conduct water and nutrients.

U

Unduliseptate – Septum with wavy margins.

Uniaxial – Plant axis made of single filament or its branches.

Unilocular – Pertainting to sporangia borne on sporophyte generation in Phaeophta; unilocular sporangia not divided into separate chambers but produce numerous spores, by meiosis, in the single chamber of sporangium, spores produced being haploid e.g. *Sargssam*.

Uniseriate – Arranged in single now or series.

Urseolate – Shaped like pitcher.

Utricle – Small sac-shaped filament, e.g. in *Codium*.

V

Vacuolar apparatus – Includes two kinds of organelles, viz. contractile vacuole and cytopharyngeal apparatus consisting of gullet, reservoir, and a number of small contractile vacuoles surrounding reservoir.

Vacuole – Small, usually spherical space within cell, bounded by membrane and containing fluid sap.

Valve – One of the two parts of a diatom theca; the other is the connecting band.

Valve view – Top surface of apitheca or hypotheca of diatom.

Vegetative – Formation of plant body lacking reproductive structures or organs.

Ventral – Lower or inner side.

Verrucose – Knobby or warty.

Vesicle- Small bladder-like sac filled with gas.

W

Water bloom – Luxuriant growth or one or a few species of algae in water, often imparting colour to water.

Whorl – Group of similar organs radiating from mode.

Whorled – Surrounding axis or branch in a ring.

Z

Zonate – Division of tetrasporangium in three parallel planes so that tetraspores apper arranged in linear row.

Zoogamete– Motile flagellate gamete.

Zoospore – Motile flagellated asexual cell.

FUNGI

A

Acervulus (pl. **acervuli; L. acervus** - heap, dimin form) – a mat of hyphae giving rise to short conidiophores closely packed together forming a bed-like mass; characteristic of the Melanconiales.

Achlorophyllous – (Gr. a = not + chloros + green + phyllon + leaf); lacking chlorophyll.

Adventitious septum – a septum formed independently of nuclear division; especially associated with changes in the concentration of protoplasm in parts or a hypha.

Aeclospore (Gr.*aikia* = injury + spora = seed. spore) – a binucleate spore produced in an aecium.

Aecium (pl. **aecia;** Gr. aikia = injury) – a structure consisting of binucleate hyphal cells, with or without a peridium, that produce spore chains consisting of aeciospores alternating with disjunctor cells, following the successive, conjugate division of the nuclei.

Aethallum (pl. aethalia -Gr. aethalos = soot) – a rather large, sometimes massive. generally cushion-shaped fructification of some Myxomycetes.

Akinete (Gr. a = not + kinetos = mobile) – a thick walled; nonmotile spore in algae; see reproduction of phycobionts in lichens.

Amatoxins – cyclopeptides present in certain mushrooms; alpha-amanitin and beta-amanitin are the principal amatioxins.

Anisogamous planogametes (Gr. a = not + isos = equal + gamos = marriage; planetes = wanderer + gameter = husband); motile gametes that are morphologically similar but differ in size.

Annellide (L. annelus = ring) – a type of conidiogenous cell that produces blastic conidia in a basipetal fashion; it typically elongates with the production of each conidium and has ring-like scars on its outer surface near the conidiogenous locus.

Antheridiol (antheridium = male sex organs) – a hormone liberated by the female thallus of *Achyla bisexualis* that induces formation of antheridia in the male thallus.

Antheridium (Pl. antheridia; Gr. *antheros* = flowery + idion, a dimin. suffix) - a male gametangium.

Antherozoid (Gr. antheros = flowery + bios = life) – a substance produced by a living organism. which injures or kills another living organism.

Aphanoplasmodium (Pl. **aphanoplasmodia;** Gr. *aphaners* = invisible = plasmodium) – a plasmodium consisting, in its early stages, of a network of very fine transparent strands that are not conspicuously differentiated into ecto-and endoplasm, and in which the protoplasm is not coarsely granular. Characteristic of *Stemonitis* and other similar genera.

Aplanetic (Gr. a = not + planetes + wanderer) - non motile.

Aplanospore (Gr. a = not + *plancetes* = wanderer + spore = seed, spore) –

a nonmotile spore.

Apothecium (Pl. **apothecia**; Gr. apotheke store house) – an open ascocarp.

Appressorium (Pl. appresoria; L. apprimere = to press against) – a flattened. hyphal, pressing organ, from which a minute infection peg usually grows and enters the epidermal cell of the host.

Archicarp (Gr. arche = beginning + karpos = fruit) – the initial stage of a fruiting body.

Arthrospore (Gr. arthron = joint + spora = seed. spore) – a spore resulting form the fragmentation of a hypha.

Ascigerous (Gr. askos = sac) the ascus stage of an ascomycete.

Ascocarp (Gr. askos = sac + karpos = fruit) – a fruiting body containing asci.

Ascogenous hypha (Gr. askos = sac + gennao = 1 give birth + hyphe = web) – a specialized hypha that gives rise to one or more asci.

Ascogonium (p). ascogonia; Gr. askos = sac + gennao = 1 give birth) – the female gametangium of the Ascomycetes.

Ascospore (Gr. askas = sac + spora = seed, spore) – a meiospore borne in an ascus.

Ascostroma (pl. **ascostromata**; Gr. askos = sac + stroma =mattress. cushion) – a stromatic ascocarp bearing asci directly in locules within the stroma.

Ascus (pl. asci; Gr. askos = sac) – a sac-like cell generally containing a definite number of ascospores (typically eight) formed by free cell formation usually after karyogamy and meiosis; characteristic of the class Ascomycetes.

Ascus mother cell – the binucleate crook cell in the Ascomycetes in which karyogamy occurs and which develops into the ascus.

Aseptate (L. ab = away + septum = hedge) – lacking cross-walls.

Asexual (L. ab = away + sexus = sex) – reproduction not involving karyogamy and meiosis.

Aspergillosis (*Aspergillus* = a genus of Deuteromycetes) – one of a group of diseases of animals and humans caused by various species of *Aspergillus.*

Asporgogenous (Gr. a = not +spora + seed, spore + gennao = I give birth) – nonspore-forming.

Autoecism (Gr. autos = self, i.e. the same + oikos = home) – the ability of a parasitic fungus to complete its entire life cycle on a single host species; used particularly for certain rusts.

Axenic (Gr. a = not + xenos = stranger) – without another organism being present.

Azotodesmic (Gr. a = not + zygos + yoke + spora + seed, spore) – refers to lichens in which the phycobiont fixes nitrogen.

Azygospore - (Gr. a = not + zygos = yoke + spora = seed, spore) – a zygospore that develops parthenogenetically.

B

Ballistospore (Gr. *ballo* = to throw + spora = seed, spore) – any spore that is focrcibly discharged.

Basidiocarp (Gr. basidion = small base. basidium + *karpos* = fruit) - a fruiting body that bears basidia.

Basidiospore (Gr. *basidion* = small base + spora = seed, spore) – a spore borne on the outside of a basidium, following karyogamy and meiosis.

Basidiole or basidiolum (L. dimin. of basidiun) – a type of sterile element in the hymenium of certain Basidiomycetes; resembles a basidium without basidiospores.

Basidium (pl. basidia; Gr. basidion = a small base) – a structure bearing on its surface a definite number of basidiospores (typically four) that are usually formed following karyogamy and meiosis.

Binding hyphae – thick-walled. typically aseptate, highly branched hyphae present in the basidiocarps of some fungi.

Binomial (L. Bi = two + nomen ≈ name) – the scientific name of an organism: it is composed of two names: the first designating the genus, the second the species.

Bipolar heterothallism (L. Bi = two = Gr. polos = pole) – a condition of sexual compatability in which there are only two mating types; also known as unifactorial heterothallism. (See also **Heterothallism and Tetrapolar heterothallism**).

Bitunicate (L. bis = twice, two + tunica = coat, mantle) – an ascus in which the inner wall is elastic and expands greatly beyond the outer wall at the time of spore liberation.

Blastic conidium (Gr. blastos = a sprout) – a conidium arising form only a portion of a pre-existing conidiogenous cell.

Blepharoplast (Gr. blepharis = eyelash + plastid) – see kinetosome.

Budding (ME, budde = bud): the production of a small outgrowth (bud) from a parent cell; a method of asexual reproduction.

C

Capillitium (pl. **capillitia;**L. capillus = hair) – sterile, thread-like structures present among the spores in the fruiting bodies of many Myxomycetes and Gasteromycetes.

Centrum (pl. **centra;** Gr. kentron = center) – the totality of structures enclosed by the ascocarp wall.

Cephalodium (pl. **cephalodia;** Gr. kephale = head) L: internal or external swellings in the thalli of diphycophilous lichens, in which the blue-green algae are segregated.

Chitosome (chitin + Gr. *soma* = body) – a small cytoplasmic vesicle containing chitin synthetase, the enzyme involved in chitin synthesis in fungi.

Chlamydospore (Gr. *chiamys* = mantle + *spora* = seed, spore) – a thick-walled thallic conidium that generally functions as a resting spore.

chlorophycophilous (Chlorophyceae = green algae + Gr. *philein* = to love): a lichen with green alga pohycobionts.

Cirrhus (pl. **cirhi;** L. *cirrhu*s = curl) – a ribbon-like cylinder of spores held together by mucous as it issues frm an ostiole.

Clamp connection – a bridge-like hyphal connection characteristic of the secondary mycelium of many Basidiomycetes.

Cleistothecium (pl. cleistothecia; Gr. kleistos = closed + theke = case) – a completely closed ascocarp.

Coenocytic (Gr. koinos = common + Kytos = a hollow vessel) – nonseptate; referring to the fact that nuclei are embedded in the cytoplasm without being separated by cross-walls. that is, the nuclei lie in a common matrix.

Colony (L. *colonia* = a settlement) – a group of individuals of the same species living in close association; in fungi, the term usually refers to many hyphae growing out of a single point and forming a round or globose thallus.

Columella (pl. **columellae;** L columen = column) – a sterile structure within a sporangium or other fructification often an extension of the stalk.

Compound oosphere (Gr. oon = egg + sphaira = sphere) – an oosphere with many functional gamete nuclei.

Concentric body – a round, perfectly concentric body found in the hyphae of a great many lichenized fungi and in a few nonlichenized species.

Conidiophore (conidium, q. v. + Gr. Phoreus = bearer) – a simple or branched hypha arising form a somatic hypha and bearing at its tip or side one or more conidiogenous cells; sometimes used inter-changeably with conidiogenous cell.

Conidium (pl. **conidia**; Gr. konis = dust + idion. dimin. suffix) – a non-motile asexual spore usually formed at the tip or side of a sporogenous cell; in some instances a pre-existing hyphal cell may be converted to a coidium.

Conjugate nuclear division (L. con = with + jugum = yoke) the simul-taneous division of the two nuclei in a dikaryon, giving rise to four daughter nuclei these generally be-come separated by a septum into two cells, with the sister nuclei migrating into different daughter cells.

Context (L. contexere = to weave together) – the fibrous tissue that makes up the body of the pileus in the Basidiomycetes.

Coprogen (Gr. kopios = dung + gen-nao = I give birth) – a factor in dung necessary for the growth of *Pilobolus*.

Coprophilous (Gr. kopros = dung + philein = to love) – growing on dung.

Cortex (L. cortex = bark) – the upper and lower (outer) layers of a lichen thallus.

Cortina (pl. **cortinae**; L. cortina = curtin) – a curtain-like, cobwebby veil hanging form the margin of the cap of certain mushrooms.

Cruciform division (L. crus = cross) – intranuclear division in which the chromosomes are arranged in a ring around a dumbbell-shaped nucleolus as they divide.

Crustose thallus – a crust-like lichen thallus.

Cyanophycophilous (Cyanophyceae = blue green algae + Gr. Philein = to love:) – lichens with blue-green alga phycobionts.

Cystidium (pl. cystidia; Gr. kystis = bladder + idion, dimin. suffix) – a sterile element occurring in the hymenium of certain Basidiomyceter; cystidia are generally larger

than other hymenial elements and protrude beyond them.

D

Damping-off – a disease of seedlings causing them to rot at the soil level and to fall over.

Demicyclic (L. demi = half + Gr. kyk-los = circle, cycle) – a rust fungus that lacks a uredinial stage yet typical has stages O, I, III, and IV.

Dermatomycosis (Gr. derma = skin + mykes = mushroom) – a fungus infection of animal or uman skin.

Dermatophyte (Gr. derma = skin + phyton = plant) – any one of several fungi that cause skin dis-eases.

Diaspore (Gr. diaspora = dispersal, scattering) – a propagule produced by a lichen thallus such as an isidium or soredium.

Dictyospore (Gr. dictyon = net + spora = seed, spore) – a spore with both vertical and horizontal septa.

Dikaryon (NL. di = two + Gr. karyon = nut, nucleus) – a pair of closely associated nuclei, each usually derived from a different parent cell.

Dikaryotic (NL. di = two + Gr. karyon = nut, nucleus) – pertaining to a cell that contains a dikaryon.

Dimitic (Gr. dis = twice + mitos = thread, i.e. hypha:) a basidiocarp composed of generative hyphae and either binding or skeletal hyphae.

Dimorphic (Gr. dis = twice + morphe = form): (1) producing two types of zoospore: (2) a fungus able to grow either in yeast form or in mycelial form.

Dioecious (Gr. dis = twice + oikos = home) – refers to species in which the sexes are segregated in dif-ferent individuals: the use of this-term is often restricted to plants.

Diphycophilous (Gr. dis = twice + phykos = sea weed. alga = philein = to love) – lichens with both green and blue-green phycobionts.

Diplanetic (Gr. dis = twice + planetes = wanderer) – refers to a dimorphic species in which two swarming periods occur.

Diploid (Gr. diplaus = double) – containing the double (2n) number of chromosomes.

Dolipore septum (L. dolium = large jar + pore) – a septum with a central pore surrounded by a barrel-shaped swelling of the septal wall, and coverd on both sides by a perforated membrane termed the septal pore cap; common in many Basidiomycetes.

E

Ectal excipulum (Gr. ektos = outside + L. excipulum = receptacle) – the outer layer of the apothecium.

Ectomycorrhiza (pl. **ectomycorrhizae**; Gr. ektos = outside + mycorrhiza q. v.) – a mycorrhiza in which the fungal hyphae grow only intercellularly, never penetra-ting the plant cells.

Egg (Icel. egg = egg) – female gamete.

Endobiotic (Gr. endos = inside + bios = life) – an organism that lives within its substratum. usually the cells of its host.

Endomycorrhiza (pl. **endomycorrhizae**; Gr. endos = inside + mycorrhiza q.v.) – mycorrhiza in which the fungal hyphae penetrate the cells of the plant; also called vesicular-arbuscular mycorrhiza.

Epibiotic (Gr. epi = on + bios = life) – an organism whose reproductive organs are on the surface of the substratum, but part or all of whose soma is within the substratum.

Epigean (Gr. epi = upon + ge = earth) – above the ground.

Epiphytotic (Gr. epi = upon + physton = plant) – a widespread occurrence of a plant disease; equivalent to epidemic on humans.

Epithecium (pl. **epithecia**; Fr. eip = upon + theke = a case) – a layer of tissue on the surface of the hymenium of an apothecium, formed by the union of the tips of the paraphyses over the asci.

Epixylous (Gr. epi = upon + xylon = wood) occurring on wood.

Epizootic (Gr. epi = upon + zoon = animal) – a wide-spread occurrence of an animal disease; equivalent to epidemic on humans.

Eucarpic (Gr. cu = good + karpos = fruit) – forming reproductive structures on certain portions of the thalus, with the thallus itself continuing to perform its somatic functions.

Excipulum (pl. **exipula**; N.L. excipu lum = receptacle) – the outer layer of the hypothecium.

F

Facultative parasite (L. facultas = ability: Gr. parasitos = table mate) – an organism capable of infecting another living organism or of growing on dead organic matter. according to circumstances.

Facultative saprobe (L. facultas = ability: Gr. sapra = rotten + bios + life) – an organism capable of growing on dead organic matter. or of infecting another living organism. according to circumstances.

Fairy ring – a ring of mushrooms on the ground representing the periphery of mycelial growth of a basidiomycete.

Fertilization tube (L. fertilis = fertile) – a tube originating form the male gametangium and penetrating into the female through which the male gametes (nuclei) are transferred.

Fission (L. fission = splitting) – the splitting of a cell into two cells.

Flagellum (pl. flagella; L. flagellum = whip) – a hair. whip or tinsel-like structure that serves to propel a motile cell.

Foliose thallus – a leaf-like lichen thallus.

Fragmentation (L. frangere = to break) – the segmentation of the thallus into a number of fragments each of which is capable of growing into a new individual – a method of asexual reproduction.

Fructification (L. fructus = fruit) – any complex fungal structure that contains or bears spores.

Fruiting body – see Fructification.

Fruticose thallus – an erect or pendent usually many branched lichen thallus whose tissues tend to form cylinders but may be flattened.

Fungi (sing. **fungus;** L, *fungus* = mushroom) – achlorophyllous, spore bearing eukaryotes with either a typically walled thallus with absorptive nutrition, or with an unwalled thallus with phagotropic nutrition.

Funiculus (pl. **funiculi;** L. *funiculus* = a small cord) a thin cord by means of which the peridioles of some Nidulariales are attached to the basidiocarp that bears them.

G

Gametangial contact (Gr. gametes = husband + angeion = vessel) – a method of sexual reproduction in which two gametangia come in contact but do not fuse; the male nucleus migrates through a pore or fertilization tube into the female gametangium.

Gametangial copulation (Gr. gametes = husband + angeion = vessel) – a method of sexual reproduction in which two gametangia or their protoplasts fuse and give rise to a zygote that develops into a resting spore.

Gametanginm (p) **gametangia;** Gr. gametes = husband – angeion = vessel) – a structure that contains gametes.

Gametes (Gr. gametes = husband, sex cell) – a differentianed sex cell or a sex nucleus that fuses with another in sexual reproduction.

Gamethalli; (pl. **gametothalli;** Gr. **gametes** = husband + thallos = shoot) – a thallus that produces gametes as opposeed to a sporothallus.

Gammna particle – small membrane-bounded structures found in the zoospores of *Blastocladiella emersonii.*

Gemmae (pl. **gemmae;** L gemma = bud) – a thickwalled cell similar to a chlamydospore.

Generative hyphae – the essential components of any basidiocarp; thin-walled septate, profusely branched hyphae capable of giving rise to basidia.

Genus (pl. genera; L genus = race) – a taxonomic category that includes a number of species; the genus name (generic name) is the first name in a binomial.

Gleba (pl. glebae; L. gleba = clod) – the inner. fertile portion of the fruiting body of the Gasteromycetes.

H

Haploid (Gr. haplous = simple) – containing the reduced (in) number of chromosomes:

Hapteron (pl. **haptera;** Gr. hapto = I touch) – a mass of highly adhesive hyphae that form an attachment organ at the. nose of the funicular cord of the Nidulariaceae.

Haustorium (pl. haustoria; L. haustor = drinker) – an absorbing organ originating or a hypha of a parasite and projecting into a cell of the host; most often associaned with obligate parasites. but also produced by some facultative parasites.

Helicospore (Gr. helix = helix + spora = seed. spore) – a colled or helical spore.

Hermophroditic (Gr. Hermes = the messenger of the gods. symbol of the male sex + Aphrodite = the goddess of love. symbol of the female sex) – refers to species in which both recognizable male and female sex which are introduced by each individual.

Heterobasidium (pl. **heterobasidia;** Gr heteros = different other-basidium) - a term used by some to refer an any type of basidium other than the single celled cup shaped basidium.

Heterocyst (heteros = different + kystis = bladder. a thick walled cell in certain blue green algae believed to be involved in nitrogen fixation see reproduction of phycobionts in lichens.

Heteroecism (Gr. heteros = other, different + oikos = home, host) - the necessity of two host species for the completion of the life cycle of certain parasitic fungi.

Heterogametangia (sing. **heterogametangium**; Gr. hetero = other, different + gametes = husban + angeion = vessel) - male and female gametangia that are morphologically distinguishable.

Heterogametes (Gr. heteros = other, different + gametes = husband) - male and female gametes that are morphologically distinguishable.

Heterokaryosis (Gr. heteros = other + karyon = nut, nucleus) - a condition in which genetically different nuclei are associated in the same protoplast or the same mycelium.

Heterokaryotic (Gr. heteros = other + karyon = nia, nucleus) - an individual exbibiting heterokaryosis.

Heterokont (Gr. heteros = other + kontos + staff) - a biflagelate structure with two flagella, unequal in size.

Heteromerous (Gr. heteros = different + meros = part) - lichen thalli in which the algal cells form a distinct layer within the thallus.

Heterothallic (Gr. heteros = other, different + thallas = shoot, thallus) - according to one version: a species consisting of self-sterile (self-incompatible) individuals requiring therefore the union of two compatible thalli (mating types) for sexual reproduction regardless of the possible presence of both male and female organs on the same individual: according to another version: a species in which the sexes are segregated in separate thalli, two different thalli being required for sexual reproduction.

Heterothallism (Gr. heteros = other, different + thallos = shoot, thallus) - the condition exemplified by heterothallic species.

Holdfast - a special structure by means of which Trichomycetes become attached to their hosts.

Holobasidium (pl, **holobasidia**; Gr. holos = entire + basidion = a small base) - a single-celled basidium; although typically club-shaped. holobasidia may resemble tuning forks in some taxa. while in others the basidium may become divided by adventitious septa.

Holocarpic (Gr. holos = entire + karpos = fruit) - refers to an organism whose thallus is entirely converted into one or more reproductive structures.

Holozoic (Gr. holos = entirely + zoikos = of animals) - ingesting food in the form of solid particles.

Homoiomerous (Gr. homoios = the same + meros = part) - lichen thalit in which the algae are more or less evenly distributed throughout.

Homokaryotic (Gr: homo = the same + karyon = nut, nucleus) - an individual whose nuclei are genetically alike.

Homothallic (Gr. homo = same + thallos = shoot, thallus) - refers to fungi in which sexual reproduction takes place in a single thallus that is therefore, essentially self-compatible.

Homothallism (Gr. homo = same + thallas = shoot, thallus) - the condition exemplified by homothallic species.

Hormogonium (pl. **hormogonia**; Gr. hormao = I thrust + ganos = offspring) - a usually motile segment of a filamentious blue green alga capable of growing into a new filament; see reproduction of phycobiont in lichens.

Host (L. hospes = one who receives a stranger as hisguest) – a living organism harboring a parasite.

hyaline (Gr. hyalinos = made of glass, i,e, coloriess) – colorless, transparent.

Hymenium (pl.) **hymenia;** (Gr. hymen = membrane) – a fertile layer consisting of asci or basidia .

Hyperplasia(Gr. hyper = over + plasis = molding, formation) – excessive multiplication of cells, abnormal rate of cell division.

Hypertrophy (Gr. hyoper = over + trophe = food) – excessive enlargement of cell.

Hypha (pl. **hyphae**: Gr. hyphe = web) – the unit of structure of most fungi; a tubular filament.

Hyphal body (Gr. hyphe = web) – a fragment of the mycelium of the Entomophthorales.

Hyphopodium (pl. **hyphopodia;** Gr, hyphe = web + pous = foot) – a small appendage on a hypha; characteristic of the Meliolales.

Hypogean (Gr. hypo = under + gè = earth) – growing below the gorund.

Hypothallus (pl. **hypothalli;** Gr. hypo = under + thallos = shoot, thallus) – a thin, often transparent deposit at the base of the fructifications of many Myxomycetes.

Hypothecium (pl. **hypothecia;** Gr. hypo = under + theke = case) – a thin ,layer of interwoven hyphae immediately below the hymenium of an apothecium.

Hysterothecium (pl. **hysterothecia;** Gr. hysteros = womb + theke = case) – an elongated, boat shaped ascocarp with a longitudinal slit; characteristic of the Hysteriales and of certain lichens such as Arthoniales.

I

Imperfect stage – the asexual (usually conidial) stage of a fungus.

Indusium (Pl. **indusia;**L. indusium = undergarment) – a skirt – like structure hanging form the receptacle of the expanded fruiting body of *Dictyaphora* (one of the stinkhorns).

Inner veil – the hyphal membrane covering the gills of some young mushrooms.

Isidium (Pl. **isidia**) – a minute, corticated more or less columnar lichen propagule consisting of both fungal hyphae and algal cells. breaking off the thallus and distributed by various means.

Isogametangia (sing. **isogametangium;** Gr. ison = equal + gametes = husband + angeion = container) – gametangia, presumably of opposite sex; that are morphologically indistinguishable.

Isogametes (Gr. isos = equal + gametes = husband) – gametes, presumably of opposite sex, that are morphologically indistinguishable.

Isoplanogametes (Gr. isos = equal + planetes + wanderer + gametes = husband) – motile gametes, presumably of opposite sex. that are morphologically indistinguishable.

K

Karyogamy (Gr. Karyaon = nut, mucleus + gamos = marriage, union) – the fusion of two nuclei.

Kinetosome (Gr. kinetos = movable + soma = body) – the base of a flagellum consisting of a cylinder of nine triplet microtubules; Kinetosomes generally arise from centrioles and are also known as blepharoplasts or basal bodies,

L

Lamella (pl. lamellae; L. lamina = plate, dimin. form) – a plate-like structure (gill) on which some Basidiomycetes produce their basidia.

Lichen (Gr. lichen = lichen) – a combination of an alga and a fungus in which the two components are so interwoven as to form what appears to be a single individual.

Lichen acid – any of a number of organic compounds synthesized by lichens; not necessarily an acid.

Locule (L. loculus = a little place) – a cavity within a stoma.

Lomasome (Gr. loma = hem, fringe, border + soma = body) – membraneous structure lying between the cell wall and the plasma membrane.

M

Macroconidium (pl. **macroconidia;** Gr. makron = long + konis = dust + idion, dimin suffix)– a conidium, as distinguished from a microconidium.

Macrocyst (Gr. makros = long, large + kystis = bladder) – a thick-walled structure representing the sexual stage of the Acrasiomycetes.

Mastigoneme (Gr. mastix = whip + nema = skein, thread) – one of the numerous small, hair-like projections on a tinsel flagellum; also known as flimmers.

Mazaedium (pl. **mazaedia;** Gr. *maza* = dough + *eidos* = like) – a fruiting body in which the spores freed from the asci then form a powdery mass.

Medium (pl. **media;** L. medium = intermediate) – substratum of a balanced chemical composition employed in the laboratory for growing microorganisms; media may be used in the liquid state or solidified with agar, gelatin, or other solidifying agents.

Medulla (L. medulla = marrow) – the central layer of a heteromerous lichen thallus just below the algal layer.

Medullary excipulum (L. medulla = marrow + excipulum = receptacle) – the inner portion of the apothecium.

Meiosis (Gr. meiosis = reduction) – a series of two nuclear divisions usually in quick succession in which the chromosome number is reduced by one - half.

Meiospore (Gr. meros = less + spora = spore) – a spore formed as a result of meiosis.

Meristogenous (Gr. meros = part + gennao = I give birth) – the origin of a fruiting body from the division of a simple cell or of adjacent cells of the same hypha.

Merosporangium (pl. **merosporangia;** Gr. meros = portion + sporangium) – a cylindrical sporangium.

Metabasidium (pl. **metabasidia;** Gr. meta = between + basidium) – that portion of the basidium in which meiosis takes place; also called the promycelium in certain Basidiomycetes.

Microconidium (pl. **microconidia;** Gr. mikron = small + konis = dust + idon, dimin. suffix) – a small coridium that often acts as a spermatium,

Microcyclic (Gr. mikros = small + kyklos = circle, cycle) – a rust in which the teliospore is the only binucleate spore in the life cycle.

Microcyst (Gr. mikros = small + kystis = bladder) – a small, encysted protoplast; usually an encysted myxamoeba of the myxomycetes or Acrasiomycetes.

Mitic system – an approach used in the taxonomy of the Aphyllophorales; involves the analysis of the hyphal types comprising the basidiocarp.

Monocentric (Gr. monos = single + kentron = center) – a thallus radiating from a single point at which a reproductive organ (sporangium or resting spore) is formed.

Monokaryotic (Gr. monos = alone, single + karyon = nut, nucleus) – containing a single nucleus.

Monomitic (Gr. monos = single + mitos = a threca i.e. a hypha) – term used to describe a basidiocarp composed of only generative hyphae.

Monomorphic (Gr. monos = alone, one + morphe form) – producing one type of zoospore.

Monophyletic (Gr. monos = alone,

single + phylon stock, race) – of a single line of descent.

Monoplanetic (Gr. monos = alone, only + planetes wanderer) – refers to a species that produces only on type of zoospore and in which there is but one swarming period.

Mushroom – a fleshy, sometimes tough, umbrella-like basidiocarp of certain Basidiomycetes.

Mycelium (pl. mycella; Gr. mykes = mushroom. fungus) – mass of hyphae constituting the body (thallus) of a fungus.

Mycetismus – mushroom poisoning.

Mycobiont (Gr. mykes = fungus + bios = life) – the fungus component of a lichen.

Mycology (Gr. mykes = mushroom, fungus + logos = discourse) – the study of fungi.

Mycoparasite (Gr. mykes = fungus + parasitos = parasite) – a fungus that parasitizes another fungus; necrotrophic forms kill the host or at least a portion of the host, whereas biotrophic forms cause little or no apparent damage to the host.

Mycophagy (Gr. mykes = mushroom + phagein = to eat) – the eating of mushroom.

Mycorrhiza (pl. mycorrhizae; Gr. mykes = mushroom + rhiza = root) – a symbiotic association between the hyphae of certain fungi and the absorptive organs typically the roots - of plants.

Myxameoba (pl. **myxamoebae;** Gr. myxa = slime + amoeba = change) – an amoeboid cell, particularly one of the Myxomycetes.

N

Nuclear cap – the characteristic structure of the motile cells of the Blastocladiales; consists of the ribosomes of the cell, clustered together near the nucleus and surrounded by what appears to be an extension of the nuclear envelope.

Nutriocyte (L. nutrio = to nourish +

Gr. kystis = blader) – the inflated portion of the ascogonium of *Azoosphaera.* which eventually develops into a sporocyst.

O

Obligate parasite (L. obligare = to bind + Gr. parasitos = table mate) – an organism that can obtain food only from living protoplasm; obligate parasites cannot be grown in culture on nonliving media.

Obligate saprobe (L. obligare = to bind; + Gr. sapros = rotten + bios = life) – an organism that must obtain its food from dead organic mater, and , and is incapable of infecting another living organism.

Ocellus (pl. **ocelli;** L. occulus = eye) – an eyespot functioning as a lens and concentrating the light rays on a photosensitive spot.

Oidiophore (Gr. oidion = small egg + phoreus = ˙ bearer) – a specialized hypha that bears oidia.

Oidium (pl. **oidia;** Gr. oidion = small egg) – a thinwalled, free, hyphal cell derived from the fragmentation of a somatic hypha into its component cells or from an oidiophore; it behaves as a spore or as a spermatium: normally used in reference to such cells produced by certain Basidiomycetes.

Oidization (Gr. oidion = small egg) – the union of an oidium with a somatic hypha. resulting in the dikaryotization of the latter.

Oogamous (Gr. oon = egg + gamos = marriage, union) – refers to a type of fertilization in which two heterogametangia come in contact, and the contents of one flow into the other through a pore or tube.

Oogonium (pl. **oogonia;** Gr. oon = egg + gennao = I give birth) – a female gametangium containing one or more eggs.

Ooplast (Gr. oon = egg + plastes = molder) – a membrane bounded cellular incision in the oospore of the Saprolegniaceae.

Oosphere (Gr. oon = egg + sphaira = sphere) - a large, naked, nonmotile. female gamete.

Oospore (Gr. oon = egg + spora = seed, spore) - a thick-walled spore that develops from an oosphere through either fertilization or parthenogenesis.

Operculum (pl. opercula; L. operculum = lid) - a hinged cap on a sporangium or an ascus.

Ostiole (L. astiolum = little door) - a neck-like structure in an ascocarp; lined with periphyses. and terminating in a pore: also the opening of a pycnidium.

P

Paraphyses (sing. **paraphysis;** Gr. para = beside + physis = a being. a growth) - sterile, basally attached structures in a hymenium.

Paraphysoid (paraphysis + Gr. para = a form) - a paraphysis-like sterile threads originating in the generative tissue and often branching to form a network: found mostly in lichens.

Parasexuality (Gr. para = beside + sexuality) - a process in which plasmogamy, karyogamy, and haploidization take place in sequence, but not at specified points in the life cycle of an individual: of significance in heterokaryotic individuals that derive some of the benefits of sexuality from a parasexual cycle.

Parasite (Gr. parasitos = eating beside another : from para = beside + sitos = wheat, food) - an organism that lives at the expense of another, usually invading it and causing disease.

Parthenogenesis (Gr. parthenos = virgin + genesis = birth) - the development of the normal product of sexual reproduction from the female gamete alone.

Pellicle (L. pellis = skin. dimin. form) - a skin-like aggregation of bacteria or yeasts on the surface of liquid media: any surface, skin like growth.

Penicillus (pl. **penicilli;** L. penicillum = small brush) - the conidiophore of the genus penicillii.

Perfect stage - the sexual stage of a fungus.

Peridiole (Gr. peridion = small leather pouch + L. olum = dimin suffix) - the glebal chamber of the Nidulariales. which has a hard. waxy wall of its own: contains the basidiospores but acts as a propagating unit as a whole.

Peridium (pl. **peridia;** Gr. peridion = small leather pouch) - the outside covering or wall of a fructification.

Periphyses (sing. **periphysis;** Gr. peri = around + physis = a being. a growth) - short, hair-like growths in the form of a fringe lining the inside of an ostiole or of a pore in a stroma.

Periphysoids (Gr. peri = around + physis = a growth) - lateral periphyses.

Periplasm (Gr. peri = around + plasma = a molded structure) - a layer of protoplasm surrounding the oosphere of certain Oomycetes.

Perithecium (pl. **perithecia;** Gr. peri = around + theke = a case) - a closed ascocarp with a pore at the top, a true ostiole, and a wall of its own.

Petri dish (named after R. J. Petri. a German scientist) - a glass container consisting of a circular. flat dish with vertical sides, and a similar but slightly larger cover that fits over it; standard equipment for the growth of microorganisms in pure culture.

Phallotoxins - cyclopeptides present in certain mushrooms: the principle phallotoxin is phalloidin.

Phaneroplasmodium (pl. **phaneroplasmodia;** Gr phaneros = visible + plasmodium) - a plasmodium consisting of a well-differentiated advancing fan and conspicuous thick strands in which ecto, and endoplasmic regions are well differenfiated and in which the protoplasm is coarsely granular: characteristic of the Physarales.

Phialide (Gr. phialis = phial) – a type of conidiogenous cell that produces blastic conidia in a basipetal fashion without detectably increasing in length.

Phialoconidium (Gr. phialis = phial + conidiun) – a conidium formed from a phialide.

Phragmobasidium (pl. **phragmobasidia;** Gr. phragma = a fence, i.e. a septum + basidium) – a basidium typically divided into four cells by either transverse or vertical septa.

Phycobiont (Gr. phykos = alga + bios + life) – the algal component of a lichen.

Pileus (pl. **pilei;** L. pileus = cap) – upper portion or cap of certain types of ascocarps and basidiocarps.

Planogamete (Gr. planetes = wandere + gametes = husband, sex cell) – a motile gamete.

Planogametic (Gr. planetes = wanderer + gametes = husband, sex cell) – a motile gamete.

Planogametic copulation (Gr. planeters = wanderer + gametes = husband: + L. copulare + to couple) – fusion of naked gametes, one or bóth of which are motile.

Plasmodiocarp (Gr. plasma = a molded object + karpos = fruirt) – a curved or branched. vein-like fruiting structure of some of the Myxomycetes.

Plasmodium (pl. **plasmodia;** Gr. plasma = a moided object) – a naked. multinucleate mass of protoplasm moving and feeding in amoeboid fashion: the somatic phase of the Myxomycetes. of some Protosteliomycetes and of the Plasmodiophoromycetes.

Plasmogamy (Gr. plasma = a molded object + gamos = marriage. union) – the fusion of two protoplasts.

Plectenchyma (Gr. pleko = I weave + enchyma = infusion . i.e. a woven tissue) – the general term employed to designate all types of fungal tissues: the two most common types of tissues are prosenchyma and pseudoparenchyma.

Podetium (pl. **podetia;** Gr. pous, podos = root + idion = diminutive suffix) – an erect, columnar or branched structure originating on a lichen thallus and bearing apothecia.

Polycentric (Gr. poly = much, many + kentron = center) – a thallus radiating from many centers at which reproductive organs (sporangia or resting spores) are formed.

Polyphyletic (Gr. poly = much, many + phylon = stock; race) – of several lines of descent.

Polyplanetic (Gr. poly = much, many + planetes = wanderer) – refers to a species in which there are several swarming periods but only one type of zoospore.

Primary septum – a septum formed in association with a nuclear division: laid down between daughter nuclei.

Primordium (pl. **primordia;** L primordium = beginning) – the beginning stage of any structure.

Probasidium (pl. **probasidia;** Gr. pro. = before + basidium) – that portion of the basidium in which karyogamy takes place.

Promycelium (pl. **promycella;** Gr. pro = before + mycelium) – a germ tube issuing from the teliospore in which meiosis takes place and which bears basidiospores: technically. the metabasidium.

Progametangium (pl. **progametangia;** Gr. pro = before + gametangium) – a cell that gives rise to a gametangium.

Prosenchyma (Gr. pros = toward + enchyma + in fusion , i.e. approaching a tissue) – a type of plectenchyma in which the component hyphae lie parallel to one another and are easily recognized as such.

Prosorus (Pl. **prosori;** Gr. pro = before + soros = heap) – a structure that eventually divides to give rise to a sorus.

Protista (Gr. protiston = the very first) – a kingdom proposed by Haeckel in an attempt to classify organisms showing characteristics of both plants and animals.

Protoperithecium (pl. **protoperithecia;** Gr. protos = first + perithecium) – a perithecial initial that develops into a perithecium after fertilization has occurred.

Protoperithecium (pl. **protoplasmodia;** Gr protos = first + plasmodium) – a microscopic plasmodium, with no differentiated fan-shaped region or strands, that exhibits slow and irregular streaming and gives rise to only a single, minutes fruiting body; typical or the Echinosteliales. but occurring in other Myxomycetes as well.

Pseudocapillitium (pl. **pseudocapillitia;** Gr, pseudo = false + capillitium) – irregular threads, plates, or other structures present among the spores withing the fructifications of some Myxomycetes; resembles capillitium.

Pseudomycelium (pl. **pseudomycelia;** Gr. pseudo = false + mycelium) – a series of cells adhering end to end to form a chain; produced by some yeasts.

Pseudoparaphyses (sing. **pseudoparaphysis;** Gr. pseudo = false + paraphysis) – sterile threads attached both to the roof and to the base of an ascocarp.

Pseudoparenchyma (pl. **pseudoparenchymata;** Gr. pseudo = false + parenchyma = a type of plant tissue) a type of plectenchyma consisting of oval or isodiametric cells; the component hyphae have lost their individuality.

Pseudoperithecium (pl. **pseudoperithecia;** Gr. pseudo = false + perithecium) – an uniloculate ascostroma.

Pseudoplasmodium (pl. **pseudoplasmodia;** Gr. pseudo = false + plasmodium, dq. v.) – a sausage-shaped amoeboid structure consisting of many myxamoebae and behaving as a unit; the result of myxamoebal aggregation in the cellular slime molds (Acrasiomycetes) – also called grex or slug.

Pseudopodetium (**pseudopodetia;** Gr. pseudo = false + podetium) – a vertical lichen structure that develops from somatic thallus tissue and forms an ascocarp primordium at its tip.

Pseudoseptum (pl. **pseudosepta;** Gr. pseudo = falso + L. septum = hedge) – a plug-like partition of cellulin or other substance in a hypha, resembling a septum.

Pseudothecium – contraction of pseudoperithecium.

Pycnidiospore (pycnidium + Gr. spora = seed, spore) – a conidium borne in a pycnidium.

Pycnidium (pl. **pycnidia;** Gr. pyknon = concentrated + idion, dimin. suffix) – an asexual, hollow fruiting body , lined inside with conidiophores.

Pycniospore (Gr. pyknos = concentrated = spora = seed, spore) – the old designation for the spermatium of the rusts, used before the true function of the spermatia was discovered.

Pycnium (pl. **pycnia;** Gr. pyknow = concentrated) – the old designation for the spermogonium of the rusts.

Pycnosclerotium (pl. **pycnosclerotia;** pycnidium + sclerotiun from Gr. skleron = hard) – a more or less hard-walled structure resembling a pycnidium but containing no spores.

Q

Quellkorper (Ger. quellen = to swell + korper + body) – a gelatinous mass of cells inside. just below the apex of the ascocarp of the Coronophorales; functions in producing an opening for the escape of the spores.

R

Reindeer moss – various species of lichens-especially *Cladonia rangifera*-native to the far-north regions of Europe and North America. which serve as important winter food for caribou and reindeer.

Reproduction (L re = prefix for again + producere = to bring foth) – the production of new individuals having all the Characteristics typical of the species.

Resupinate (L. resupinatus = inverted) – lying flat on the substratum with the hymenium on the free surface.

Reticulate (L. reticulum = a small net) – having the form of a net; covered with net-like ridges.

Rhizoid (Gr. rhiza = root + oeides = like) – a short, thin branch of thallus. superficially resembling a root.

Rhizomorph (Gr. rhiza = root + morphe = shape) – A thick strand of somatic hyphae in which the hyphae have lost their individuality, with the whole mass behaving as an organized unit; the structure of the growing tip of the rhizomorph somewhat resembles that of a root tip, hence the mame.

Rhizomycelium(pl. **rhizomycella**; Gr. rhiza = root + mycelium) – an enucleate rhizoidal system extensive enough to resemble mycelium superficially.

Rhizoplast (Gr. rhiza = root + plastid) – term used to describe the system of cytoplasmic microtubules and filaments associated with the kinetosomes and the nucleus of motile cells; also known as the rootlet.

Rohr – a long. tubular cavity within the encysted zoospores of *Plasmodiophora* in which the rod-like stachel lies.

Rumposome (sw. **rumpa** = rump, tail + Gr. soma = body) – a complex structure composed of interconnecting tubules that lies at the posterior end of the zoospore of *Monoblepharella* and some other Chytrids.

S

Saprobe (Gr. sapros = rotten + bios = life) – an organism that utilizes dead organic matter for food.

Sclerotium (pl. **selerotia; Gr.** skleron = hard) – a hard resting body resistant to unfavorable conditions. which may remain dormant for long periods of time and germinate on the return of favorable conditions.

Scolecospore (Gr. skoles = worm + spora + seed, spore) – an elongated, needle or worm-like spore.

Self-compatible (L. compati = to suffer with) – self fertile; refers to a thallus that reproduces sexually by itself.

Self-incompathible (L. in = not + compati = to suffer with) – self-sterile; refers to thallus that cannot reproduce by itself sexually.

Septate (L. septum = hedge) – with more or less regularly occurring cross walls.

Septum (pl. septa; l. septum = hedge, partition) – a cross-wall in a hypha.

Seta (pl. **setae**; L. seta = bristle) – a bristle-like hair.

Sexual reproduction– reproduction involving nuclear fusion and meiosis.

Side body – a structure found in the motile cells of the Blastocladiales; lies just beneath the cell membrane at the posterior end of the cell and consists of a double membrane system with which are associated microbodies and lipid bodies.

Sirenin (Gr. sirein = siren) – a reproductivbe hormone secreted by the female gametes of *Allomyces* that attracts the male gametes.

Skeletal hyphae – thick-walled, typically aseptate, unbranched hyphae found in the basidiocarps of some fungi.

Slime mold – a common term for Acrasiomycetes and Myxomycetes.

Soma (pl. **somate**; Gr. soma = body) – the body of an organism as distinguished from its reproductive organs or reproductive phase.

Somatic (Gr. soma = body) – refers to the body phase in plants. the vegetative phase—structure, or function as distinguished from the reproductive.

Somatogamy (Gr. soma = body + gamos = marriage, union) – fusion of somatic cells during plasmogamy.

Soralium (pl. soralia; Gr. soros = heap) – one of several types of pustules on the thalli of certain lichens.

Soredium (pl. **soredia**; Gr. soros = heap) – a microscopic. noncorticated powdery mass of algal cells enveloped by fungal hyphae and produced in soralia on some lichen thalli; soredia are propagative units.

Sorocarp (Gr. soros = heap + karpos = fruit) – the fructification of the Acrasiomycetes,

Sorus (pl. **sori**; Gr. soros = heap) – a mas of sporangia or spores.

Species (sing. and pl. **species**; L. species = kind) – the unit of classification; a group of closely related individuals resembling one another in certain inherited characteristics; it is designated by a binomial consisting of the generic name and the specific epithet.

Spermatiophore (Gr. spermation = litle seed + phopreus + bearer) – a specialized hypha that produces spermatia.

Spermatium (pl. **spermatia**; Gr. spermation = litt seed) – a nonmotile, uninucleate, spore-like male structure that empties its contents into a receptive female structure during plasmogamy; spermatia are variously regarded as gametes or

gametangia.

Spermatization (Gr. sperma = seed) – plasmogamy by the union of a spermatium with a receptive structure.

Spermogonium (pl. **spermogonia**; Gr. sperma = seed, sperm +gennao + I give birth) – a structure resembling a pycnidium and containing minute, rod-shaped, or oval spore-like bodies that in some cases have proved to be functional spermatia.

Sphaerocyst (Gr. sphaira = sphere + kystis = bladder) – spherical cells present in the trama of the Russulaceae.

Spindle pole body – a small. electron opaque. nucleus associated organelle present in certain nonflagellate fungi.

Spitzenkorper (Germ. spitze = point + korper = body) – a refractive region near the hyphal apex in certain fungi: when viewed with transmission electron microscopy the cytoplasm in this region is typically very granular and often contains many small vesicles.

Sporangiolum (pl. **Sporangiola**; (Gr. spora = seed. spore + angeion = vessel + L. olum, dimin, suffix) – a small sporangium containing few spores.

Sporangiophore Gr. spora = seed, spore + angion = vessel + phoreus = bearer) – a hypha that bears a sporangium.

Sporangiospore (Gr. spora = seed, spore + angion = vessel + spora): a spore borne within a sporangium.

Sporangium (pl. **sporangia**; Gr. spora = seed. spore + angeion = vessel): a sac-like structure, the entire protoplasmic contents of which become converted into an indefinite number of spores.

Spore (Gr. spora = seed, spore) – a minute propagative unit functioning as a seed, but differing from it in that a spore does not contain a preformed embryo.

Sporidium (Gr. spora = seed, spore +

L. idium. dimin. suffix) – term sometimes used to designate the basidiospore of a rust or smut.

Sporocladium (pl. **sporocladia**; Gr. spora = seed. spore + lkados + branch) – a special type of fertile branch of a sporangiophore that bears merosporangia.

Sporodochium (pl. **sporodochia**; Gr. spora = seed, spore + docheion + container): a cushion-shaped stroma covered with conidiophores.

Sporophore (Gr. spora = seed, spore + phoreus = bearer) – any structure that bears spores.

Sporothallus (pl. **sporothalli**; Gr. spora = seed, spore + thallos = shoot, thallus) – a thallus that produces spores. as oposed to a gametothallus.

Squamulose thallus – a scale-like lichen thallus or a foliose thallus with scales.

Stachel – a rod-like structure present in the encysted zoo spore of plasmodiophora.

Sterigma (pl. **sterigmata**; Sterigma = support) – a small hyphal branch or structure, which supports a sporangium. a conidium, or a basidiospore.

stipe (L. stipes = stalk) – the stalk of a stipitate basidiocarp or ascocarp.

Stroma (pl. **stromata**; Gr. stroma = mattress) – a compact somatic structure, much like a mattress, on which or in which fructifications are usually formed.

Stylospore (Gr. stylos = pilar + spora + seed, spore) – an elongated or cane-shaped pycnidiospore of unknown function.

Subiculum (pl. **subicula**; L. dimin, of subex = underlayer) – a loose hyphal mat on or in which fruiting bodies are formed.

Suboperculate (L. sub = under + operculum = small door) – an ascus with a thick apical ring capped by a plug or hinged operculum.

Swarm cell – a flagellated cell; usual-ly applied to the motile cells of the Myxomycetes and the plasmodiophoromycetes.

Symphogenous (Gr. symphyein = to grow together + gignesthai = to be born, i.e. originating from structures that grow together) – the origin of a fruiting body from a number of interweaving hyphae.

Synaptinemal complex – a tripartite structure visible with transm; ssion electron microscopy in meiotic nuclei during prophase I; it binds homologous chromosomes together throughout their length.

Synnema (pl. **synnemata**; Gr. syn = together + nema = yarn) – a group of conidiophores cemented together and forming an elongated spore-bearing structure.

T

Taxonomy (Gr. taxis = order, arrangement + nomos = law) – the science of classification.

Teleutospore – see **Teliospore**.

Teliospore (Gr. telos = end + spora = seed, spore) – a thick- walled resting spore of the rusts and smuts in which karyogamy occurs; it is a part of the basidial apparatus.

Telium (pl. **telia**; Gr telos = end) – a group of binucleate cells that produce teliospores.

Tetrapolar heterothallism – (See also Heterothallism and Bipolar heterothallism.)

Thallic condidium(Gr. thallos = a young shoot, scion) – a conidium formed by the transformation of an entire pre-existing hyphal cell; also termed an **arthroconidium**.

Thallophyte (Gr. thallos = shoot, thallus + phyton = plant) – a plant whose somatic phase is devoid of stems, roots, or leaves, and which propagates by means of spores.

Thallus (pl. **thalli**; Gr. thallos = shoot) – a relatively simple plant body devoid of stems, roots, and leaves; in fungi, the somatic phase.

Trama (pl. **tramae;** L. trama = woof)
– the fungal tissue composing the
pileus or bearing the hymenium or
the Holobasidiomycetidae.

Trichogyne (Gr. thrix = hair + gyne
= woman, female) – the receptive
neck of the ascogonium, which is
often long and hair-like.

Trichospore (Gr. thrix, trichos =
hair + spora = seed, spore) – a
unispored sporangium of certain
Trichomycetes; has one to several
filamentous appendages at its
base.

Trimitic (Gr. tris = trice + mitos =
thread, i.e. hypha) – term used to
describe a basidiocarp composed of
generative, binding, and skeletal
hyphae.

Trophocyst (Gr. trophe = food + kystis = bladder) – an enlarged cell;
the swollen portion of the sporangiophore of *Pilobolus.*

U

Unitunicate (L. unus = one + tunica
= coat, mantle) – an ascus in which
both the inner and outer wall are
more or less rigid and do not
separate during spore ejection.

Universal veil – a thin , veil-like
membrance that covers certain
types of young mushrooms; upon
expansion of the mushroom, the
kuniversal veil tears and its remnants may be seen in the form of
scales on the pileus and in the
form of a volva.

Urediniospore (L. urere + to burn +
Gr. spora = seed spore) – a binucle-
ate, repeating spore of the
Uredinales.

Uredinium (pl. **uredinia;** L. urere =

to burn) – a group of binucleate
cells that give rise to uredinio-
spores.

V

Volva (pl. volvae; L, volva = covering) – a cup at the base of the stipe
of certain mushrooms; a remnant
of the universal veil.

W

Woronin body – an electron dense
spherical body found in the hyphae
of Ascomycetes and many
Deuteromyceter; Woronin bodies
are usually concentrated near the
septa.

Z

Zoosporangium (Gr. zoon = animal +
spora = seed, spore) – a motile,
asexually produced spore.

Zoospore (Gr. zoon = animal + spora
= seed, spore) – a motile, asexually
produced spore.

Zygophore (Gr. zygos = yoke =
phoreus = bearer) – a special hypha
capable of developing into a progametangium in the Zygomycetes.

Zygosporangium (pl. zygosporangia.
Gr. zygos = yoke + sporangium) – a
sporangium containing a zygospore; develops following the fusion
of two gametangia.

Zygospore (Gr. zygos = yoke spora =
seed, spore) – a resting spore that
results from the fusion of two
gametangia in the Zygomycetes.

Zygote (Gr. zygos = yoke) – a diploid
cell resulting from the union of two
haploid cells.

Key to symbols

**Ger = German; Gr = Greek; Icel = Icelandic; I. = Latin; ME = Middle English;
Nl = New latin: Sw = Swedish.**

PLANT PATHOLOGY

A

Abscission – The shedding of leaves or fruits.

Anthracnose – Lesions on stems, leaves and fruits caused by acervulus fungi like *Colletortrichum* and *Gloeosporium*.

Antibiosis – Antagonism (toxicity) between two organisms, particularly microorganisms in soil or between one organism and a metabolic product of another organism, harmful to one of them.

Antibody – Specific protein produced in the blood of mature animals against any macromolecule as a defense mechanism (called antigen) produced by pathogen (or chemical), which is foreign particle to the host.

Antigen – A substance capable of inducing antibody formation. Most macromolecules, proteins, nucleic acids and even polysaccharides act as antigens.

Appresorium (Pl. Appresoria) – An enlarged, usually disc shaped, structure formed at the tip of hypha of some parasitic fungi through which entry is made into the host.

Autographic – An organism capable of utilizing energy from light or from oxidation of inorganic compounds for metabolic synthesis.

B

Benign (Virus) – A mild or advertent type, not producing visible symptoms.

Biological control – Control of a disease by the activity of organisms.

Biotype – A subdivision of races of fungi.

Blight – A non-restricted, necrotic symptom on leaves which is characterized by sudden and serious leaf damage.

Blotch – A necrotic spot having fibrillose margins of visible mycelial strands.

C

Callus – Parenchymatous tissue formed by cambial activity for protection after wounding.

Canker – A necrotic symptom of woody parts. Spreading wounds (usually on bark) surrounded by a raised tumor-like margin.

Chemotherapy – Treatment of disease by chemicals which act internally without damaging the host seriously.

Clone – A group of plants derived from an individual by vegetative propagation.

Chlorosis (pl. Chloreses) – Symptom of chlorophyll deficiency.

Control – The prevention of, or reduction in the development of disease.

D

Damping off – Disease of germinating seeds (pre-emergence damping off) and seedlings (post-emergence damping off) characterized by their toppling due to rotting of stem tissues at soil level.

Dieback – Symptom of disease in which apical shoots die first followed by lower shoots.

Disease – A malfunctioning of physiological processes due to continuous irritation from the presence or absence of some agent.

Disinfect – To kill a pathogen within plant tissues.

Disinfest – To kill a pathogen or organisms on the surface of plants or on within non-living substances.

E

Epidemic – Widespread and destructive development of a disease in a large human population. For plants, the equivalent word is epiphytotic; and epizootic for animal diseases.

Epidemiology – The study of epidemics.

Epinasty – downward curvature of leaf.

Epiphyte – A non-parasitic plant attached to another plant for mechanical support, e.g. orchids, lichens, mosses etc.

Epiphytology – The study of epiphytotics; analogous to epidemiology.

Epiphytotic – Widespread and destructive development of a plant disease in large populations.

Eradication – Complete removal of the disease.

Etiology – The study of the causal agent of disease and its relation to the susceptible plant.

F

Forma specials (P. *formae specials, abb. f. sp)*: – Special form; and group of biotypes of species.

Fructification – Spore production in fungi or the structure in which spores are formed.

Fumigant – A chemical that is used in the vapour phase to kill microorganisms, insects etc.

Fungiside – A chemical that kills or inhibits fungi.

Fungistat – A chemical that reversibly inhibits the growth of a fungus as long as it is in contact with fungus.

G

Gall – A tumor; formed by hyperplastic growth of unorganized cells.

H

Heterokaryosis – The state of a fungus having genetically different nuclei in its cells.

Heterotopy – Hyperplastic symptom in which an organ grows at an unusual place (e.g, the development of ears in the tissues of corn plants).

Heterotrophic – Organism deriving food from other living or dead (parasitic or saprophytic) organisms; are unable to synthesize organic materials from inorganic sources (Cf. autotrophic).

Host – A living organism providing sustenance to a parasite.

Host-indexing – Determination of whether a given plant is carrier of a pathogen.

Hypersensitive plant – A plant that shows extreme resistance against a pathogen by quick death of cells around the point of attack, thus preventing spread.

Hyperparasite – An organism that parasitizes another parasite.

Hyperplasia – Outgrowth due to abnormal increase in size of cells.

Hypertrophy – Outgrowth due to abnormal increase in size of cells.

I

Immune – Free from disease by virtue of inherent properties of the plant.

Infect – To establish a pathogenic relationship with the host plant (Cf. Disinfect).

Infest – To introduce a pathogen into the environment of the host (Cf. Disinfest).

Infection court – The site of initiation of infection.

Infectious – Refers to a pathogen that can transmit from one host to another by an external agent.

Injury – Transient irritation by some causal agent. Different from disease.

Inoculum – Portion of pathogen capable of dissemination and initiation of disease.

L

Lesion – A localized area of injured or diseased tissue (e.g, a spot or Serb or functional change (biochemical lesion).

M

Micro-environment – Any localized environment.

Mosaic – Irregular dark and light areas in leaves: a type of virus disease.

N

Necrosis (Pl. necroses) – Death or disintegration of cell.

P

Parasite – An organism that derives its food from a different species of plant.

Pathogen – Causal agent of a disease.

Psychokinesis – The chain of events, from origination to development of disease.

Pathogenic – Causing or capable of causing disease.

Pathogenicity – The capabilities of the pathogen to cause disease.

Phyllody – Transformation of petals into leafy structures.

Protectant – A substance, usually a fungicide that prevents infection.

Physiological race – One or more biotypes or clones of a pathogenic species producing similar reactions in selected varieties of host plants.

Pustule – A pimple like eruption of host tissue formed by fruiting body such as uredium or telium of a rust.

R

Resistance – The inherent capacity of a plant to prevent, restrict or withstand disease.

Rhizosphere – The zone around roots of a plant under its influence.

Rogue – To remove unwanted plants from a planting of the same species.

Rugose – Rough, crinkled leaves e.g. rugose mosaic.

Russetting – Brownish, rough areas on fruits or tubers due to formation of abundant cork-cells.

S

Scab – rough, crusty lesions formed by excessive cork-cells.

Sign – Visible structures of pathogens appearing on or in diseased tissue.

Symptom – A visible expresion of disease in the host plant.

Syndrome – The pattern of symptoms and signs of a disease.

Suscept (= Host) – A plant suceptible to diseases.

Symptomatology – Study or symptoms of diseases.

Systemic – 1. A disease in which the pathogen is spread throughout the plant body.
2. A pesticide which spreads throughout the plant body.

T

Tolerance – The ability of a host to survive and give satisfactory yield when other varieties of the same species suffer great loss.

V

Veriety – Subdivision of a species that differs from the type species in certain characteristics.

Vector – An agent that causes dissemination of a pathogen.

Vein-banding – A symptom of virus infected leaves in which tissue close to veins are darker-green than laminar tissues.

Vein-clearing – A symptom of virus

infected leaves in which tissues
are lighter-green, clearer than tissues develop green pigmentation.

Virulence – Relative ability to cause
disease; a measure of pathogenicity.

Viruliferous – An insect vector containing virus and capable of inoculating it into a host.

W

Witche's-broom – A symptom of disease due to clustering of shoots.

Y

Yellowing – Green tissues turning
yellow.

BRYOPHYTES

A

Abaxial surface – The side of a lateral organ, *e.g.*, leaf, away from the axis. (Dorsal).

Acrocarpous – Of mosses, bearing the archegonia and hence the sporogonia, at the tip of stem or main branches.

Acropetal – Development of organs successively towards the apex, the oldest at the base, youngest nearest the apex.

Adaxial – The side of a lateral organ nearest to the main axis. (Ventral).

Androcyte – Antherozoid mother cell. The cell which later develops into the antherozoid.

Androgonial cell – Any cell within an autheridium other than androcyte or androcyte mother cell.

Antheridial cell – A cell which divides to form the antheridium.

Antheridium – The male sex organ of the cryptogams.

Antherozoid (Spermatozoid) – Small, motile male gamete with flagella.

Antical – The term applied, in a leafy liverwort, to that part of a stem or leaf that lies in front (or above, if the stem is creeping), hence away from the substratum and nearer to the observer.

Anticlinal – Perpendicular to the surface.

Apogamy – The production of a sporophyte from a gametophyte directly, without syngamy or sexual fusion.

Apophysis – The more or less swollen sterile tissue at the base of capsule, of some mosses, next to the seta and below the spore sac.

Apospory – The production of a gametophyte from a sporophyte directly, without the formation of spores.

Archegoniatae – Plants having archegonia, applied to bryophytes, pteridophytes and gymnosperms.

Archegonium – The female sex organs of bryophytes, pteridophytes and gymnosperms, conaining the egg inside a cellular jacket.

Archesporium – The first cell generation of sporogenous tissue, or the cell of group of cells from which the spoes of a sporangium or sporogonium are ultimately derived.

Atavistic – Reverting to an older type of structure, or recurrence in descendants of a character which had been possessed by an ancestor.

Autoicous – Having the male and female organs on the same plant but in separate branches.

Autotrophic – Self-feeding from inorganic material.

B

Basipetal – An order of development of organs in which the youngest structures are at the base and the oldest at the apex.

Benthic – Pertaining to sea floor.

Benthos – Attached aquatic organisms which live on sea floor.

Blepharoplast – A minute extranuclear granule which gives rise to the motile cilia of the antherozoid.

Bract – A modified leaf.

Branch-trace – Vascular supply from stem to a branch.

C

Cainozoic (Cenozoic) – Geological era lasted approximately from 70 till : million years ago.

Calyptra – A covering developed from the venter of the archegonium in bryophytes and pteridophytes, which surrounds the young sporophyte.

Capsule – The part of the sporogonium containing the spores.

Carboniferous – Geological period lasted approximately from 270 till 220 million years ago.

Cauline – Belonging to a stem.

Cespitose – Growing in tufts.

Chlorophyllose (Chlorophyllous) – Containing chlorophyll.

Cleistocarpous – Applied to a capsule opening by irregular splitting, not by a lid or by some other regular method of dehiscence such as slits or valves.

Cortex – The region of a vascular plant that lies between the epidermis and the stele.

D

Dichotomy – The state of equal division of the growing point to form two equivalent branches.

Dimorphic – Existing under two forms upon the same plant or upon other plants of the same species.

Dimorphism – The state of existing under two distinct forms.

Dioecious – Having unisexual male and female sexual reproductive organs borne on different individuals.

Diploid   generation – The sporophyte.

Diploid – Of nuclei having double the typical number of chromosomes (2n), so that twice the haploid number is present. This is characteristic of the nuclei of the sporophyte generation.

Distal – Situated remote from the point of attachement or the part farthest from the main axis.

Distichous – Having two rows, two-ranked.

Distribution – Geographical spread of a form.

Dolabrate apical cell – An apical cell with two cuttings faces.

Dorsal – In lateral organs—the side away from axis. In dorsiventral organs—the upper surface.

Dorsiventral – An organ having distinct dorsal and ventral surfaces which usually show difference in structure.

E

Environment – The sum total of all the external conditions (temperature, light, water, other organisms) which act upon an organism.

Egg – A female gamete.

Elaters – Gells or part of cells forming hygroscopic spirally twisted structures helping to disperse the spores. (i) In *Equisetum* four clubbed arm-like appendages attached to the spore, (ii) In many Hepaticopsida—spirally thickened sterile cells in the capsule.

Embryo – Young plant developed within the archegonium of cryptogams, or inside the seed in the phanerogams, after sexual or parthenogenetic reproduction from an ovum.

Embryophyta – Plant possessing multicellular embryo. The group includes bryophytes, pteridophytes and spermatophytes.

Endemic – Native to particular restricted region.

Endophyte – A plant that grows inside another plant of different species not parasitically but symbiotically.

Endothecium – The inner layer of a young sporogonium in bryophytes.

Eocene – Geological period, subdivision of Tertiary, lasted approximately from 60 till 45 million years ago.

Evolution – Irreversible hereditary change in the characteristics of organisms occurring in successive generations and producting new types of living things.

Epibasal – Forming the upper part of the embryo.

Epidermis – The outermost layer of cells of a plant body of vascular plants.

Epiphragm – A membrane which closes the opening of the theca in some mosses, *e.g.*, *Polytrichum*.

Epiphyllous – Growing on leaves.

Epiphyte – A plant attached to another plant, not growing parasitically upon it, but merely using it for support.

Exine – The outer layer of the cell wall of spore.

Exotic – Non-native.

F

Fertilization – The union or fusion of two similar or dissimilan gametes to form a diploid zygote.

Foliar – Belonging to a leaf.

Foot – A specialized organ of attachment and temporary nutrition of the embryo sporophyte of bryophytes and pteridophytes which absorbs food from the gametophyte.

Fossil – The remains of organisms, that existed in prehistoric time, or direct evidence of their presence in rocks.

Frond – The foliage leaf of ferns and other cryptogams.

G

Gametangium – Organ in which gametes develop, a sex organ.

Gamete – A unisexual haploid nucleated protoplasmic reproductive cell. It unites with another gamete to form a diploid zygote.

Gametophyte – The haploid or gamete-producing generation of any plant.

Gemma – Organ of vegetative reproduction, consisting of one to many cells, that becomes detached from parent plant and develops into new plant.

Genotype – The hereditary properties of an individual as represented by genetic constitution.

Germ tube – The first tubular outgrowth or process from a germinating, spore.

Guard cells – Specialized paired chlorophyllous cells which bound the stoma and regulate its opening.

H

Habit – General form and aspect of a plant.

Habitat – Place with particular kind of environment, in which an organism lives.

Halophyte – Plant that tolerates very salty condition.

Haploid generation – Gametophyte generation.

Haploid – Having a single set of unpaired or reduced number of chromosomes in each nucleus, characteristic of gametophyte generation.

Haustorium – The absorbing organ of a parasite.

Heteromorphic – When the two phases of the life cycle are morphologically dissimilar.

Heterosporous – Producing two kinds of spores, usually of different size.

Heterospory – The condition of producing two kinds of spores differing in size, i.e., the smaller microspores and the larger megaspores, the former producing the male and the latter the female plants.

Homologous – Of organs having a similar orgin, though varying in form and function.

Homosporous – Producing one kind of spores, which are all alike in appearance and behaviour.

Homospory – The condition of producing one kind of spores only.

Hygrophilous – Pertaining to plants which need a large supply of moisture for their growth.

Hypobasal – Forming the lower

part of the embryo.

I

Infloreseence – Group of antheridia, or archegonia and associated structures.

Innovation – A young offshoot from the stem.

Intercalary – Occurring between apex and base. Used of meristem which is not apical, but situated between regions of permanent tissue.

Intine – The inner coat of spore wall.

Involucre – A protective envelope enclosing the reproductive organs.

J

Jurassic – Geological period, lasted approximately from 170 till 140 million years ago.

Juvenile – Early forms.

L

Lacuna – A space, especially in the midst of tissue.

Lamellae – Any thin sheets or plates of tissue, *e.g.*, the longitudinal plates of green tissue in the leaves of *Polytrichum*.

Lysigenous – Applied to a cavity formed by the disintegration and dissolution of cells.

M

Medulla – The pith, tissue inside the vascular strands.

Meiosis – A double mitosis in which the nucleus divides twice but the chromosome only once, and the chromosome complement of the nucleus is reduced qualitatively and quantitatively from the diploid (2n) number to haplod (n).

Meristem – Localized region of active cell division from which permanent tissue is derived.

Mesozoic – The geological era, which lasted approximately from 190 till 70 million years ago.

Metabolism – The chemical processes occurring within the protoplasm of an organism.

Micron – 1/1000 millimeter.

Miocene – Geological period, subdivision of Tertiary, lasted approximately from 35 till 15 million years ago.

Mitosis – Ordinary nuclear division in which the chromosomes are duplicated.

Monoecious – Having both male and female reproductive organs on the same plant.

Mutant – An organism that differs from its parents in one or more characteristics that are heritable.

Mutation – A sudden and relatively permanent heritable change in an individual gene which results in a new hereditary variation.

Mycorrhiza – A symbiotic association of a fungus with the sub-terranean organ of a higher plant. They may be *endotrophic* when the fungus is present within the cells; or *ectotrophic* when the mycelium forms a mantle that completely inveats the subterranean organ.

N

Neck canal cells – The cells that occupy the neck canal of an archegonium.

O

Obcordate – Inversely heart-shaped, the notch being apical.

Oligocene – Geological period, subdivisionof Tertiary lasted approximately from 40 till 30 million years ago.

Ontogeny – The whole course of development of an individual organism throughout its successive growth stages.

Oogamy (Anisogamy) – Sexual reproduction by gametes differentiated into a large female non-motile egg and a small motile male gamete or the antherozoid.

Oogonium – Female sexual organ of

many algae and fungi containing one or more oospheres or eggs.

Oosphere (Egg) – A naked spherical, non-motile female gamete which is haploid.

Oospore – Fertilized oosphere.

Operculum – The covering or lid of moss-capsule.

P

Palaeozoic – Geological era lasting approximately from 500 till 190 million years ago.

Palea – The ramenta or chaffy scales of many ferns.

Palisade tissue – The elongated, perpendicular chlorenchyma comprising the upper mesophyll of many leaves.

Paraphyllia – Thread-like or minute leaf-like appendages on the stem of mosses and some Hepaticopsida.

Paraphyses – Sterile slender filaments accompanying reproductive organs.

Parasite – An organism living on another organism, from which it obtains food.

Paroicous – In mosses, when the antheridia are below the archegonia on the same stem or branch.

Parthenogenesis – The development of an embryo from an unfertilized egg.

Peat – Accumulated dead plant material that remains incompletely decomposed.

Peltate – Shield-shaped, with the support attached to the lower surface, instead of at the base or margin.

Periclinal – Of cell division, whose walls run parallel to the surface of a plant organ.

Perigonial leaves or bracts – Leaves surrounding the antheridia.

Perichaetial leaves or bracts – Leaves surrounding the archegonia.

Pericycle – The outermost layer of stele in the vascular plants.

Perigynium – Tissue surrounding the archegonia of liverworts.

Perimetes – The distance all round a plane figure.

Peristome – The ring of hygroscopic teeth round the mouth of dehiscent capsule in mosses, concerned in spore dispersal.

Phenotype – The observed characters manifested by an organism, produced by the interaction of genotype with environment. Individuals of the same phenotype appear alike but may not breed alike.

Phloem – A tissue in vascular plants, consisting of sieve tubes and often of companion cells, phloem parenchyma and fibres, for the conduction of food.

Photoperiod – The favourable length of day, or more correctly, relative length of day and night for an organism.

Phylogeny – The ancestral or evolutionary history of the group to which an organism belongs.

Pith – The central portion of a siphonostelic stem.

Plagiotropic – Having the direction of growth oblique or horizontal.

Plankton – Floating organisms of sea or lake.

Planktonic – Of Floating organisms.

Plasmodium – A mass of naked, multinucleate protoplasm.

Pleurocarpous – Bearing the archegonia and hence the sporogonia on a small side branch and not on the tip of the main stem or principal branches (cf. acrocarpous).

Polyploid – Having more than two complete sets of chromosomes in each nucleus.

Prothallus (Prothallium) – The sexual or gametophyte generation in pteridophytes.

Protonema – The early filamentous stage produced on germination of the asexual spore in some bryophytes.

Protoxylem – The first-formed elements of primary xylem.

Proximal – That part of an organ nearest the main axis.

Pseudoperianth – A delicate gametophytic sheath, usually late to develop, that surrounds the young sporogonium in some Hepaticopsida, *e.g., Marchantia.*

Pyrenoid – A small rounded protein body surrounded by a starch sheath found embedded in chloroplasts of various algae, *Anthoceros* and some pteridophytes, *e.g. Selaginella.*

Receptacle – The apex of a reproductive branch on which the reproductive organs are borne.

Retort cells – Special enlarged cortical cells of the young branches of *Sphagnum* having recurved apices with a terminal pore.

Rhizoids – Single or many-celled hair-like structures which attach the gametophytes of bryophytes and pteriodophytes to the substratum, and perform the function of roots.

Rhizome – An elongated underground, horizontal stem.

Rosette – Rose-like in pattern.

S

Saprophyte – An organism obtaining its nutrient food material from dead or decayed organic matter.

Scalariform – Ladder-like. In some plants applied to internal thickenings of xylem vessels or tracheids, in the form of more or less transverse bars, stiggestive of ladder rings.

Schizogenous – Of intercellular spaces which originate by separation of cells.

Sclereid – An irregular highly lignified cell occurring singly or in groups.

Sclerenchyma – Thick-walled fibrous tissue for mechanical support.

Seta – The stalk of sporogonium in liverworts and mosses which connects the foot and capsule.

Sieve plate – The perforated portion of a transverse, oblique, or sometimes lateral wall of sieve tubes.

Sieve tube – Characteristic elements of phloem consisting of thin-walled elongated cells arranged in a longitudinal row and concerned with transport of food material.

Spermatozoid – (Antherozoid)

Spermatocyte – (Androcyte).

Spore – A specialized haploid asexual reproductive body usually consisting of a single cell and capable of direct development independently into a new gametophyte.

Sporeling – A young plant from a germinated spore.

Spore mother cell – Diploid cell giving rise by reduction division to four haploid spores.

Sporogonium – The spore-producing structure of bryophytes that develops after fertilization; diploid generation of these plants.

Sporophyte – The asexual generation of the life-cycle of a plant which has diploid nuclei, and reproduces asexually by spores.

Stegocarpous – With the capsule operculate.

Stele – The central vascular cylinder taken as a whole.

Stoma – Microscopic air pores in the epidermis of land plants bounded usually by guard cells.

Substrate – The surface to which a fixed organism is attached.

Suspensor – A filament of cells forming a temporary organ, in some pteridophytes and most spermatophytes, that forces the growing embryo deep into the tissue of the female gametophyte.

Symbiosis – Association of two dissimilar organisms with joint physiological action to their mutual advantage.

Syngamy – (Fertilization).

Synoicous – In mosses when the antheridia and archegonia are mixed together in the same involucre.

T

Tertiary – Geological period, lasted approximately from 70 till 1 million years ago.

Tetrad – Group of four spores resulting from reduction division of one spore mother cell.

Tetraploid – Having four sets of chromosomes in each nucleus.

Thallose – Having the form of a thallus.

Theca – The capsule of mosses and liverworts.

Triassic – Geological period, lasted approximately from 190 till 170 million years ago.

Trilete – Possessing a triradiate tetrad scar.

Tuber – Swollen end of an underground stem bearing buds in axils of scale-like rudimentary leaves, serving for propagation.

V

Vacuole – A cavity in the cytoplasm of a cell filled with cell sap.

Vegetative reproduction – Asexual reproduction in plants by the detachment of some part of the plant body and its further development into a complete separate plant.

Venter – The expanded basal region of an archegonium containing the egg cell, and when young the ventral canal cell.

Ventral – In lateral organs—the side nearest to the main axis. In dorsiventral organs—the lower surface.

Ventral canal cell – A small cell above the oosphere in the upper part of the venter of an archegonium.

Vestigial organ – Organ, the size, structure and function of which have diminished in the course of evolution, and of which only a trace remains.

W

Whorl – A group or circle of similar plant parts all inserted at the same level.

X

Xerophyte – Plants which can endure conditions of drought, or which live in very dry places.

Xylem – Water conducting tissue of vascular plants.

Z

Zygote – The fertilized egg before it undergoes further differentiation.

TAXONOMY

The following abbreviations will be used to indicate parts of speech:
a.- adjective, n.- noun, v-verb.

A

Abaxial – a. Said of an embryo which is out of the axis of the seed as the result of one-sided development of the albumen, or of the side of a lateral organ away from the axis.

Aberrant – a. Differing from the type of a species, genus, or higher taxon in one or more characters, but not readily as signable to another taxon.

Abortion – n. The arrested development of an organ.

Abortive – a. Imperfectly developed, not fully developed at maturity, as abortive stamens with filaments only.

Abrupt –a. Changing suddenly rather than gradually, as in a pinnately compound leaf without a terminal leaflet.

Absciss-layer – n. A layer of separation especially with reference to the phenomena of defoliation.

Acantha – n Thorn, spine, prickle.

Acarpic – a. Without fruit.

Acarpothropic – a. Not throwing off its fruits.

Acaulescent – a. Stemless or apparently stemless.

Acclimation – n. The process of becoming inured to a climate at first harmful.

Acclimatization – n. See Acclimation.

Accrescent – Enlarging with age, as the budscales of some hickories or sepals of some flowers.

Accumbent – a. Lying against another body.

Accumbent cotyledons – n. Cotyledons with edges lying against the radicle.

Acephalous – a. Headless.

Aceriform – a. Like a maple leaf.

Acerose – a. Needle-shaped; having a sharp rigid point, as the leaf of the pine.

Acetose a. Acetic sour.

Acicular – a. Slenderily needle shaped.

Achene (akene) – n.A small, dry, indehiscent, one -seeded fruit in which the ovary wall is free form the seed.

Achenodium – n. A double achene, as the cremocarp of Umbelliferae.

Achilary – a, Without a lip, as in some orchids.

Achlamydeous – a. without calyx or corolla. as in willows.

Acicula – n. The bristle continuation of the rachilla of a grass; a needle-like spine.

Acies – n. the edge or angle of certain stems.

Acquired character – n. A nonheritable environmental variation.

Acropetal – a. Produced in a succesion toward the apex, as applied to development of organs (the antithesis of basipetal).

Acrophilous – a. Dwelling in the alpine region.

Actinomorphic – a. With radial symmetry.

Aculeate – a. Prickley; beset with prickles or sharp points.

Aculeolate – a. Beset with small prickles.

Acuminate – a. Tapering to a prolonged point; attenuate.

Acute – a. distinctly and sharply pointed, but not drawn out.

Acyclic – a. Said of flowers whose parts are arranged spirally, not in whorls.

Adaxial – a. With side or face next to the axis; ventral.

Adenophorous – a. Gland-bearing.

Adherent – a. Attached or joined, though naturally or normally separate; adnate.

Adhesion n. The union or fusion of unlike parts.

Adhesive disc – n. The disclike tip of some tendrils such as is found on Virginia Creeper.

Adnate – a. With unlike parts congenitally grown together.

Aduncate – a. Hooked.

Adventitious – a. Said of plants recently introduced, or of organs arising from abnormal positions, as buds from a root or roots from the stem or leaf.

Adventive – a, Imperfectly naturalized.

Adynamogyny – n. Loss of function in the female organ of a flower.

Aelophilous – a. Disseminated by wind.

Aestival – a. belonging or peculiar to summer.

Aestivation – n. The arrangement of the parts of a flower in the bud.

Affinity – n. The closeness of relationship between plants as shown by similarity of important organs.

Afoliate – a. Aphyllous, leafless.

Agad – n. A beach plant.

Agamandroecism – n. In composites, the state of having female and neuter flowers in the same individual.

Agamogynaecism – n. In composites, the state of having female and neuter flowers in the same individual.

Agamohermaphroditism – n. A condition in which hermaphrodite and neuter flowers appear on the same plant.

Agamospermy – n. Seed production without fertilization.

Age and area hypothesis – n. The older the species the greater the area occupied.

Agglomerate – a. Crowded into a dense cluster, but not cohering.

Agglutinate – a. Stuck together, as the pollen -masses of asclepiads or orchids.

Aggregate – a. Assembled; collected together.

Aggregate fruit – n. A cluster of ripened ovaries traceable to separate pistils or the same flower and inserted on a common receptacle.

Aggregate species – n. A super species which may be compounded of more than one true species.

Agonisis – n. Certation; competition, as between pollen grains of different genotypes, in the rapidity with which they can grow down the style.

Agrophilous – a. Growing in grain fields.

Agrostography – n. The description of grasses.

Agrostology – n. The study of grasses- their description, identification, classification, distribution and habitat.

Agrotype – n. An agricultural race.

Agynic – a. Said of stamens which are free from the ovary; pistils wanting; destitute of pistils.

Aianthous – a Constantly flowering; having everlasting flowers.

Aigialophilous – a. Beach-dwelling.

Aigicolous – a. Inhabiting a stony strand or beach.

Akene – n. See Achene.

Ala – n. A wing, a lateral petal of a Papilionaceous flower; a membranous expansion of any kind, as in the seed of Bignoniaceae; the outer segment of the corolla lobes in some asclepiads.

Alate – a. Winged.

Albinism – n. The absence of pigmentation in organisms normally pigmented.

Albino – n. Any animal with congenital deficiency of pigment in skin, hair eyes, etc; a plant with colorless chromatiophores, due to the absence of chloroplasts or undeveloped chromoplasts.

Albinotic – a. Affected with albinism.

Albumen – n. Any deposit of nutritive material within the seed coats and not in the embryo.

Albuminous – a. Having albumen. See also Exalbuminous.

Aliferous – a. Having wings .

Allautogamia – n. The state of having two methods of pollination, one usual, and the other facultative.

Alliaceous – a. With the smell or taste of garlic; pertaining to the genus *Allium*.

Allochronic species – pl. n. Species which do not belong to the same time level, as opposed to contemporary, or synchronic.

Allogamous – a. Reproducing by cross-fertilization.

Allogamy – n. The pollination of a flower with pollen form another flower. See also Geitonogamy; Xenogamy.

Allogenous flora – n. Relic plants of an earlier prevailling flora and environment; epibiotic plants.

Allopatric – a. Inhabiting distinct separate areas.

Allotropous flower – n.a flower so shaped that its nectar is easily available to insects.

Alluring glands – pl.n.Glands in the pitchers of pitcher plants which tempt insects down the tube.

Alpestrine – a.Pertaining to the Alps or high mountains.

Alpine – a. Pertaining to the Alps or to the Arctic zone of a mountain; above timberline.

Alsad – n.A grove plant.

Alternate – a. Any arrangement of leaves or other parts not opposite or whorled; placed singly at different heights on the axis or stem.

Alveolate – a. With pits or depressions suggesting honey-comb.

Alveolation – n. A honey -combed condition.

Amathicolous – a. Growing on sandy plains.

Amathophilous – a. Growing in sandy plains or in sandy hills.

Ambiparous – a. Producing two kinds, as a bud which contains both flowers and leaves.

Ament – n. A catkin, a spike of flowers unsually bracteate,pendulous, and deciduous.

Amentiferous – a. Bearing aments.

Amentum – n. catkin.

Amethystine – a. Violet-colored.

Amixia – n. Cross-sterility.

Ammochthad – n. Asand-bank plant.

Ammophilus – a. Sand-loving.

Amphibious – a. Capable qf living on land or in water.

Amphicarpous – a. Producing two kinds of fruit.

Amphichromy – n. The ' abnormal production of two different colors of flowers on the same stem.

Amphigean – a. Native around the world.

Amphimixis – n. Cross-fertilization.

Amphitropous – a. Turned both ways applied to an ovule with hilum intermediate between the micropyle and chalaza.

Amphora – n. The lower part of a pyxis, as in henbane.

Amplectant – a. Embracing, clasping by the base.

Amplexicaul – a. clasping or embracing the stem, as a leaf.

Ampliate – a. enlarged.

Ampulla – n. The flasks found on aquatics such as *Utricularia*.

Anadromous – a. Said of ferns in which the first set of nerves in each segment of the frond is given off on the upper side of the midrib toward the apex, as in *Aspidium* and *Asplenium*.

Anametadromous - a. Said of ferns in which the weaker pinnules are anadromous and the stronger are catadromous.

Anastomosing - a. Netted; interveined; said of leaves marked by cross-veins forming a network; interlacing.

Anastomosis - n. The union of one vein with another, the connection forming a reticulation.

Anatropous - a. The ovule reversed with micropyle close to the side of the hilum and the chalaza at the opposite end.

Ancipital - a. Two-edged.

Ancophilous - a. Loving mountain glens or valleys.

Androecium - n. The stamens of a flower (a collective term).

Androgynous - a. Hermaphroditic; having both male and female flowers in the same inflorescence. Occasionally used with meaning of monoecious.

Androphore - n. A supporrt or column on which the stamens are raised.

Anemochore - n. An organism that is disseminated by the wind.

Anemochoroyus - a. Distributed by wind.

Anemogamous - a. Wind-pollinated.

Anemophilous - a. Said of flowers pollinated by wind.

Anemotropism - n. The tropic response of organisms to wind and air currents.

Angiosperm - n. A plant with seeds enclosed in an ovary or pericarp.

Angiospermous - a. Having the seeds borne within a pericarp.

Annotinous - a. A year old, or in yearly growths.

Annual - a. Of one year's duration; completing its life cycle in one year.

Annular - a. Said of any organs disposed in a circle.

Annulus - a. In ferns, the elastic

organ which partially invests the theca, and at maturity bursts it; in Equisetaceae. The imperfectly developed foliar sheath below the fruit spike; the fleshy rim of the corolla in Asclepiads, as the genus *Sta-pelia.*

Anomalous - a. Not equal; unlike its allies in certain points; contrary to rule; unusual; out of the ordinary.

Anterior - a. Front ; on the front side: away form the axis; toward the subtending bract.

Anthecology - n. The study of the flower and its environment.

Anthela - n. The panicle of *Juncus*, in which the lateral axes exceed the main axis.

Anthelate - a. With elongate flower bearing branches, as in some *Junci*.

Anthemia - n. See Anthemy.

Anthemy - n. A flower-cluster of any kind.

Anther - n. That portion of the stamen which bears the pollen.

Antheridium - n. In Cryptogams, an organ or receptacle in which male sex cells are produced.

Antheriferous - a. Anther – bearing.

Antheroid - a. Anther – like.

Antherozoid - n. A male motile cell provided with cilia and produced in an antheridium, a sperm cell.

Anthesis - n. The act of flowering; strictly, the time of expansion of a flower, but often used to designate the flowering period.

Anthesmotaxis - n. The arrangement of the different parts of a flower.

Anthopocarpous - a. Said of fruits with accessories, sometimes termed pseudocarps as the strawberry and pineapple.

Anthracine - a. Coal-black.

Anthropochorous - a. Said of plants which follow man, or cultivation.

Antrorse - a. Directed upward or forward; opposite of retrorse.

Apetalous - a. Without petals, or with a single perianth.

Aphaptotropism - n. The state of not being influenced by touching stems or other surfaces.

Apheliotropism – n. The act of turning away form the sun; negative phototropism.

Aphercotropism – n. The act of turing away from an obstruction.

Aphototropism – n. The act of turning away from light.

Aphyllous – a. Without leaves.

Apical – a. Pertaining to the apex or tip.

Apicula – a. A short, sharp, but not stiff point.

Apiculate – a. Having a minute pointed tip.

Apiculation – n. A short, sharp, but not stiff point, in which a leaf, petal, or other organ may end.

Apocarpous – a. With carpels separate, not united. See also syncarpous.

Apocarpy – n. The condition of having the carpels separate.

Apogamous – a. Developed without fertilization, parthenogenetic.

Apomixy – n. The phenomenon of limited or not cross-fertilization. See also panmixy.

Apophysis – n. An enlargement or swelling of the surface of an organ; the part of a cone scale that is exposed when the cone is closed.

Apendiculate – a. Furnished with an appendage.

Aplanate – a. Flattened.

Appressed – a. Lying flat against an organ.

Approximate – a. Drawn close together but not united.

Apterous – a. Wingless.

Apyrene – a. Said of fruit which is seedless.

Aquatic – a. Living in water.

Aquila – n. Eagle.

Arachnoid – a. Cobwebby; composed of soft, slender entangled hairs; spider – like.

Araneose – a. Like a spider – web.

Arboreous – a. Treelike or pertaining to trees.

Arborescent – a. Attaining the size or character of a tree; treelike.

Arbuscula – n. A small shrub with the aspect of a tree.

Archegonium – n. The organ or receptacle in which the female sex cells are produced in the higher Cryptogams and some Gymnosperms.

Arctic – alpine – a. Used for plants of arctic and alpine distribution but found only south of the Arctic zone.

Arcuate – a. Moderately curved; bent like a bow; descriptive of leaf venation of *Cornus, Ceanothus*, etc.

Arenaceous – a. Of or pertaining to sand: sandy; growing in sand.

Arenicolus – a. Growing in sand or sandy places.

Areolate – a. Marked with areoles, divided into distinct spaces; reticulate.

Areole – n. A space marked out on a surface.

Argenteoguttate – a. With silvery spots.

Argillaceous – a. Clayey, pertaining to clay, or clay-colored.

Argillaceous – a. With silver colored nerves or veins.

Aril – n. An appendage or an outer covering of a seed growing out from the hilum or funiculus; sometimes it appears as pulpy covering.

Arillate – a. With an aril.

Aristate – a. Awned; provided with a bristle at the end, rarely on the back or edge.

Armed – a. Provided with any kind of strong and sharp defense, as of thorns, spines, prickles , barbs, etc.

Aromatic – a. Fragrant, spicy, pungent.

Article – n. A segment of a constricted pod or fruit, as in *Desmodium.*

Articulate – a. Jointed; provided with nodes or joints, or places where separation may naturally take place.

Arundinaceous – a. Reedlike, haveing a culm like tall grasses.

Ascending – a. Rising up; produced

somewhat obliquely or directly upward.

Asepalous – a. Without sepals.

Asexual – a. Sexless; without sex.

Asperous–a. Rough or harsh to touch.

Assumentum (pl. assumenta) – n. The valve of a siliqua.

Assurgent – a. Ascending, rising.

Astigmatic – a. Wind - pollinated plants which do not possess stigmas, such as Gymnosperms.

Asyngamic – a. Unable to cross by reason of differences in time of flowering.

Atavism – n. Ancestral resemblance, reversion to a more primitive type.

Atavistic form – n. A reversion to the primitive form.

Atratous – a. Turning black; blackened, as in some species of *Carex*, the apex of the glumes being darkened.

Attenuate – a. Long tapering, acuminate.

Aurantiaceous – a. Orange-colored; like an orange.

Auricle – n. An ear; applied to earlike lobes at base of leaf blades and to small lobes at the summit of sheath in many species of Gramineae.

Auriculate – a. With earlike appendages.

Austral – a. Southern; occasionally applied to plants which are native to warmer countries, even if not from the southern Hemisphere.

Autocarp – n. A fruit obtained as a result of self-fertilization.

Autogamous – a. Self-fertilized.

Autogamy – a. Self-fertilized.

Autogamy – n. The fertilization of a flower by its own pollen as in an autophilous flower.

Autoorthotropism – n. The tendency of an organ to grow in a straight line forward.

Autophilous – a. Self-pollinated.

Autophytic – a. Said of a plant able to produce its own food through the presence of chlorophyll.

Autumnal – a. Of or pertaining to autumn; flowering in autumn; serotinal.

Auxiliary – a. Helping.

Awl-shaped – a. Narrow and sharp-pointed; gradually tapering from base to slender or stiff point.

Awn – n. A bristle-like appendage, especially on the glumes of grasses.

Axil – n. The upper angle formed between the axis and any organ that arises from it.

Axile – a. In the axis, said ordinarily of the placentae in the ovary.

Axillary – a. Situated in the axil.

Axis – n. The main or central line of development of any plant or organ; the main stem.

Azure – a. Sky blue.

B

Baccate – a. Berry-like; pulpy or fleshy.

Badious – a. Dark reddish-brown, chestnut-brown.

Balausta – n. The fruit of pomegranate with firm rind, burried within, crowned with the lobes of an adnate calyx.

Balsamiferous – a. Balsam-bear-ing.

Banner – n. The topmost petal in the corolla of a member of the pea family; standard; vexillum.

Barbed – a. With rigid points or short bristles, usually reflexed like the barb or a fishhook.

Barbellate – a. Finely barbed.

Barbulate – a, Finely bearded.

Barotropism – n. The response of an organism to changes in barometric pressure.

Barrier – n. Any obstacle that limits the distribution of a species; any condition that reduces or prevents crossbreeding.

Basifixed a. Attached or fixed by the base, as an ovule that is affixed to its support by its bottom rather than by its side.

Basinerved – a. Veined from the base.

Basipetal – a. Growing in the direction of the base (the antithesis of a Cropetal).

Basonym – n. The specific or sub-specific epithet which has priority and is retained when transferred to a new position.

Bast – n. Phloem; fibrous tissues serving for mechanical support.

Bay – a. Reddish-brown.

Beak – n. A long, prominent, and substantial projection; allied particularly to a prolongation or a fruit or carpel.

Beaked – a. Ending in a firm, prolonged, slender tip.

Beard – n. A long awn, or bristlelike hair.

Bearded – a. Bearing or furnished with long or stiff hairs.

Bellying – a. Swelling on one side, as in the corolla of many Labiatae.

Berry – n. Any simple fruit having a pulpy or fleshy pericarp, as the grape, gooseberry, tomato, or banana.

Betaceous – a. Of the beet; beet like.

Bicarpellary – a. Composed of two carpels.

Bicolored – a. Two-colored.

Bicruris – a . Two-legged, as the pollen masses of asclepiads.

Bicuspidate – a. Having two sharp points.

Bidentate. – a. Having two teeth.

Biennial – a. Of two seasons'duration from seed to maturity and death.

Biferous – a. Producing two crops of fruit in one season.

Bifid – a. Forked.

Bifurcate – a. Forked or two pronged.

Bijugous – a. Yoked, two together.

Bilabiate – a. Two-lipped.

Bilateral – a. Arranged on opposite sides.

Bilobate – a. With two lobes.

Bilocular – a. Two-celled, with two compartments.

Binomial – a. The generic and specific name of an organism.

Biological races or species-pl. n. Races or species which differ only in their physiological behavior, being morphologically identical.

Biosystematics – n. Taxonomic studies involving cytology and genetics.

Biotype – n. A group of individuals all of one genotype.

Bipinnate – a. A condition in which both primary and secondary divisions of a leaf are pinnate.

Bisexual – a. Having both sexes on the same individual; a hermaphrodite.

Bivalvular – a. With two valves.

Bladdery – a. Inflated; empty with thin walls like the bladder of an animal.

Blade – n. Lamina; the expanded portion of a leaf or petal.

Blastochore – n. A plant distributed by offshoots.

Bloom – n. The white, waxy. or pruinose covering on many fruits, leaves, and stems.

Blossom – n. A flower, especially of fruit trees.

Bole – n. The main trunk of a tree.

Bolochore – n. A plant distributed by propulsion.

Boreal – a. Northern.

Boss – n. A knoblike or rounded protruberance.

Bossed – a. With a rounded surface having a projection in its center.

Brachiate – a, Spreading with branches suggesting arms.

Bract – n. A modified leaf subtending a flower or belonging to an inflorescence.

Bracteate – a. With bracts.

Bracteody – n. The replacement of the floral whorls by bracts.

Bracteolate – a. With small bracts or bractlets.

Bracteole – n. A bractlet, or small bract.

Bracteose – a. Havifng conspicuous or numerous bracts.

Bractlet – n. Bract borne on secondary axis, as on the peducle or even on a petiole.

Bradycarpic – a. Fruiting after the winter, in the second season after flowering.

Bradyspore – n. A plant which disperses its seed slowly.

Branch – n. A lateral division of the stem, or axis of growth.

Branchlet – n. The ultimate divisions of a branch.

Bristle – n. A stiff hair.

Bristly – a. Bearing stiff , strong hairs.

Brotochore – n. A plant dispersed by man.

Brunescent – a. Brownish; becoming brown.

Brusque variation – n. A sudden, heritable deviation from type; mutation.

Bud – n. An embryonic axis with its appendages.

Bulb – n. A modified bud, usually underground;   imbricated – with-complete enveloping coats, as in the onion.

Bulbiferous – a. Bulb-bearing.

Bulbil – n. A bulb arising from the mother bulb.

Bulblet – n. A little bulb produced in the leaf axils, inflorescence, or other unususal places.

Bulbose – a. Having bulbs or the structure of a bulb.

Bulbous – a. Having the character of a bulb.

Bullate – a Blistered or puckered on the surface, as the leaf of a savoy cabbage.

Bulliform – a. Applied to large thin-walled epidermal cells of most Gramineae and Cyperaceae.

Bumble - bee flowers – pl. n. See Humble-bee flowers.

Bur, burr – n. Any rough or prickly envelope, as of a pericarp, a persistent calyx, or an involucre; any plant which bears burs.

Bursicle – n. A pouchlike receptacle.

Bursicule, burshicula – n. The pouchlike expansion of the stigma

into which the caudicle of some orchids is inserted.

Bush – n. A low shrub, branching from the ground.

C

Caducous – a. Falling off early, or prematurely, as the sepals in some plants.

Caerulescent – a. Bluish; becoming blue.

Caespi tose – a. Growing in tufts.

Calathiform – a. Cup-shaped.

Calcarate – a. Spurred.

Calcareous – a. Of or pertaining to calcium carbonate (limestone), as a calcareous soil.

Calciforma. Shoe-shaped.

Calcicolous – a. Growing best in a soil with a high lime content.

Callosity – n. A hardened thickening.

Callus – a. Having the texture of a callus.

Callus – n. A hard prominence or protuberance; in a cuting or on a servered or injured part, the role of new covering tissue; an extension of the flowering-glume below its point of insertion and grown to the axis or rachilla of the spikelet.

Calycanthemy – n. Petaloidy of the calyx; the formation of colored petal-like structures in place of a normal calyx.

Calyciflorous – a. Having calyx, corolla, and stamens adnate.

Calyculate – a. Calyx-like; bearing a part resembling a calyx; particularly, furnished with bracts against or underneath the calyx resembling a supplementary or outer calyx.

Calyptra – n. A hood or lid; particularly, the hood or cap of the capsule of a moss or lid in the fruit of a *Eucalyptus*.

Calyx – n. The outermost circle of the floral envelopes.

Cambium – n. A layer, usually regarded as one cell thick, of persistent meristematic tissue (referring to vascular and cork cambia); or a persistent meristematic layer which gives rise to secondary wood and secondary phloem (Vascular cambium).

Campanula – n. Small bell.

Campanulate – a. Bell-shaped.

Campestrian – a. Of plains or open country.

Camptodromous – a. Said of venation in which the secondary veins curve towards the margins, but do not form loops.

Campylodromous – a. Said of venation with its primary veins curved in a more of less bowed form towards the leaf apex.

Campylotropous – a. Said of an ovule or seed which is curved in its formation so as to bring the micropyle or true apex down near the hilum.

Canaliculate-a. Longitudinally channeled.

Cancellate – a. Latticed; resembling lattice-work.

Candelabra hairs – n. Stellate hairs in two or more tiers.

Canescence – n. Hoariness, usually with gray pubescence.

Canescent – a. Becoming hoary, usually with a gray pubescence.

Cantharophilous – a. Said of plants that are pollinated by beetles.

Capillary – a. Hairlike; very slender.

Capitate – a. Headed; in heads; formed like a head; aggregated into a very dense or compact cluster.

Capoe – n. A palm thicket (Brazil).

Capreolate – a. Having tendrils.

Capsella – n. A small seed vessel.

Capsular – a. Pertaining to a capsule; formed like a capsule.

Capsule – n. A simple dry fruit, the product of a compound pistil splitting along two or more lines of suture.

Cardinal – a. Of cardinal-red color.

Carina – n. A kell; used either for the two combined lower petals of a Papilionaceous flower or for a salient longitudinal projection on the center of the lower surface of an organ, as on the lemmas of many grasses.

Carneous – a. Flesh-colored.

Carpel – n. A simple pistil; one unit of a compound pistil; in conifers, the cone scale of the female cone.

Carpellate – a. Possessing carpels.

Carphospore – n. A plant whose seeds are disseminated by means of a scaly or chaffy pappus.

Carpography – n. Description of fruits.

Carpophore – n. A portion of receptacle prolonged between the carpels, as in Umbelliferae.

Caruncle – n. An excrescence or appendage at or about the hilum of the seed.

Carunculate – a. With a caruncle.

Caryopsis – n. The grain or fruit of most grasses. with the seed coat grown fast to the pericarp.

Castaneous – a. Chestnut-colored; dark brown.

Catadromous – a. Said of ferns in which the first set of nerves in each segment of the frond is given off on the basal side of the midrib, as in *Osmunda*.

Catkin – n. A flexible , usually pendulous scaly spike bearing apetalous, unisexual flowers; ament.

Caudate – a. With a tail or tail-like appendage.

Caudex. – n. The woody base of a perennial plant.

Caudicle – n. A cartilaginous strap which connects certain pollen masses to the stigma. as in orchids.

Caulescent – a. More or less stemmed or stem-bearing; having an evident stem above ground.

Cauline – a. Pertaining or belonging to the stem.

Caulis(pl. Caules) – n. The stalk or stem of a plant.

Cecidium – n. A gall produced by fungi or insects, in consequence of infection; an abnormal growth.

Cell – n. Any structure containing a cavity. as the cell of an anther or ovary; locule; a unit of plant structure.

Cellular – a. Pertaining to cells.

Cement-disk – n. The retinaculum in orchids.

Cenanthy – n. The retinaculum in orchids.

Cenanthy – Censer-action-n. The action of capsules that, like censers (in-cense-burners), partially open by valves, the seeds being gradualy shaken out by the wind, as in papaver and *Cerastium*.

Centrifugal – a. In inflorescences, bloomibng from top to base.

Centripetal – a. In inflorescences, blooming from the outside inward, or from the base upward.

Centrospore – n. A plant with spurred fruits.

Centrum – n. The central portion,as the large central air space in hollow stems, as in *Equisetum*.

Cerasiferous – a. Cherry-bearing.

Cereal – n. Any grass whose seeds serve as food (from Ceres, the goddess of agriculture).

Ceriferous – a. Wax-bearing; waxy.

Cernuous – a. Drooping; inclining somewhat from the perpendicular; nodding.

Certation – n. Competition, as between pollen grains of different genotypes, in the rapidity with which they can grow down the style.

Cespitose, Caespitose – a. Matted; growing in tufts; in little dense clumps; said of low plants that take tufts or turf of their basal growths.

Chaff-n. Small membranous scales, degenerate bracts in many Compositae; the outer envelopes of cereal grains.

Chalaza – n. That part of the ovule or seed in which the nucellus joins the integuments; the base of the nucellus. always opposite the upper end of the cotyledons.

Chalicard n. A gravel slide plant.

Chalicophilous – a. Dwelling in gravel slides.

Channeled – a. Grooved longitudinally.

Chartaceous – a. Having the texture of writing paper.

Chasmogamous – a. With pollination taking place while the flower is open (the opposite of cleistogamous).

Chasmophilopus – a. Having a fondness for crannies.

Chemotropism – n. Curvature , in response to a chemical stimulus.

Cheradad – n. A set sandbar plant.

Cheradad n. **Cheradophilous** – a. Loving dry habitats; dwelling in dry places.

Chersed – n. A plant of a dry waste.

Chersophilous – a. Dwelling in dry places.

Chionad – n. A snow-plant.

Chionic – a. Of snow fields.

Chiropterophilous – a.Said of plants which are pollinated by bats.

Chledocolous – a. Dwelling in waste places.

Chledophilous – a. Preferring waste places.

Chloranthous – a. Having green, usualy inconspicuous flowers.

Chloranthy – n. The reversion of petals to green leaves.

Chlorophyll – n. The green coloring matter in the cells of autophytic plants.

Chlorophyllous – a. Containing chlorophyll.

Chlorosis – n. A yellowing of the plant due to chlorophyll deficiency.

Chlorotic – a. Lacking chlorophyll.

Choripetalous –a. Polypetalous, with petals separate.

Chorisis – n. Separation of an organ (leaf, petal, stamen, etc)into more than one.

Chorology – n. The geographic study of the distribution of organisms.

Chromosome – n. One of the small bodies, ordinarily definite in number in the cells of a given species and often more or less definite in shape, into which the chromation of the cell nucleus resolves itself previous to the mitotic division of the cell.

Chrysanthine – a. Yellow-flowered.

Chrysophyllous – a. Golden-leaved.

Cicatrice – n. Scar, the mark left by the separation of one from another, as by the leaf from the stem.

Cicatrix – n. See cicatrice.

Ciliate – a. Said of a margin fringed with hairs.

Ciliolate – a Said of a margin fringed with small hairs.

Cilium (pl. cilia) – n. Used generally in the plural to designate marginal hairs.

Cincinnus – n. A one-branched scorpoid cyme.

Cineraceous – a. Somewhat ashy in tint.

Cinereous– a. Ash-colored; light gray.

Circinate – a. Coiled from the top downward; coiled into a ring, or partially so.

Circumscissile – a. Opening or dehiscing along a horizontal line around the fruit or anther, the valve usualy coming off like a lid.

Cirriferous – a. Curl-bearing tendril-bearing.

Cirrus – n. A curl, a tendril.

Citreous – a. Lemon yellow.

Cladode – n. A branch of a single internode simulating a leaf; a cladophyll.

Cladophyll – n. A branch assuming the form and function of a leaf; a cladode.

Class – n. The name of the taxon which is next higher than order.

clastotype – n. A fragment from the original type.

Clathrate – a. Latticed

Clavate – a. Club-shaped; said of a long body thickened toward one end.

Clavellate–da. Diminutive of clavate.

Clavicle – n. A tendril, cirrus.

Claviculate – a. Furnished with tendrils or hooks.

Claviform – a. Club-shaped.

Claw – n. The long narrow spetiole-like base of the petals or sepals in some flowers; the modified auricle of some grass leaves. such as wheat and barley.

Cleft – a. Divided into lobes separated by narrow or acute sinuses which extend more than half way to the midrib.

Cleistogamous – a. Having fertilization occur within the unopened flower.

Cleistogamy – n. The state of being cleistogamous.

Cleistogene – n. A plant which bears cleistogamous flowers.

Cleistogenous – a. Cleistogamous.

Cleistogeny – n. The state of bearing cleistogamous flowers.

Climbing – a. Ascending by using other objects as supports.

Clinandrium – n. The anther bud in orchids, that part of the column in which the anther is concealed.

Clinanthium – n. The receptacle in Compositae.

Cline – n. A series of form changes; a gradient of biotypes along an environmental transition.

Clinium – n. The receptacle of a composite flower.

Clip – n. The seizing mechanism in the flowers of asclepiads.

Clitochore – n. A plant that is distributed by falling or sliding.

Clockwise – a. In the same direction as the hands of a clock, dextrorse.

Clon – n. See Clone

Clone – n. The vegetatively produced progeny of a single in dividual.

Close fertilization – n. Fertilization by its own pollen.

Coalescence – n. The union of like parts or organs.

Coarctate – a. Crowded together.

Cob – n. Rachis of the pistillate corn (maize) spike.

Coccus – n. A berry; in particular, one of the parts of a lobed fruit with one-seeded cells.

Cochlea – n.a Closely coiled legume.

Cochlear – a. Spoon-shaped; said of a form of imbricate aestivation with one piece exterior.

Cochleate – a. Spiral, like a snail shell,

Coelospermous – a. Hollow-seeded; said of the seedlike carpels of Umbelliferae. with ventral face incurved at the top and bottom as in Coriander.

Coenocarpium – n. The collective fruit of an entire inflorescence, as a fig or pineapple.

Coenospecies – n. The total sum of possible combinations of a variable hybrid of two Linneans or ecospecies.

Coerulescent – a. See Caerulescent.

Coherent – a. Two or more similar parts or organs joined.

Cohesion – n. Union of like parts between primary stem and root; the back side of the union of the blade and sheath in grasses.

Collateral – a. Descriptie of accessory buds arranged on either side of a lateral bud.

Colliculose – a. Covered with little round elevations or hillocks.

Colonial – a. Forming colonies; used chiefly for plants with a sexual reproduction.

Column – n. A combination of stamens and styles into a solid central body as in orchids; the lower, twisted portion of an awn of grasses., not always present.

Coma – n.The hairs at the end of some seeds; the tuft at the summit of the inflorescence as in the pineapple; the entire head of a tree.

Comal tuft – n. A tuft of leaves at tip of a branch.

Combination nova (eomb, nov.) – n. New combination, i.e, a hitherto unpublished scientific plant name based on a rearrangement of name already published.

Comb – shaped – a. Pectinate.

Commissure – n. The place of joining or meeting, as the face by which one carpel joins another.

Comose – a. Bearing a tuft or tufts of hair.

Complanate – a. Folded upon itself.

Compound – a. Similar parts aggregated into a common whole.

Compound inflores: :e – n. An inflorescence composeu of secondary ones.

Compound leaf – n. One leaf consisting of two or more blades (leaflets).

Compound pistil – n. Two or more carpels coalesced into one body.

Compressed – a. Flattened; especially, flattened laterally.

Concave – a. Hollow, as the inside of a saucer.

Concolor – a. Of the same color.

Conduplicate – a. Folded together lengthwise with the upper surface within, as in the blades or many grasses.

Cone – n. The fruit of a pine, cycad, or fir-tree with scales forming a strobile; an inflorescence or fruit with overlapping scales.

Conelet – n. The diminutive of cone, applied to a cone of the first year in hard pines.

Conferted – a. Closely packed, or crowded.

Confluent – a. Blended into one, passing by degrees one into another.

Congested – a. Crowded.

Conglomerate – a. Clustered.

Conical – a. Having the form of a cone, as the carrot.

Conifer – n. A cone-bearer.

Coniferous – Producing or bearing cones.

Conjugate – a. Coupled, or in pairs.

Connate – a. United congenitally or subsequently.

Connate-perfoliate – a. United at the base in pairs around the supporting axis.

Connivent – a. Coming together or converging but not organically connected.

Conocarpium – n. An aggregate fruit

consisting of many fruits on a conical receptacle, as the strawberry.

Conoidal – a. Cone-shaped.

Conopodium – n. A conical floral receptacle.

Constipate – a. Crowded, or massed together.

Contorted – a. Twisted or bent; in aestivation, the same as convolute.

Contortuplicate – a. Twisted and plaited or folded; twisted back upon itself.

Contracted – a. Said of the inflorescences that are narrow and dense, the branches short or appressed.

Convergent – a. Applied to veins which run from the base to the apex of a leaf in curved manner.

Convergent evolution – n. The evolution of similar structures produced by different means in different lines of descent.

Convex – a. Having a more or less rounded surface.

Convolute – a. Said of floral envelopes in the bud in which one edge overlaps the next part, as sepal or petal or lobe, while the other edge or margin is over laped by a preceding part; rolled by a preceding part; rolled up from the sides longi-tudinally.

Copious – a. Abundant.

Coppice – n. A small wood which is regularly cut at stated intervals, the new growth arising from the stools.

Coracoid – a. Shaped like a crow's beak.

Cordate – a. Heart-shaped; said of leaves having the petiole at the broader and notched end.

Cordiform – a. Shaped like a heart.

Coriaceous – . Like leather.

Cork – n. Protective tissue replacing the epidermis in older superficial parts of plants; the outer cell contain air, and are elastic and spongy in texture, but impervious to liquids.

Corm – n. A solid bulblike stem, usually subterranean, as the "bulb of *Crocus*, or *Gladiolus*.

Cormatos – a. Producing corms.

Cormel – n. A corm arising from a mother corm.

Corneous – a. Horny, with a horny texture.

Cornet – n. A hollow hornlike growth.

Corniculate – a. Bearing or terminating in a small hornlike protuberance or process.

Corolla – n. The inner floral envelope, composed of separate or connate petals.

Corolline – a. Seated on a corolla; corolla-like; petaloid, or belonging to a corolla.

Corolloid – a. Corolline; corolla-like; petaloid.

Corona – n. Crown, coronet; any appendage or intrusion that stands between the corolla and staments, or on the corolla, as the cup of a daffodil, or that is the outgrowth of the staminal part or circle, as in the milkweed.

Coronate – a. Crowned; with a corona.

Coroniform – a. Shaped like a crown.

Corrugate – a. Wrinkled.

Cortex – n. Rind or bark.

Corymb – n. Short and broad, more or less flat-topped indeterminate flower cluster, the outer flowers opening first.

Corymbiform – a. Shaped like a corymb.

Corymbose – a. Arranged in corymbs.

Coryphad – n. An alpine meadow plant.

Costa – n. A rib, as a midrib.

Costate – n. A Ribbed; with one or more longitudinal ribs or nerves.

Cotyledon – n. Seed leaf; the primary leaf or leaves in the embryo.

Cotype – n. An additional or associate type specimen from which a species is described.

Counterclockwise – a. Sinistrorse, turning the reverse way of clock-hands.

Crampon – n. A hook or adventitious root which acts as a support, as in ivy.

Craspedodromous – a. A condition in which the lateral veins of a leaf run from midrib to margin without dividing.

Crateriform – a. Saucer – or cup-shaped; shallow.

Creatospore – n. A plant with nut fruits.

Creeper – n. A trailing shoot that roots throughout most of its length; sometimes said of a tightly clinging vine.

Creeping – a. Running along on the ground and rooting.

Cremocarp – n. A dry, seedlike fruit composed of two one-seeded carpels invested by an epigynous calyx, separating when ripe into mericarps.

Crena –n. DA rounded tooth or notch.

Crenad – n. A plant growing near a spring.

Crenate – a. Said of a margin with rounded or blunt teeth.

Crenicolious – a. Dwelling in brooks fed by springs.

Crenulate – a. Finely crenate.

Creophagous – a. Carnivorous, as applied to plants.

Crested – a. With elevated and irregular toothed ridge.

Crisp – a. Curled.

Cristate – a. Crested.

Cristulate – a. With small crests.

Cross – pollination – n. The pollination of the stigma by pollen derived from another plant not in the same clone.

Crown – n. Corona; the base of a tufted, herbaceous, perennial grass; the hard ring or zone at the summit of the lemma of some species of *Stipa*; the part of a stem at the surface of the ground; a part of a rhizome with a large bud, used in propagation.

Crosier – n. Any plant structure with a curled end, as the young leaves of most ferns.

Cruciate – a. Cross-shaped, said especially of the flowers of Cruciferae.

Crucifer – n. A plant with four petals and tetradynamous stamens; a member of the family Cruciferae.

Cruciform – a. Cross-shaped.

Crustaceous – a. Of hard and brittle texture.

Crymophilous – a. Dwelling in polar regions.

Cryotropism – n. Movement induced by cold or frost.

Cryptanthous – a. With hidden flowers; cleistogamous; the stamens remaining enclosed in the flower.

Ctenoid – a. Comblike, pectinate.

Cucullate–a. Hooded or hood shaped.

Culm – n. The jointed stem of grasses and sedges.

Cultigen – n. Plant or group known only in cultivation; presumably originating under domestication; contrast with indigen.

Cultivar – n. A variety or race that has originated and persisted under cultivation, but not necessarily referable to a botanical species.

Cultrate – sa. Having the shape of a knife blade.

Cuneate – a. Wedge-shaped' triangular, with the narrow end at the point of attachment, as of leaves or petals.

Cuneifoliate – a. With wedge-shaped leaves.

Cup – n. An involucre, as of an acorn.

Cupule – n. The cup of such fruits as the acorn; an involucre composed of bracts adherent by their base at least.

Cupuliform – a. Cup-or cupuleshaped.

Cuspidate – a. Tipped with a sharp, rigid point.

Cutin – n. A substance present as a thin continuous external layer on the outer wall of the epidermis or a leaf or stem.

Cyamium – n. A kind of follicle resembling a legume.

Cyanochrous – a. Having a blue skin.

Cyanthiform – a. Cup-shaped.

Cyanthum – n. The ultimate inflorescence of *Euphorbia*, consisting of a cuplike involucre bearing the flowers from its base.

Cycle – n. Said of foliar structures arranged in whorls; coiled into a cycle or relating to a cycle.

Cylindrical – a. Elongated with a circular cross section.

Cyme – n. A broad, more or less flat-topped determinate flower cluster, with central flowers blooming first.

Cymose – a. Cyme-like.

Cymule – n. A small cyme.

Cynarrhodium – n. A fruit like that of the rose, fleshy, hollow, and enclosing achenes, as a rose hip.

Cypsela – n. An achene invested by an adnate calyx, as the fruit of Compositae.

D

Dactyliferous – a. Finger-bearing.

Dasycarpous – a. Thick-fruited.

Dasyphyllous – a. Thick – leaved.

D. B. H. – n. Diameter breast-high.

Decamerous – a. In tens.

Decandrous–a. Having ten staments.

Decapetalous – a. Having 10 petals.

Deciduous – a. Not persistent; said of leaves falling in autumn or of floral parts falling after anthesis.

Decompound – a. More than once compound.

Decumbent – a. Reclining or lying on the ground, but with the ends ascending.

Decurrent – a. Said of a leaf or leaf scar, part of which extends in a ridge down the twig below the point of insertion.

Decussate – a. Said of a leaf or leaf scar, part of which extends in a ridge down the twig below the point of insertion.

Decussate – a. In pairs alternately crossing at right angles.

Definite – a. Precise; of a certain number, as of stamens not exceeding twenty; applied to inflorescence, it means cymose.

Definite inflorescence – n. An inflorescence in which the axis terminates in a flower, cymose, determinate.

Deflexed – a. Bent or turned abruptly downward.

Deflorate–a. Past the flowering state.

Defoliation – n. The act of shedding leaves.

Dehiscence – n. The method or process or opening of a seed-pod or an anther.

Dehiscent – a. That which dehisce, as the opening of an anther or fruit along regular lines or suture.

Dehisce – v. To open spontaneoulsy when ripe, as seed capsule.

Deliquescent – a. Dissolving or melting away; said of a stem which loses itself by repeated branching; opposed to excurrent.

Deltoid – a. Triangular, delta-like.

Deme – n. Any specified assemblage of taxonomically closely related individuals.

Dendrocolous – a. Dwelling on trees.

Dendroid – a. Treelike; shaped like a tree.

Dendrology – n. The study of trees their description, classification, identification, and distribution.

Dendrophilous – a. Dwelling on or among trees; tree-loving.

Dentate – a. Said of a margin with sharp teeth pointing outward.

Denticulate – a. Minutely or finely dentate.

Dentoid – a. Tooth-shaped.

Denudate – a. Stripped, made bare, or naked.

Depdauperate – a. Reduced or undeveloped, impoverished, dwarfed.

Depurlation – n. The act of throwing off bud-scales in leafing.

Deplanate –a. Flattened or expanded.

Depressed – a. More or less flattened endwise or from above; pressed down.

Derma (pl. dermata) – n. Surface or an organ, bark, rind, or skin.

Descending – ad. Tending gradually downward; as the branches of some trees or as the roots.

Desmobrya – n. A group of ferns in which the fronds are adherent to the caudex.

Determinate – a. Said of an inflorescence in which the terminal flower blooms slightly in advanve of its nearest associates; limited in number or extent.

Dextrorse – a. Turning to the right, clockwise.

Diadelphous – a. Said of stamens formed in two groups through the union of their filaments.

Diadromous – a. Said of a venation shaped like a fan, as as in *Ginkgo biloba*.

Dialycarpic – a. Having a fruit composed of distinct carpels.

Dialypetalous – a. Plypetalous.

Diandrous – a. Possessing two stament.

Dianthic – a. Fertilized by the pollen from the same plant.

Diaphototropism – n. The act of placing itself at right angles to incident light.

Diaphragm – n. A dividing membrane, or partition, as in the pith of Juglans.

Diaspore – n. A disseminule; any spore, seed fruit, or other portion of a plant capable of producing a new plant.

Diatropism – n. The act of organs placing themselves crosswise to an operating stimulus.

Dicarpellary – a. Composed of two carpels.

Dichasium – n. A cyme with two lateral axes.

Dichlamydeous – a. Having double perianth, calyx and corolla.

Dichogamous – a. Hermaphrodite with one sex maturing earlier than the other , stamens and pistil not synchronizing.

Dichogamy – n. A condition in perfect flowers in which the sexes do not mature simultaneously.

Dichotomous – a. Branching by constantly forking in pairs.

Diclinism – n. The separation of the anther and stigma in space, as dichogamy is in time.

Diclinous – a. Having staminate and pistillate flowers either on the same plant or on different plants.

Dicotyledons – pl. n. A class of angiosperms differentiated by possession of two cotyledons.

dicotyledonous – a. Having two cotyledons.

Dictyodromous – a. With reticulate venation.

Dicymose – a. Doubly cymose.

Dicyclic – a. Having a series of organs arranged in two whorls, as a perianth; biennial.

Didymous – a. Found in pairs, as the fruits of Umbelliferae; divided into two lobes.

Didynamous – a. Said of four-stamend flowers with stamens in pairs, two long,two short, as in some Labiatae.

Diffuse – a: Loosely branching or spreading; of open growth.

Digitate – a. Finger-like; compound with the members arising from one point, as the leaflets of horse chestnut.

Digonous – a. Two-angled, as the stems of some cacti.

Dimerous – a. Flowers with the parts in twos.

Dimidiate – a. Halved, as a condition which half an organ is so much smaller than the other as to seem wanting.

Dimorphic – a. Occurring in two forms.

Dimorphous – a. Occurring in two forms.

Dioecious – a. Unisexual, the male and female elements in different plants.

Diphotic – a. With two surfaces equally lighted.

Diplobiont – n. A plant flowering or fruiting twice each season.

Dipterid – n. Fly flowers, visited chiefly by dipterous flies.

Dipterous – a. Two-winged.

Disarticulate – v. To separate at a joint, as the leaves in autumn.

Disc, disk – n. Development of the torus within the calyx or within the corolla and stamens; the central part of a capitulum in Compositae as opposed to the ray; the base of a pollinium; the expanded base of the style in Umbelliferae; in a bulb, the solid base of the stem around which the scales are arranged.

Disc flowers – n. The tubular flowers in the center of the heads of Compositae, as distinguished from the ray flowers.

Dischisma (pl. dischismate) – n. The fruit of *Platystemon*, which divides into logitudinal carpels, each of which again divides transversely.

Disciform – a. Depressed and circular like a disk.

Discoid – a. With a round thickened lamina and rounded margins.

Disepalous – a. With two sepals.

Disk – n. See disc.

Disk flowers – n. See Disc flowers.

Dispermous – a. Two-seeded.

Dispersal – n. The act of dispersing or scattering.

Dissemination – n. The act of dispersing or scattering such objects as seed, fruit, pollen, etc.

Disseminule – n. See Diaspore.

Dissepiment – n. A partition in an ovary or pericarp caused by the adhesion of the sides of the carpellary leaves.

Dissilient – a. Bursting asunder.

Distant – a. Said of similar parts not closely aggregated; opposed to approximate. remote.

Distichous – a. Conspicuously two-ranked, in two rows.

Distinct – a. Separate; not united with parts in the same series.

Diurnal – a. Occurring in the day time; sometimes used meaning ephemeral.

Divaricate – a. Widely divergent.

Divergent – a. Inclining away from each other.

Divided – a. Characterized by a lobing or segmentation which extends to the base.

Dodecagynous – a. Possessing twelve pistils or distinct carpels.

Dodecamerous – a. In twelve parts, as in a cycle.

Dodecandrous – a. Normally possessing twelve staments, occasionally extended to more than twelve.

Dolabriform – a. Axe-shaped or hatchet-shaped.

Doliform – a. Barrel-shaped.

Domesticated – a. Thriving under cultivation.

Dormant – a. Said of parts which are not in active life.

Dorsal – a. Relating to the back, or attached thereto. the surface turned away from the axis, which in a leaf is the lower surface; opposed to ventral.

Dorsifixed – a. Attached by the back.

Dorsiventral – a. With a distinct upper and lower surface.

Down – n. Soft pubescence; the pappus of such plants as thistles.

Drepanioform – a. Sickle-shaped.

Drepanium – n. A sickle-shaped cyme.

Drimyphilous – a. Salt-loving.

Driodad – n. A plant of a dry thicket.

Dromotropism – n. The irritability of climbing plants which results in their spiral growth.

Drupaceous – a. Resembling a drupe, possessing its character, or producing similar fruit.

Drupe – n. A fleshy one – seeded indehiscent fruit, with seed enclosed in a stony endocarp.

Drupelet – n. One drupe in a fruit made up of aggregate drupes, as in a raspberry.

Duct – n. A tube or canal which carries resin, latex or oil.

Dubious – a. Doubtful, said of plants whose structure of affinities are doubtful.

Duetose – a. Bushy; relating to bushes.

Dumose – a. Full of bushes, or of shrubby aspect.

Dysteleology – n. The supposition that nature (and especially organic evolution) lacks any fore-ordained direction or purpose.

Dystropous – a. Said of an insect whose visit is injurious to the flower.

E

Ebeneous – a. Black as ebony.

Ebracteate – a. Without bracts.

Eburneopus – a. Ivory-white, white more or less tinged with yellow.

Ecad – n. A form arising by adaptation to environment.

Ecblastesis – n. The appearance of buds within a flower; proliferation of an inflorescence.

Echinate – a. Armed with prickles.

Echma (pl. **Echmata**) – n. The hardened hook-shaped funiculus which suppoprts the seed in most Acanthaceae.

Ecological – a. Pertaining to the relation of organisms to their environment.

Ecology – n. The study of organisms in relation to their environment.

Ecospecies – n. A habitat form of a species.

Ecotype – n. A habitat type of plant.

Ectopy – n. The abnormal position of an organ.

Edaphotropism – n. Tropic responeses to the soil.

Edoble – n. A plant whose seeds are scattered by propulsion through turgescence.

Eeltrap hairs – pl. n. Hairs which detain insect visitor, as in *Sarracenia* and *Aristolochia*.

Efflorescence – n. The season of flowering, anthesis.

Effuse – a. Patulous, expanded, loosly spreading.

Eglandular – a. Without glands.

Elater – n. In *Equisetum*, four club-shaped hygroscopic bands attached to the spores, which serve for dispersal.

Ellipsoid – n. An elliptic solid.

Elliptic – n. A flat part or body that is oval and narrowed to rounded ends.

Elliptical – a. Stretched; lengthened.

Emasculation – n. The removal of the anthers from a bud or flower.

Emarginate – a. With shallow notch at the apex.

Embracing – a. Clasping by the base, aplectant.

Embryo – n. The rudimentrary plant formed in the seed.

Emersed – a. Raised above stand out of the water.

Enaulophilous – a. Dwelling in sand draws.

Endemic – a. Indigenous or native to.

Endocarp – n. The inner layer of a pericarp.

Endogenous – a. Growing by internal accessions, as monocotyledonous stems.

Endosperm – n. The albumen of a seed in Angiosperms; in Gymnosperms the prothallium within the embryo sac.

Enneanderous – a. With nine stamens.

Ensiform – a. Sword-shaped, as in the leaf of *Iris*.

Entire – a. Without toothing or division, with even margin.

Entomogamous – a. Insect-pollinated.

Entomogamy – n. The pollination of flowers by insects.

Entomophilous – a. Said of a plant whose flowers are pollinated by insects.

Envelope – n. The surrounding part.

Environment – n. The aggregate of surrounding conditions.

Epharmonic convergence – n. Morphological and anatomical simi-

larity between taxonomically unrelated, or distantly related plants.

Epharmosis – n. Organic adaptation
to a changed environment.

Ephemer – n. A flower that closes
after a short term of expansion.

Ephemeral – a. Persisting for one
day only, as flowers of spider wort.

Epibiotic species – n. Endemic
species that are relics of a past
flora.

Epicalyx – n. A series of bracts close
to and resembling the calyx.

Epicarp – n. The external layer of a
pericarp, exocarp.

Epicormic – a. Said of adventitious
buds which develop on the trunks
of trees, and of branches which
develop on the body of a forest tree
from which surrounding trees have
been removed.

Epiderm – n. The true cellular skin
or covering of a plant below the
cuticle.

Epigeotropism – n. Tropism resulting in growth on the surface of the
soil.

Epigeous – a. Growing upon or above
the ground; on land as opposed to
water; said of seed-lings which
bring their cotyledons above
ground.

Epigynous – a. Gynoeclum occupying
a lower position to sipals, petals
and stamens.

Epipthyte – n. A plant which grows
on other plants, but not parasitically.

Epither – n. A single descriptive word
or single descriptive phrase; in
taxonomy, it is applied to the subdivision of genera, to the second
component of the name of species,
and to the subdivisions of species.

Equal – a. Alike as to length, size, or
number.

Equitant – a. Folded over as if
astride; used for conduplicate in
which the leaves are folded
together lengthwise in two ranks,
as in *Iris*.

Eremophilous – a. Desert-loving.

Ericoida. Erica – like, like the heath.

Eriophorous – a. Wool-bearing, densely cottony.

Erose – a. Irregularly toothed or
eroded as though bitten or gnawed.

Erubescent – a. Blushing red.

Erythrosorus – a. With red sori.

Esculent – a. Suitable for human
food, edible.

Essential – a. The necessary constituent of an existing object.

Essential organs – pl. n. Organs
which are absolutely necessary, as
staments and pistil.

Estipitate – a. Without a stipe.

Estival – a. Of or pertaining to summer. See also Aestival.

Ethnobotany – n. The study of the
relations between man (especially
primitive man)and his surrounding
vegetation.

Etiolated – a. Blanched.

Eutropic – a. Twining in the direction of the sun, clockwise,
dextrorse; said of flowers to which
only a restricted class of specialized insects can gain access.

Evergreen – a. Remaining green
during its dormant season; said of
plants that are green through out
the year.

Exalbuminous – a. Destitute of albumin, used only of seed in which
the embryo occupies the whole
cavity within the testa.

Exannulate – a. Said of ferns which
do not possess an elastic ring
around their sporangia.

Exannulate – a. Rough, with hard
projecting points.

Excentri – a. One-sided, off -center,
abaxial.

Excurrent – a. Running through. to
the apex and beyond, as a mucro; a
stem that remains central, the
other parts being regularly disposed around it.

Exfoliate – v. To come off in scales or
flakes, as the bark of the sycamore.

Exfoliating – a. Coming off in thin
layers.

Exine – n. The outer coat of a pollen grain.

Exocarp – n. The outer layer of a pericarp.

Exogenous – a. Said of growth by the addition of layers on the outside, as with dicotyledons.

Exotic – a. Foreign, not native, from another region.

Explanate – a. Spread out flat.

Explosive Speciation-n. The rapid production, within one locality, of a number of new species from a single species.

Exserted – a. Sticking out; projecting beyond, as stamens from a perianth; not included.

Exsiccate (sing. exsiccatum) – pl. n. Dried plants, usuually in sets for sale or for subscribers.

Exsiccated – a. Dried.

Exstipulate – a. Without stipules.

Extravaginal – a. Beyond or out-side the sheath; said of branches springing from buds, which break through the sheath of the subtending leaf, chiefly in grasses.

Extrorse – a. Facing or opening outward.

Exuviae – pl.n. Cast-off parts, as shed scales.

F

Facultative – a. Having the power to live under different conditions, as a facultative parasite, a plant which is normally saprophytic, but which may exist wholly or in part as a parasite; opposed to obligate.

Falcate – a. Sickle-or scythe-shaped.

Family – n. The taxon next higher than the genus.

Farinaceous – a. Mealy, like meal or pertaining to meal.

Farinose – a. Covered with a meal.

Fasciated – a. Much flattened; an abnormal widening and flattening of the stem as though several stems had coalesced in one plane.

Fascicle – n. A close cluster or bundle of flowers, leaves, stems, or roots.

Fascicled – a. In clusters or bundles.

Fasciculate – a. In close bundles or clusters.

Fastigiate – a. Parallel, clustered, and erect, as the branches of *Populus fastigiata*.

Faucal – a. Pertaining to the throat of a gamopetalous corolla.

Fauces – n. The throat of a gamopetalous corolla.

Faveolate – a. Honey – combed, alveolate.

Favose – a. Honey-combed. as the receptacles of many Compositae.

Feather – veined – a. With secondary veins proceeding from the midrib; penninerved.

Fecundation – n. Fertilization.

Felted – a. Matted with intertwined hairs.

Female – n. The fruiting element in plants, the pistil and its analogues.

Fenestra – n. Lattice, window, an opening through a membrane.

Fenestrate – a. Pierced with holes, as the septum in some Cruciferae.

Feral – a. Wild; not cultivated.

Fertile – a. Said of pollen-bearing stamens and seed-bearing fruits; capable of producing fruit.

Ferruginous – a. Rust-colored.

Fertile flowers – pl.n. Female flowers, those which possess pistils.

Fetid – a. Having a disagreeable odor.

Fiber, fibre – n. A thread, or thread-like structure. a long, slender, thick-walled cell, as in sclerenchyma; the fusiform cells of the inner bark; the ultimate rootlerts.

Fibriform – a. Fiber-shaped.

Fibril, fibrilla – n. A small fiber.

Fibrillose-a. Having numerous woody fibers, as the rind of a coconut.

Fibro – vascular – a. Composed of woody fibers, xylem vessels, and sieve tubes.

Ficoid – a. Figlike.

Fiddle-shaped – a. Panduriform.

Filament – n. The part of a stamen that supports the anther; thread-like structures.

Filamentous – a. Formed of filaments or fibers.

Filical – a. Fernlike, relating to the Filicineae.

Filicoid – a. Fernlike.

Filiferous – a. With filaments or threads.

Filiform – a. Threadlike, long and very slender.

Filipendulous – a.Hanging from a thread.

Filmbria – n. A Fringe.

Fimbriate – a. Fringed, the hairs longer or coarser as compared with ciliate.

Fimbrillate – a. With a minute fringe.

Fimbriolate – a. Very finely fimbriate.

Fimetarious – a. Growing on or among dung.

Fimicolus – a. Growing on manure.

Fissile – a. Tending to split, or easily split.

Fistula–n. A pipe, or hollop cylinder.

Fistular – a. Hollow-cylindrical.

Fistulous – a. Hollow-cylindrical, as the leaf and stem of an onion.

Flabellate – a. Fan-shaped, dilated in a wedge-shape, sometimes plaited.

Flabelliform – a. Fan-haped.

Flaccid–a. Withered and limp, flabby.

Flagelliform – a. Whip-shaped.

Flask – n. The utricle of *Carex*.

Flattening – n. The fasciation of a stem; the production of a cladodium.

Flavescent – a. Yellowish, becoming yellow.

Fleshy – a. Succulent.

Flexible – a. Capable of being bent, but elastic enough to be able to resume its original shape,.

Flexuous – a. Bent alternately in different directions.

Floccose – a. Bearing tufts of woolly hairs.

Flocculent – a. The diminutive of floccose.

Flocculose – a. Bearing small tufts of wooly hairs,

Flora– n. The aggregate of plants of a country or district, or a work which contains the enumeration of them.

Floral –a. Of or pertaining to flowers.

Floral diagram – n. A drawing to show the relative position and number of the constituent parts.

Floral envelope – n. The perianth leaves, calyx, and corolla.

Floral formaula – n. A formula compossed of letters, figures, and symbols arranged to show number, union, sex shape, elevation, etc, of the floral parts.

Floral glume – n. The lower glume of the floret in grasses; the lemma.

Florepleno – a. With full or double flowers.

Florescence – n. Anthesis, the peroid of flowering.

Floret – n. The lemma and palea with included flower (stamens and pistil); a small flower; one of a cluster, as in Compositae.

Floricane – n. The flowering cane usually the second year's development of the primocane, in *Rubus* and other such genera.

Floriferous – a. Flower-bearing.

Flos (pl. flores) – n. A perfect flower with some protecting envelope.

Floscule – n. A little flower, a floret.

Floss – n. The down of certain compositae. as thistle – down.

Floer – n. (See Flos)A modified plant structure concerned with the producation of seeds in the Agniosperms.

Flower bud – n. An unexpanded flower, as distinct from a leafbud or mixed bud.

Flower – glume – n. The lower of the two scales which subtend the flower of grasses, the lemma.

Fluminal – a. Said of a plant which grows in running water.

Fluted – a. Regularly marked by alternating ridges and groovelike depressions.

Fluvial – a. Applied to plants growing in streams.

Foliaceous – a. Leaflike. said particularly of sepals, calyx-lobes, and of bracts that in texture, size, or color look like leaves.

Foliage – n. The leafy covering, especially of trees.

Foliar – a. Pertaining to a leaf.

Foliose – a. Closely clothed with leaves, leafy.

Follicle – n. A single carpellate dry fruit dehiscing along one line of suture.

Follicular – a. Of or pertaining to a follicle.

Foramen (pl. **foramina**) – n. An aperture, especially that in the outer integuments of the ovule; micropyle.

Foraminose – a. Perforatd by holes.

Forcipate – a. Forked liked pincers.

Forest – n. Land covered with trees exclusively, or with under -growth of shrubs or herbs.

Fornicate – a. Provided with arched scalelike appendages in the corolla tube, as in *Myosotis*.

Fornix(pl.fornices) – n. A small arched acale.

Fovea – A depression or pit,as in the upper surface of the leaf-base in Isoetes, which contains the sporanҫium; the seat of the pollinium in orchds.

Foveolate – a. Marked with small pittings.

Free – a. Not joined to anther organ.

Frond – n. The foliage of ferns and some other Cryptogams; the leaves of palms.

Frondose – a. Leafy; frondlike or bearing fronds.

Fructescence – n. The time of maturity of fruit.

Fructiferous – a. Producing or bear.ing fruit.

Fructification – n.The act of fruiting.

Fruit – n. A mature ovary or ovaries with or without closely related parts.

Fruit dots – n. The sori of ferns.

Frustraneous – a. Said of Compositae with disk flowers hermaphrodite, and those of the ray neuter or im-

perfect.

Frutescent – a. Becoming shrubby.

Frutex – n. A woody plant destitute of a trunk.

Frutical – a. Shrubby with a soft, woody stem, such as shrrby species of *Geranium*.

Fruticose – a. Shrubby or shrub like in the sense of being woody.

Fugacious – a. Soon falling or fading; not permanent.

Fulcrum (pl. fulcra)-n. An appendage of leaves, as a prickle, tendril, stipule, etc.

Fuliginous – a. Sooty, or soot-colored.

Fulvous – a. Yellow, tawny.

Funicle – n. A stem or thread which connects the ovule or seed to the placenta; funiculus.

Funiculus – n. A stem or thread which connects the ovule or seed to the placenta; funicle.

Funnel – n. A space below the thick outer coats of the macro-spore in Marsiliaceae, into which the apical papilla projects.

Funnel form – a. With tube gradually widenbing upward and passing insensibly into the limb as in many flowers of *Convolvulus*; infundibuliform.

Furcate – a. Forked with terminal lobes which are like prongs.

Furrowed – a. With longitudinal channels or grooves; sulcate; striate on a large scale.

Fuscous – a. Frayish-brown.

Fusiform – a. Spindle-shaped; narrowed both ways from a swollen middle, as *Dahlia* roots.

G

Galbulus – n. The fruit of the *Cypress*, a modified spherical cone, the apex of each carpellary scale being enlarged and somewhat fleshy.

Galea – n. A petal shaped like a helment placed next to the axis, as in *Aconitum*.

Galcate – a. Hollow and vaulted, as

in many Labiate corollas.

Galericulate – a. Covered, as with a hat.

Galochrous – a. Milk-white.

Gamopetalous – a. With corolla of one piece; petals united.

Gamophyllous – a. With leaves united by their edges.

Gamosepalous – a. With sepals united into one piece.

Geitonogamy – n. Pollination by pollen from another flower on the same plant.

Gelatinous – a. Jelly-like.

Geminate – a. In pairs, binate, twin.

Gemma(pl. **gemmae**) – n. A bud or a body analogous to a bud capable of produciong a new plant.

Gemmiferous – a. Bearing buds.

Gemmiparous – a. Bearing gemmae.

Gene – n. A unit of inheritance, which occupies a fixed place on a chromosome.

Gene – flow – n. The spread of genes which takes place within a group (variety. subspecies, or species)as a result of outcrossing followed by natural crossing within the group.

Generitype – n. The type species of a genus.

Geniculate – a. Abruptly bent so as to resemble the knee-joint, as of awns and the lower nodes of some culms.

Genitalia – pl.n. The stamens and pistils; reproductive organs.

Genospecies – n. A. group,all the members of which are genotypically identical.

Genotype – n. The type of a genus, the species upon which the genus was established.

Gens – n. A tribe in botany.

Genus – n. The smallest natural group containing distinct species; large genera are frequently divided for the sake of convenience into subgenera, but the generic name is applied to all species.

Geocarpy – n. The subterraneous ripening of fruits which have

developed from a flower above ground.

Geodiatropism – n. The tendency to place an organ at right angles to the force of gravity.

Geographic speciation – n. The gradual formation of new species by reason of spatial isolation of different stocks of the original species.

Geonasty – n. The act of curving toward the earth.

Geophilous – al. Earth-loving; said of plants which fruit under-ground.

Geotaxis – n. Orientation of organisms with reference to gravity.

Gibbous – a. Swollen on one side as the glume ln *Sacciolepis*; a pouchlike enlargement of the base of an organ, as of a calyx.

Glabrate – a. Nearly glabrous, or becoming glabrous with maturity or age.

Glabrous – a. Smooth, devoid of pubescence or hair or whatso ever form.

Gladiate – a. Flat, straight, or slightly curved, with acute apex and approximately parallel edges, ensiform, swordlike.

Gland – n. An acorn, or acorn-like fruit;a definite secreting structure on the surface, embedded, or ending a hair; any protruberance of like nature which may not secrete, as the warty swellings at the base of the leaf in the cherry and peach.

Glandular – a. Furnished with glands, or of the nature of glands.

Glandule – n. A viscid gland in orchids and asclepiads, which holds the pollen-masses in their place; the retinaculum.

Glanduliferous – a. Bearing glands.

Glandulose – a. Glandular.

Glans (pl.glandes) – n. A fruit one-seed by abortion, or a few-seeded, dry, inferior, indehiscent pericarp seated within a cupular involucre, as the nut of an acorn as distinguished from the cup.

Glareose – a. Frequenting gravel.

Glaucescent – a. Becoming seagreen; somewhat glaucous.

Glaucophyllous– a. Glaucous leaved.

Glaucous – a. Covered with a "bloom" or a whitish substance that rubs off, as of a plum or cabbage leaf.

Glittering – a. With luster from a polished surface which is not uniform.

Globose – a. Spherical, globular.

Globular – a. Spherical.

Glochid – n. A barbed hair or bristle.

Glochideous – a. Pubescent with barbed bristles.

Glochidiate – a. Pubescent with barbed bristles.

Glome – n. A rounded head of flowers, as *Echinops*.

Glomerate – a. In a dense, compact cluster or clusters.

Glomerule – n. A cluster of capitula in a common involucre.

Glumdaceous – a. With glumes; resembling the glumes of grasses.

Glume – n. The chaffy two-ranked members of the inflorescence of grasses and similar plants; one of the two empty bracts at the base of a grass spikelet.

Glume, empty – n. Glume which subtends a spikelet, and does not include a flower.

Glume, flowerifng – n. the glume in grasses which includes a flower; the lemma.

Glutinous – a. Covered with a sticky exudation.

Gnaurs – pl. n. Burls or knotty excrescences on tree trunks or roots, probably form clusters of adventitious buds; burls.

Gnesiogamy – n. Fertilization between different individuals of the same species.

Gorge – n. The throat of a flower.

Gossypine–a. Cottony, flocculent, like the hairs on the seeds of *Gossypium*.

Gourd – n. A fleshy, one-celled, many-seeded fruit, with parietal placentation, as a melon; a pepo.

Gramineous – a. Relating to grass, or grain-bearing plants.

Graminicolus – a. Grass-inhabiting.

Granular – a. Coverd with very small grains; minutely or finely mealy.

Granulose – a. Granular.

Gray – a. A cold, neutral tint.

Gregarious – a. Growing in groups or masses, but not matted.

Grumose – a. Crumby.

Guttation – n. The act of forming drops.

Gymnospermous – a. Bearing naked seeds.

Gynandrous – a. With the stamens adnate to the pistil, as in orchids.

Gynecandrous – a. With staminate and pistillate flowers in the same spike, the pistillate at the apex; used chiefly in reference to the Cyperaceae.

Gynodioecious – a. Dioecious, with some flowers hermaphrodite, others pistillate only on separate plants.

Gynoecium– n. The pistil or pistils of a flower; the female part of a flower.

Gynomonoecious – a. With pistillate and perfect flowers on one plant.

Gynophore – n. Stipe of an ovary prolonged within the calyx.

Gynospore – n. One of the larger reproductive bodies (female)in the Isoetaceae.

Gynostegium – n. A sheath or covering of the gynoecium of what ever nature as in *Calotropis*.

Gynostemium – n. The compound structure resulting from the union of the stamens and pistil in Orchidaceae.

Gypsophilous –a. Growing on gypsum soils.

H

Habit – n. The general appearance of a plant, whether erect, prostrate, climbing etc.

Habitat – n. The kind of locality in which a plant grows.

Hair – n. An outgrowth of the epidermis consisting of one to several cells.

Halberd-shaped – a. Hastate; sagittate (arrow-shaped). with the lobes turned out.

Halophilous – a. Salt-loving.

Halophyte – n. A plant which grows in saline soil.

Hamate – a. Hooked at the tip.

Hamous – a. Hooked.

Haplocaulus – a. Having a simple-unbranched stem.

Haplochlamydeous – a. Monochlamydeous, having a single perianth.,

Haplostemonous – a. Monochlamydeous, having a single perianth.

Heplostemonous – a. Possessing only one whorl of stamens.

Hastate–a. Halberd-shaped, sagittate, with the basal lobes turned outward.

Haustorium – n. A sucker of ecto parasitic plants.

Head – n. A dense spherical or flat-topped inflorescence of sessile flowers clustered on a common receptacle.

Heart-shaped – a. Cordate; broadly ovate with two rounded lobes at the base.

Hebecarpous – a. Having a fruit covered with downy pubescence.

Helad – n. A marsh plant.

Heliad – n. A heliophyte, or sun-loving plant.

Helicoid – a. Curved or spiraled like a snail-shell.

Heliotropic – a. Turning in response to sunlight.

Heliotropism – n. The act of turning in response to the sun.

Helohylophilous – a. Dwelling in wet forests.

Hemeranthous – a. Day-flowering.

Hemicarp – A half-carpel, a mericarp.

Hepaticous – a. Liver-colored, dark purplish-red.

Herb – n. A plant naturally dying to the ground at the end of the growing season, without persistent stem above ground, and lacking definite woody, firm structure.

Herbaceous – a. Not woody; dying to the ground each year; said also of soft branches before they become woody.

Hercogamy – n. The condition of a bermaphrodite flower in which some structural peculiarity prevents self-fertilization, requiring some other form of pollination.

Hermaphrodite – a. With stamens and pistil in the same flower.

Hesperidium – n. A superior, polycarpellary, syncarpous berry, pulpy within, and externally covered with a tough rind.

Heterandrous – a. With two sets of stamens; said of flowers whose stamens vary in size or length.

Hypanthium – n. An enlargement or development of the torus under the calyx.

Hyperboreal – a. Of the far North.

Hyperdromy – n. A condition in which anadromous and catadromous venation occurs on one side of a fern-frond.

Hyphodromous – a. Said of a condition in which the veins are sunk in the substance of the leaf, and thus are not readily visible.

Hypocarpogenous – a. Having flowers and fruit placed underground.

Hypochil – n. The (often fleshy or otherwise modified) basal portion of the labellum or lip in *Orchidaceae.*

Hypocotyl – n. The axis of an embryo below the cotyledons, but not passing beyond them.

Hypocrateriform – a. Salver-shaped, with a salver-shaped corolla.

Hypogeous – a. Under the earth or soil.

Hypogynium – n. The perianth-like structure subtending the ovary in *Seleria* and some other *Cyperaceae.*

Hypogynous – a. Corolla and calyx present below the pistil or gynoecium.

Hypogyny – n. The condition of possessing hypogynous flowers.

Hyponastic – a. Said of a dorsiventral organ in which the ventral surface grows more actively than the dorsal, as shown in flower expansion; employed by Van Tieghem for anatropous or campylotropous ovules when the curvature is in an upward direction.

Hyponym – n. A name to be rejected for want of an identified type.

Hysteranthous – a. Said of leaves which are produced after the flowers, as in the almond.

I

Icon (pl. icones) – n. A plate, engraving, picture, or other form of image; usually used in the plural in botany as in *icones plantarum.*

Icosahedral – a. Having twenty sides, as the pollen grains of *Tragopogon.*

Icosandrous – a. With twenty or more stamens.

Imbricated – a. Overlapping as shingles on a roof; in aestivation, said of a calyx or corolla in which one piece must be wholly internal and one wholly external, or overlapping at the edge only.

Immersed – a. Entirely under water; embedded in the substance of a leaf.

Immobile – a. Immovable, as many anthers (opposed to versatile).

Imparipinnate – a. Unequally pinnate, odd pinnate, with a single terminal leaflet.

Imperfect – a. Said of a flower with one of the sexes wanting.

Implexed – a. Entangled, interlaced.

Implicated – a. Entangled, interlaced.

Implicated – a. Bent inward, hollowed, or furrowed as if by pressure.

Incanescent – a. Becoming gray, canescent.

Incarnate – a. Flesh-colored.

Incised – a. Cut sharply and irregularly, more or less deeply.

Included – a. Not protruding beyond the surrounding organ; not exserted.

Incompatibility – n. The inability of pollen to effect fertilization.

Incompletae – pl. n. Usually synonymous with *Monochlamydeae,* but variously circumscribed by different authors.

Incomplete – a. Said of flowers with one or more of the four floral organs wanting.

Incrassate – a. Made thick or stout, as the leaves of house-leek.

Incubous – a. With the leaves inserted obliquely so that the base of each is covered by the upper portion of the next lower, as in *Bazzania.*

Incumbent – a. Resting or leaning upon.

Incumbent auther – n. An anther attached to the inner face of its filament.

Incumbent cotyledon – n. A cotyledon with its back lying againt the radical.

Indefinite – a. Uncertain or not positive in character too many for easy enumeration, as abundance of stamens; in an inflorescence, indeterminate.

Indefinite growth – n. Continuous growth until cold weather kills back the immature terminal bud as well as the outer end of the twig as in certain species of *Rubus.*

Indefinite inflorescence – n. An inflorescence that is indeterminate or centrifugal, acropetal according to some authors, one that blooms progressively from outside inward or from the bottom upward.

Indehiscent – a. Not opening by valves or along regular lines.

Indeterminate – a. Said of an inflorescence in which the flowers open progressively from the base upward or from the outside inward.

Indigen – n. A native, not introduced.

Indigenous – a. Native to the

country, not introduced.

Indigo – n. A deep blackish-blue obtained from various species of *Indigofera*.

Indument – n. Any hairy covering or pubescence.

Induplicate – a. With the margins bent inwards, and the external face of these edges applied to each other without twisting.

Indurated – a. Hardened.

Indusial – a. Pertaining to indusia.

Indusial flaps – n. A false indusium as in *Woodwardia*.

Indusiate – a. Possessing an indusium.

Indusium (pl. indusia) – n. The epidermal outgrowth covering the sori or "fruit-dots" in ferns; a ring of collecting hairs below the stigma.

Induviae – pl. n. Persistent portions of the perianth or leaves which wither but do not fall off; scale-leaves.

Induviate – a. Clothed with withered remnants.

Inequilateral – a. Asymmetrical, unsymmetrical.

Inermous – a. Without spines or prickles, unarmed.

Inferals – n. A division of gamopetalous dicotyledons proposed for *Rubiaceae, Compositae, Campanulaceae, etc.*

Inferior – a. Said of one organ when below another, as an inferior ovary with an adnate or superior calyx.

Inferior ovary – n. Ovary with the perianth located on top.

Inflated – a. Bladder-like, swollen, puffed up.

Inflexed – a. Turned in at the margins.

Inflorescence – n. Mode of flower-bearing; technically less correct but much more common in the sense of a flower-cluster.

Inflorescence, definite – n. A condition in which each axis in turn is terminated with a flower, as in a cyme.

Inflorescence, indefinite – n. A condition in which the floral axis is capable of continuous extension, as in a raceme.

Infra-axillary – a. Below the axil, sub-axillary.

Infundibuliform – a. Funnel-shaped.

Infundibular – a. Funnel-shaped.

Innate – a. Borne at the apex of the supporting part, as some anthers.

Innovation – n. The basal shoot of a perennial grass; a newly formed shoot which becomes independent from the parent stem by dying off behind.

Inrolled – a. Rolled inwards.

Insectivorous – a. Said of those plants which capture insects and presumably absorb nutriment from them.

Insect pollination – n. The transfer of pollen from the anther to the stigma by insects; entomophylly.

Inserted – a. Attached to, or growing out of.

Insular – a. Pertaining to an island.

Integument – n. The covering of an organ or body; the envelope of an ovule.

Intercalary inflorescence – n. A condition in which the main axis continues to grow vegetatively after giving rise to the flowers.

Intercostal – a. Between the ribs or nerves of a leaf.

Interfoliaceous – a. Between the leaves of a pair, as the stipules of many *Rubiaceae*.

Internerves – pl. n. The space between the nerves.

Internode – n. The part of the stem between two successive nodes.

Interrupted – a. Having any symmetrical arrangement destroyed; not continuous.

Interruptedly plante – n. Pinnate with small leaflets interposed with those of larger size.

Intramarginal – a. Within and near the margin.

Introduced – a. Brought from another region.

Introrse – a. Turned or faced inward

or toward the axis, as an anther facing toward the center of the flower.

Intumescent – a. Tumid, swollen, enlarged, distended.

Invaginated – a. Enclosed in a sheath.

Inverted – a. Turned over; endforend; top side down.

Involucel – a. A secondary involucre; a small involucre about the parts of a cluster.

Involucral – a. Pertaining to an involucre.

Involucrate – a. With an involucre.

Involucre – n. A cluster of bracts subtending a flower or inflorescence.

Involute – a. Rolled in from the edges, the upper surface within.

Irregular – a. Wanting in regularity of form; asymmetric, as a flower which cannot be halved in any plane, or one that is capable of bisection in one plane only, zygomorphic.

Irregular flower – n. A flower with some parts different from other parts in the same series.

Isadelphous – a. Characterized by equal brotherhood, the number of stamens in each group being equal.

Isogenous–a.Having the same origin.

Isolation – n. The separation of one group from another within a species so that crossing between the groups is prevented.

Isomerous – n. With the members of successive circles of equal numbers.

Isotype – n. A specimen believed to be a duplicate of the holotype.

Iteology – n. The study of the genus *Salix*, willows.

J

Joint – n. An articulation, as a node in grasses or other plants.

Jointed – a. With nodes or points of real or apparent articulation.

Jordanon – n. A microspecies; a small species of slight variablity.

Jugum – n. A pair of leaflets; the ridges on the fruits of Umbelliferae.

Julaceous – a. Bearing catkins, amentaceous.

Juvenile – a. Young, early forms.

Juxtaposition – n. The relative position in which organs are placed; a placing or being placed side by side.

K

Keel – n. A central dorsal ridge; the united petals of a Papillionaceous flower.

Kernel – n. The nucellus of an ovule, or of a seed, that is the whole body within the coats.

Kettle-trap – n. A flower such as that of *Aristolochia*, which imprisons insects until fertilization is effected.

Key – n. A short statement of the contrasted characters of a genus or other group.

Key fruit – n. The samara of ash.

Kingdom – n. One of the highest groups of organic nature; the Vegetable Kingdom includes all plants.

Kieistogamous – a. See Cleistogamous.

Kione – n. See Clone.

Knee – n. An abrupt bend in a stem or tree-trunk; an out-growth of some tree-roots.

Knight-Darwin Law – n. "That no organic being fertilizes itself for an eternity of generations"; "Nature abhors perpetual self-fertilization."

L

Label – n. The pinnule or ultimate segment of a fern-frond; labellum.

Labellum – n. The third petal of orchids, usually enlarged, and by torsion of the ovary having become anterior from its normal posterior position; a similar petal in other flowers; lip.

Lablate – a. Lipped; a member of the Labiatae.

Labium – n. The lower lip of a lablate flower the lip subtending the ligule in *Isoetes*.

Labyrinthiform–a. With complicated sinuous lines or winding passages.

Lacerate – a. Torn at the edge or irregularly cleft, as in some ligules.

Laciniate – a. Cut into lobes separated by deep, narrow, irregular incisions.

Lactiferous – a. Latex-bearing.

Lacuna – n. An air space in the midst of tissue; said of the vallecular canals of *Equisetum*; a hole or cavity.

Lacuno-rugose – a. Marked with irregular wrinkles, as the stone of a peach.

Lacustrine – a. Belonging to, or inhabiting lakes or ponds.

Ladaniferous – a. Ladanum-bearing.

Laevigate – a. Smooth, as if polished.

Lagenarious – a. Of a bottle or flask.

Lamella – n. A thin plate.

Lamellate – a. See Lamellose.

Lamellose – a. Having thin plates.

Lamina – n. The limb, blade, or expanded part of a leaf or petal.

Lanate – a. Clothed with wooly and interwoven hairs.

Lanceolate – a. Lance-shaped, rather narrow, tapering to both ends with the broadest part below the middle.

Lanose – a. Wooly.

Lanuginous – a. Cottony or wooly.

Lapidose – a. Growing amongst stones.

Lasiocarpous – a. With pubescent fruit.

Lateral – a. On or at the side.

Laterally compressed – a. Flattened from the sides as certain spikelets, glumes, and lemmas.

Latericious – a. Brick-red.

Latex – n. The milky juice of such plants as spurge or milkweed; the moisture of the stigms.

Latifoliate – a. Broad-leaved.

Latisquamate – a. Broad-scaled.

Laurine – a. Laurel-like.

Lax – a. Loose, distant.

Layering – a. Rooting, said of side branches.

Leader – n. The primary or terminal shoot of a tree.

Leaf – n. A lateral projection on a stem at a node and subtending a bud; it is usually expanded and concerned with the manufacture of food (photosynthesis).

Leaf bud – n. A bud which develops into a leafy branch; opposed to "flower bud."

Leaf, compound – n. A leaf with two or more blades, called leaflets.

Leaf cycle – n. In phyllotaxis, a spiral which passes through the insertion of intermediate leaves until it attains the next leaf exactly above its starting point.

Leafing – n. The unfolding of leaves.

Leaflet – n. A single division of a compound leaf.

Leaf scar – n. The mark or cicatrice left by the acticulation and fall of a leaf.

Leaf sheath – n. The lower part of the petiole, which more or less invests the stem.

Leaf, simple – n. A leaf with a single blade; not compounded.

Leaf stock – n. The stem of a leaf; petiole, foot-stalk.

Leaf tendril – n. A tendril which is a transformed leaf.

Leaf trace – n. All the common bundles in a stem belonging to one leaf.

Lecotropal – a. Shaped like a horseshoe, as certain ovules.

Lectotype – n. A specimen or other element selected from the original material to serve as the nomenclatural type, when the holotype was not designated at the time of publication, or when the holotype is missing.

Legume – n. A member of the legume family, Leguminosae; a superior, one-celled monocarpellary fruit usually dehiscent into two valves and having the seeds attached along the ventral suture; indehiscent legumes are usually constricted between the seeds and break crosswise into one-seeded segments at maturity.

Leguminous – a. Pertaining to a legume, or to the Leguminosae.

Lemma (pl. lemmata) – n. In grasses, the flowering glume, the lower of the two bracts immediately enclosing the flower.

Lenitie – a. Pertaining to, or living in, quiet or still water.

Lenticel – n. A lenticular corky spot on young bark, corresponding to an epidermal stoma.

Lenticellate – a. Having lenticels.

Lenticular – a. Shaped like a double convex lens.

Lepanthium – n. A petal which contains a nectary.

Lepidoid – a. Scalelike; said of leaves, as Thuja.

Lepidophyllous – a. With scaly leaves.

Lepidopterid – n. A flower adapted to lepidopterous (butterfly and moth) pollination.

Lepis (pl. lepides) – n. A scale usually attached by its center.

Lepisma (pl. lepismata) – n. A membranous scale in some Ranunculaceae, an apparently aborted stamen in Paeonia papaveracea.

Leucanthous – a. White-flowered.

Liana – n. Luxuriant woody climbers in the tropics with anomalous structure.

Lid – n. Operculum.

Ligneous – a. Woody.

Lignified – a. Converted into wood.

Ligulate – a. With a ligule; strap-shaped or straplike.

Ligule – n. A strap-shaped body, such as the limb of the ray florets in Compositae; the lobe of the outer corona in Stapelia; the thin scarious projection from the top of the leaf-sheath in grasses; a narrow membranous, acuminate structure, internal to the leaf-base in Isoetes and Selaginella; an appendage to certain petals as those of Silene and Cuscuta; the ovuliferous scale in Araucaria, united with the bract, and resembling the ligule in Isoetes; the envelope which protects the yound leaf in palms, as in Chamacrops and Rhaphis.

Liguliform – a. Strap-shaped.

Limb – n. The border or expanded part of a gamopetalous corolla, as distinct from the tv   or throat; the lamina of a leaf c_   f a petal.

Limicolous – a. Growing in mud, as on the margins of pools, lakes, and ponds.

Limnodophilous – a. Dwelling in marshes.

Limnophilous – a. Dwelling in lakes.

Linear – a. Long and narrow with margins parallel, or nearly so.

Linete – a. Lined; bearing thin parallel lines.

Lineolate – a. Marked with fine lines.

Lingulate – a. Tongue-shaped.

Linnaeun System – n. The artificial classification devised by Linnaeus, based upon the number and position of the stamens and pistils.

Linneon – n. A large species, usually polymorphic but with well characterized limits; species according to Linnaeus.

Lip – n. One of te two divisions of a bilabiate corolla or calyx, that is, a bilabiate corolla or calyx, that is, a gamopetalous or gamosepalous organ cleft into an upper (superior or posterior) and a lower (inferior or anterior) portion; the labellum of orchids.

Littoral – a. Beloging to, or growing on the seashore.

Livid – a. Pale lead-colored.

Lobate – a. Divided into, or bearing lobes.

Lobe – n. Any part or segment of an organ; specifically, a part of petal, calyx, or leaf that represents a division to about the middle.

Lobulate – a. Divided into small lobes.

Lobule – n. A small lobe.

Lochmophilous – a. Dwelling in thickets.

Locule – n. Compartment or cell of a pistil or anther.

Loculicidal - a. With dehiscence on the back, between the partitions into the cavity.

Lodger-arrangement - n. The retention by certain flowers of their insect visitors.

Lodicule - n. A small scale outside the stamens in the flowers of grasses.

Loment - n. A flat legume which is constricted between the seeds, falling apart at the constrictions when mature into one-seeded joints.

Lomentaceous - a. Bearing or resembling loments.

Lorate - a. Strap-shaped, ligulate.

Lucid - a. Shining, referring to the surface.

Lunate - a. Crescent-shaped, half-moon shaped.

Lurid - a. Pale yellow.

Lustrous - a. Glossy, shiny,

Lutescent - a. Becoming yellow.

Luticolous - a. Growing in miry places.

Lyrate - a. Lyre-shaped, pinnatifid wth the terminal lobe large and rounded, the lower lobes small.

M

Mace - n. The aril of the nutmeg.

Marcrocladous - a. With long branches.

Macrophyllous - a. Having elongated or large leaflets or leaves.

Macrosporangium - n. The receptacle in which macrospores are produced.

Macrospore - n. The larger of the two kinds of spores in *Selaginella* and related plants.

Macrostylous - a. Long-styled.

Macula - n. A spot.

Maculate - a. Blotched or spotted.

Major - a. Larger.

Malacophilous - a. Said of plants that are pollinated by snails or slugs.

Malacophyllous - a. With soft leaves.

Malicorin - n. The rind of a pomegranate.

Malleolus - n. A layer; a shoot bent into the ground and half divided at the bend, whence it emits roots.

Malpighiaceous hairs - pl. n. Hairs which are straight and appressed ut attached by the middle, frequent in Malpighiaceae.

Malvaceous - a. Malva-like, mallow-like.

Mammiform - a. Breast-shaped, conical with rounded apex.

Mammilla - n. A nipple or teat.

Mammillate - a. With little teat-shaped processes.

Mammose - a. With teat-shaped processes.

Manicate - a. Covered with a pubescence so thick and inter-woven that it can e striped off like a sleeve.

Marled - a. Stained with irregular streaks of colour.

Marcescent - a. Withering without fallng off.

Marginal - a. Of, pertaining to, or attached in the edge.

Marginate - a. Broad-brimmed, furnished with a margin of distinct character.

Marine - a. Growing within the influence of the sea, or immersed in its waters.

Maritime - a. Pertaining to the sea.

Massula - n. A group of cohering pollen-grains produced by one primary mother cell, as in orchids; also styled pollen-mass.

Mast - n. The fruit of such trees as beech, and other Cupuliferae.

Mattula - n. The fibrous material surrounding the petioles of palams.

Matutinal - a. Pertaining to morning; plants flowering early, as *Impomoea purpurea*.

Median - a. Pertaining to the middle.

Megaphyllous - a. With large leaves or leaf-like expansions.

Megasporangium - n. The sporangium which produces the megaspores.

Megaspore - n. The more correct form of macrospore; the larger spore of heterosporous plants.

Megasporocarp – n. The product of the development of the megasporangium in *Azolla*, finally containing a single perfect mega-spore.

Melangeophilous – a. Dwelling in loam.

Melanophyllous – a. Having leaves of a dark color.

Melanoxylon – n. Black wood.

Membranaceous – a. See Membranous.

Membranous – a. Thin, more or less flexible, and translucent; like a thin membrane.

Meniscoidal – a. Thin and concave-convex, like the crystal of a watch.

Mentum – n. An extension of the foot of the column in some orchids, in the shape of a projection in front of the flowers.

Mericarp – n. One of the two seed-like carpels of an Umbelliferous fruit.

Meristem – n. Embryonic or undifferentiated tissue the cells of which are capable of active division.

Meristematic – a. Pertaining to the meristem.

Merotype – n. A specimen collected from the original type in cultivation by means of vegetative reproduction.

Mesocarp – n. The middle layer of a pericarp.

Mesocotyl – n. An interpolated node in the seeling of some grasses, so that the sheath and cotyledon are separated by it.

Mesophyte – n. A plant intermediate between hydrophytes and xerophytes; a plant of medium moisture requirement.

Mesophytic – a. Growing under medium moisture conditions.

Metandry – n. A condition in which the female flowers mature before the male; protogyny.

Metatype – n. A specimen from the original locality, recognized as authentic by the describer himself.

Metoecious – a. Existing on different hosts; heteroecious.

Metonym – n. A synonymous name rendered invalid by the existence of an earlier valid name for the same species or other plant group.

Micropyle – n. The aperture in the skin of a seed formerly the foramen of the ovule; it marks the position of the radicle.

Microsorus – n. The male sorus in *Azolla*.

Microsporangium – n. The receptacle in which the microspores develop.

Microspore – n. The smaller of the two kinds of spores in such Pteridophytes as *Selaginella*.

Midrib – n. The main rib or central vein of a leaf or leaflike structure.

Migration – n. Any movement by which the range of a species is extended. (Strictly speaking, it means moving under its own powder)

Migrule – n. The unit of migration, as seed, fruit, runner, bulb, etc.

Minlate – a. The color of red lead; more orange and duller than vermillion.

Minor – a. Smaller.

Minute – a. Very small, inconspicuous.

Mitriform – a. Shaped like a mitra or cap.

Mixed-inflorescence – n. One in which partial inflorescences develop differently from the main axis, as centrifugal and centripetal together.

Molendinareous – a. Furnished with large, winglike expansions.

Monadelphous – a. With stamens united by their filaments into a tube or column.

Monandrous – a. With one stamen.

Monanthous – a. With one flower.

Moniliform– a. Necklace-shaped; like a string of beads.

Monocephalous – a. Bearing a single head or capitulum.

Monochasium – a. A one-branched cyme, either pure or resulting from the reduction of cymes.

Monochlamydeae – pl. n. A large

division of Angiosperms which have only one set of floral envelopes.

Monochlamydeous – a. Having only one set of floral envelopes.

Monoclinous – a. Having both stamens and pistils in the same flower; said of the capitula of composites which have only hermaphrodite flowers.

Monocotyledon – n. A plant having but one cotyledon or seed-leaf.

Monocotyledoneae – pl n. Plants of the class identified by the possession of only one cotyledon.

Monocotyledonous – a. With one cotyledon or seed-leaf.

Monocyclic – a. With the members of a floral series in only one whorl, as with the calyx only.

Monodynamous – a. With one stamen much longer than the others.

Monoecious – a. Having unisexual flowers with both sexes born on the same plant.

Monogynous – a. With one pistil.

Monolocular – a. One-celled, unilocular, applied to ovaries.

Monopetalous – a. One-petaled; gamopetalous, with the corolla composed of several petals laterally united.

Monophyllous – a. One-leaved as an involucrum of a single piece; said of a leaf-bud in which a single leaf is subtended by an investing stipule; gamosepalous or gamopetalous.

Monopterous – a. One-winged.

Monosepalous – a. With one sepal.

Monospermous – a. One-seeded.

Monostachous – a. Arranged in one spike.

Monostichous – a. In a single vertical row.

Monostylous – a. Having a single style.

Monosymmetrical – a. Said of a flower which can be bisected equally in one plane only; zygomorphic; bilaterally symmetrical.

Monotrichous – a. Having one bristle.

Monotrophic – a. With nutrition confined to one host-species.

Monotropie – a. Said of bees which visit only one species of flower.

Monotype – n. A genus that contains only one species; the term is applicable to other categories.

Montane – a. Pertaining to mountains, as a plant which grows on them.

Moschate – a. Musky or musk-scented.

Moth-flowers – pl. n. Flowers adapted for moths as pollinating visitors; they are usually white, night-blooming flowers.

Motion-dichogamy – n. A condition in which the sexual organs vary in length or position during flowering.

Mucilaginous – a. Slimy, composed of mucilage.

Mucro – n. A sharp terminal point.

Mucronate – a. Furnished with a mucro (bristle-tipped).

Multiciliate – a. With many cilia.

Multicipital – a. With many heads, referring to the crown of a single root or to several caudices.

Multicostate – a. Many-ribbed, as the ribs runing from the base of a leaf towards its apex.

Multifarious – a. Many-ranked, as leaves in vertical ranks.

Multifid – a. Cleft into many lobes or segments.

Multifoliate – a. Many-leaved.

Multipartite – a. Divided or cut many times.

Multiple fruit – n. A cluster of ripened ovaries traceable to the pistils of separate flowers, as the mulberry and the pineapple.

Multiplicate – a. Folded often or repeatedly.

Multiradiate – a. With numerous rays.

Multiseptate – a. With many partitions.

Muricate – a. Rough with short, hard points.

Muriculate – a. Very finely muricate.

Muriforma. With brick-like markings, pits, or reticulations, as on some seed-coats.

Muscariana. With flowers that attract files by a putrid stench.

Mutablea. Able to produce mutants.

Mutanta. That which undergoes mutation.

Mutationn. A sudden hereditary variation of an offspring from its parents.

Muticousa. Pointless, blunt, awnless, curtailed.

Myochrousa. Mouse-colored.

Myrmecochorousa. Dispersed by means of ants.

Myrmecochoryn. A plant which provides shelters in which ants live.

Myrmecophilousa. Said of plants which are inhabited by ants and offer specialized shelteres or food for them; pollinated by ants.

Myrmecophobousa. Shunning ants, said of plants which by hairs or glands repel ants.

Myrmecosymbiosisn. The mutual relation between the ants and their host plant.

N

Nacreous – a. With pearly luster.

Naked – a. Wanting its usual covering, as without pubescence, or flowers destitute of perianth, or buds without scales.

Naked bud – n. A brook plant.

Napaceous – a. Turnip-shaped or rooted.

Nascent – a. In the act of being formed.

Natant – a. Floating under water, that is, wholly immersed.

Naturalized – a. Having become thoroughly established in a region to which it is not indigenous.

Natural selection – n. The natural processes contributing to the "survival of the fittest".

Natural System – n. An arrangement according to the affinity of plants, and the sum of their characters, opposed to any artificial system, based on one set of characters.

Naucum – n. The fleshy part of a drupe; seeds with a very large hilum.

Naucus – n. Certain Cruciferous fruits which have no valves.

Nautiloid – a. Spiral-formed like the shell of *Nautilus*.

Navicular – a. Boat-shaped; like the bow of a canoe.

Nebulose – a. Cloudy, misty, said of such finely divided inflorescence as of *Eragrostis*; smoke-colored.

Neck – n. The collar or junction of stem and root; the point where the blade separates from the sheath in certain leaves; the contracted part of the corolla or calyx tube.

Necrocoleopterophilous – a. Pollinated by carrion beetles.

Necrotype – n. A form that formerly existed, but is now extinct.

Nectar – n. The sweet secretion from glands or nectaries, which act as an inducement to insect visitors.

Nectar glands – n. The secreting organs which produce nectar.

Nectar guides – n. Lines, spots, or other devices directing to the nectary.

Nectariferous – a. Nectar-bearing.

Nectarostigma (pl. stigmata) – n Some mark or depression indicating the presence of a nectariferous gland.

Nectarotheca – n. The portion of a flower which immediately surrounds a nectaripore.

Nectary – n. The organ in which nectar is secreted, formerly applied to any anomalous part of a flower, as its spurred petal.

Needle – n. A stiff linear leaf as in Pinaceae.

Neism – n. The origin of an organ on a given place, as the formation of roots on a cutting.

Nema (pl. nemata) – n. A filament, a thread.

Neotype – n. A 'specimen selected to serve as the nomenclatural type of a taxon in a situation when all' material on which the taxon was based is missing.

Nephroid – a. Reniform, kidney-shaped.

Nepionic – a. Said of the first leaves of seedlings developed immediately succeeding the embryonic stage of the cotyledons.

Nervation – n. Ventation, the manner in which the foliar nerves or veins are arranged.

Nerve – n. In botany, a simple or un-branched vein or slender rib.

Nervose – a. Full of nerves or prominently nerved.

Netted – a. Reticulated, net-veined with any system of irregularly anastomosing veins.

Neuter – a. Sexless, as a flower which has neither stamens nor pistils.

Nidulent – a. Partially encased or lying free in a cavity; embedded in a pulp, as the seeds in a berry.

Nitid – a. Smooth and clear, lustrous, glittering.

Niveous – a. Snowy-white.

Nocturnal – a. Occurring at night, as night-blooming; active in the night.

Nodal – a. Relating to the node.

Nodal diaphragm – n. Any septum which extends across the hollow of a stem at the node.

Nodding – a. Curved somewhat from the vertical.

Node – n. That point on a stem which normally bears a leaf or leaves.

Nodiferous – a. Bearing nodes.

Nodose – a. Knotty or knobby.

Nodule – n. A small knot or rounded body.

Nodulose – a. With little knobs or knots.

Nomad – n. A pasture plant.

Nomenclature – n. The names of things in any science; in botany, frequently restricted to the correct usage of scientific names in taxonomy.

Nomen conservandum – n. A name retained in biological nomenclature regardless of priority.

Nomen novum (nom. nov.) – n. New name, i.e. a name hitherto unpublished, substituted for one in general use but found to be untenable.

Nomen nudum (nom. nud.) – n. A naked name, i.e. a name only; a plant name published without any description or figure, and hence which cannot be tied in with assurance to any plant or plant group. Nomina nuda are very properly rejected by all codes.

Nomogenesis – n. The theory that the evolution of organisms is the result of certain processes inherent in them and that it follows definite laws.

Nomophilous – a. Dwelling in pastures.

Normal – a. According to rule, usual as to structure.

Notate – a. Marked with spots or lines.

Nototribal – a. With the stamens and styles turned so as to strike their insect visitors on the back.

Nox – n. Night.

Nucamentaceous – a. Having the hardness of a nut; synonym for indehiscent, monospermal fruit; also, catkin-like.

Nucamentum – n. An ament or catkin.

Nuciferous – a. Not-bearing.

Nucleus – n. A kernel of an ovule, which by fertilization becomes a seed; a dense protoplasmic structure near the center of living cells.

Nude – a. Bare, naked, uncovered.

Nut – n. A dry indehiscent, usually one-celled, one-seeded fruit (though usually traceable to a compound ovary) with a bony, woody, leathery, or papery wall and in general, partially or wholly encased in an involucre or husk.

Nutant – a. Nodding, drooping.

Nutlet – n. The diminutive of nut.

Nux – n. Nut.

Nyctanthous – a. Night-flowering.

Nyctigamous – a. Said of flowers which close by day, but open at night.

Nyctitropic – a. Turning in response to darkness.

Nyctitropism – n. The act of responding to darkness.

O

Obclavate – a. Club-shaped and attached at the thicker end.

Obcompressed – a. Compressed dorso-ventrally instead of laterally.

Obconic – a. Conical, but attached at the narrower end.

Obconical – a. Inversely conical, having the attachment at the apex.

Obcordate – a. Inversely heart-shaped, the notch being apical.

Obcordate – a. Inversely heart-shaped, the notch being apical.

Obex (pl. obices)–n. A barrier; a hindrance to plant distribution. Biological obices, as the constitution of the plant. Physical obices, as the shutting in by mountains, etc.

Oblanceolate– a. Inverted lanceolate.

Oblate – a. Flattened at the poles, as a tangerine-orange.

Obligate – a. Necessary, essential, the reverse of facultative.

Oblique – a. Slanting, unequal-sided.

Oblong – a. Longer than broad, with the margins nearly parallel.

Obovate – a. Reversed ovate, the distal end the broader.

Obovoid – a. Appearing as an inverted egg.

Obscure – a. Dark or dingy in tint; uncertain in affinity or distinctiveness.

Obsolescent – a. Becoming rudimentary or extinct.

Obsolete – a. Not evident or apparent; rudimentary; no longer used.

Obturator – n. A small body accompanying the pollen mass of orchids and asclepiads closing the opening to the anther.

Obtuse – a. Blunt or rounded at the end.

Occlusion – n. The process by which wounds in trees are healed by the growth of callus.

Ocellus – n. A small eye.

Ochraceous – a. Ochre-colored.

Ochroleucous – a. Yellowish-white, buff.

Ochthad – n. A bank plant.

Ochthophilus – a. Bank-loving.

Ocrea – n. A legging-shaped or tubular structure formed by the union of two stipules.

Ocreate – a. With sheathing stipules.

Ocreola (pl. ocreolae) – n. The smaller or secondary sheaths, as in the inflorescences of Polygonum.

Octagynia – n. A Linnean order of plants with eight-styled flowers.

Octandrous – n. Having eight stamens.

Octolocular – a. Said of an eight-celled fruit or pericarp.

Octopetalous – a. With eight petals.

Octoradiate – a. With eight rays, as in some Compositae.

Octosepalous – a. With eight sepals.

Octostemonous – a. With eight fertile stamens.

Octostichous – a. In eight rows.

Oculus – n. The first appearance of a bud, especially on a tuber, as the eyes on a potato.

Odd pinnate – a. With a terminal leaflet, imparipinnate.

Official – a. Of the shops; used in medicine or arts.

Oleaginous – a. Oily and succulent.

Oleiferous – a. Oil-bearing.

Oleraceous – a. Having the qualities of garden herbs used in cooking.

Oligandrous – a. With few stamens.

Oligocarpic – a. Few-fruited.

Oligophyllous– a. Having few leaves.

Oligospermous – a. Few-seeded.

Oligotrophic – a. Growing in poor soil and competing for the nutritive salts in it.

Oligotropic – a. Said of insects which visit only a few species of plants.

Olivaceous – a. Olive green; olive colored.

Ombrophilous – a. Rain-loving.

Ombrotropism – n. Tropic responses of organsism to the stimulus of rain.

Oncospore – n. A plant having seeds with hooks which aid in dispersion.

Ontogeny, ontogenesis – n. The developmental history of an individual from fertilized egg to adult organism.

Oospore – n. The fertilized egg in the archegonium of crytogams, from which the new plant develops directly.

Opaque – a. Dull; neither shining nor translucent.

Operculate – a. Furnished with a lid.

Operculum (pl. opercula) – n. A lid or cover which separates by a transverse line of division, as in the pyxis, also in some pollen grains.

Opposite – a. On both sides at the same level, as two leaves at a node; one part before another, as a stamen in front of a petal.

Oppositiflorous – a. Having opposite peduncles or pedicels.

Oppositifolorous – a. With opposite leaves.

Optimal – a. With opposite leaves.

Optimal – a. The most advantageous for an organism or function.

Orange – a. The fruit of *Citrus aurantium*; a secondary color, red and yellow combined, taking its name from the tint of the fruit mentioned.

Orbicular – a. Flat with a circular outline.

Orbiculate – a. Disk-shaped, round.

Orchiold – a. Orchid-like.

Order – n. In botany, a group between genus (tribe and sub-order) and class.

Orgadophilous – a. Dwelling in open woodland.

Organ – n. A group of tissues or-ganized to perform a definite function.

Orifice – n. An opening by which spores, etc. escape; any opening.

Ornithogamous – a. Pollinated by birds.

Ornithophilous – a. Pollinated by birds.

Orophilous – a. Dwelling in mountainous regions.

Orthocladous – a. Straight-branched.

Orthogenesis – n. Purposive, "predetermined" evolution toward a definite objective; a tendency to vary continously in the same direction.

Orthoosmotropism – n. The assuming of an erect position to osmotic action.

Orthopterrous – a. Straight-winged.

Orthostichous – a. Straight-ranked.

Orthotropism – n. The assumption of a vertical position.

Orthotropous – a. Said of an ovule or seed with a straight axis, chalaza at the insertion, the orifice at the other end.

Osseous – a. Bony.

Ossiculus – n. The pyrene or stone of a drupe.

Ossified – a. Becoming hard as bone, as the stones of drupes, such as the peach or plum.

Oval – a. Broadly elliptical with the width greater than half the length.

Ovary – n. That part of the pistil which contains the ovules.

Ovate – a. Shaped like a longitudinal section of a hen's egg, the broader end basal; applied to ovoid.

Ovoid – n. A solid that is oval (less correctly, ovate) in flat out-line.

Ovulate – a. Pertaining to the ovule, or possessing ovules.

Ovule – n. That which becomes a seed after fertilization.

Ovuliferous – a. Bearing ovules.

P

Pabular – a. Of fodder or pasturage.

Pachycladous – a. Thick-branched.

Pagina – n. The blade or surface of a leaf.

Painted – a. having colored streaks of unequal density.

Palate – n. In personate corollas, a rounded projection or prominence of the lower lip closing the throat or very nearly so.

Pale – n. A chaffy scale such as often subtends the fruit of Compositae.

Palea – n. The inner bract of a grass floret; the chaffy scales on the receptacle of many Compositae; the ramenta or chaffy scales on the stipe of many ferns.

Paleaceous – a. Chaffy, chafflike in structure.

Paleobotany – n. Fossil botany, the study of plants in a fossil state.

Paleola (pl. paleolae) – n. A diminutive of palea, or of secondary order, as applied to the lodicule of grasses.

Paleolate – a. With a lodicule.

Palephytological – a. Relating to the study of fossil plants.

Palet – n. See Palea.

Pallescent– a. **Becoming light in tint.**

Pallid – a. Pale.

Palmate – a. Resembling a hand with the fingers spread; having lobes radisting from a common point.

Palmately – a. In a palmate manner.

Palmatifid – a. Cut in a palmate fashion nearly to the petiole.

Paludose – a. Growing in marshy places.

Palustrine – a. Of or growing in marshes.

Pampiniform – a. Resembling the tendril of a vine.

Pampinus – n. Tendril.

Pandura – n. Violin.

Pandurate – a. Fiddle-like.

Panduriform – a. Fiddle-shaped.

Panicle – n. A compound or branched receme.

Paniculate – a. Having a panicle type of inflorescence.

Panmixy – n. Free and more or less unlimited cross-ferilization.

Panniform–a. Having the appearance or texture of felt or woolen cloth.

Pannose – a. Having the appearance or texture of felt or woolen cloth of very close texture.

Papaveraceous – a. Belonging to or resemblir.g the poppy.

Papery – a. Having the texture of paper, charactaceous.

Papilionaceous – a. Descriptive of the flower of many legumes having a standard, wings, and keel; with a pealike flower; like a butterfly.

Papilla (pl. papillae) – n. A minute nipple-shaped projection.

Papillary – a. Resembling papillae.

Papillose – a. Bearing papillae.

Pappiferous – a. Bearing pappus.

Pappus – n. Thistledown; the various tufts of hairs on achenes or fruits; the limb of the calyx of Compositae florests.

Papyraceous – a. Papery; white as paper.

Paracarpium – n. An abortive pistil or carpel; the persistent portion of some styles or stigmas.

Paracarpons – a. Said of ovaries whose carpels are joined together by their margins only.

Parachute – n. Sometimes applied to a fruit which is readily carried by wind, by means of membranous expansions or pappus, recalling the action of a parachute.

Paraheliotropiem – n. Diurnal sleep, the movement of leaves to avoid the effect of intense sun-light.

Parallel – a. Extended in the same direction, but equally distant at every point.

Parallelodromous – a. Having parallel veins, as in lilies.

Parallel-veined – a. With lateral veins straight, as in *Alnus*; the entire system straight, as in the grasses.

Paraphototropism – n. The assumption of a position at right angles to the incident light.

Parasite – n. An organism subsisting

on another, the host.

Parasitic – a. Deriving nourishment from another organism.

Parasol – n. A pecullar set of spines on some cacti.

Parastas (pl. parastades) – n. Used in the plural to designate the coronal rays of *Passiflora*.

Parastomon – n. An abortive stamen, a staminodium.

Paratype – n. A specimen belonging to the original series, but not the type selected by the author.

Paravariation – n. A modification or acquired variation developed during the life of the individual as a result of environmental causes and not heritable.

Parenchyma (pl. parenchymata) – n. Soft tissue of cells with unthickened walls, as pith cells.

Parietal – a. Borne on or belonging to a wall, as parietal placentation.

Paripinnate – a. Pinnate, with an equal number of leaflets, that is without a terminal one, abruptly pinnate.

Parted – a. Divided by sinuses which extend nearly to the mid-rib.

Parthenocarpy – n. The production of fruit without true fertilization.

Parthenogenesis – n. A form of apogamy in which the oosphere develope into a normal product of fertilization without a preceding sexual act.

Parthenogenetic – a. Developing without fertilization.

Partition – n. A wall or dissepiment; a separated part or segment; the deepest division into which a leaf can be cut without becoming compound.

Pastoral – n. Pertaining to shepherds; rural.

Patelliform – a. Disk-shaped.

Patent – a. Spreading.

Pathological – a. Diseased.

Patulous – a. Standing open, spreading.

Pectinate – a. Comblike; beset with narrow, closely inserted segments like the teeth in a comb.

Pedate – a. Palmately divided or parted with the lateral divisions cleft.

Pedatified – a. Divided in a pedate manner nearly to the base.

Pedicel – n. An ultimate flower-stalk, the support of a single flower; in grasses, the stalk of a spikelet.

Pedicellete – a. Borne on a pedicel.

Pediculus – n. Pedicel; the stalk of an apple or other fruit; the filament of an anther.

Pediophilous – a. Dwelling in uplands or level country.

Peduncle – n. A primary flower stalk supporting either a cluster or a soiltary flower.

Pedunculate – a. With a footstalk or peduncle.

Peg – n. An embryonic organ at the lower end of the hypocotyl of seedlings of *Cucumis*, *Gnetum*, etc., lasting until the cotyledons are withdrawn from the testa.

Pellicle – n. A delicate superficial membrane, epidermis.

Pellucid – a. Wholly or partially transparent.

Peloria, pelory – n. Reversion, on the part of the individual to the production of regular flowers, when the species typically has symmetrical or bilaterally symmetrical flowers.

Pelta – n. A bract attached by its middle as in peppers.

Peltafid – a. A peltate leaf cut into segments.

Peltate – a. Shield-shaped, as a leaf attached by its lower surface to a stalk instead of by its margin.

Pencilled – a. Marked with fine distinct linces.

Pendent – a. Hanging down from its support.

Pendulous – a. Hanging downward.

Penicillate – a. Like a pencil.

Pennate – a. Pinnate.

Pennatifid – a. Pinnatifid.

Penniveined – a. Veined in a pinnate manner.

Pentacamerous – a. With five locules.

Pentacarpellary – a. Having five carpels.

Pentacyclic – a. With five whorls of members.

Pentadactylous – a. Five fingered or with five finger-like divisions.

Pentadelphous – a. With five fraternities or bundles of stamens.

Pentagynous – a. With five pistils or styles.

Pentamerous – a. With parts in fives, as a corolls of five petals.

Pentandrous – a. With five stamens.

Pentapetalous – a. With five petals.

Pentapterous – a. Five-winged.

Pentasepalous – a. With five sepals.

Pentastichous – a. In five vertical ranks.

Pepo– n. The fruit of the gourd family, Cucurbitaceae; an infer for berry-like fruit with more or less rind and with lateral placentation.

Perennation – a. A lasting, or perennial state.

Perennial – a. Continuing to live from year in year.

Perfect – a. Said of flowers having both sex organs present and functioning.

Perfoliate – a. Having the stem apparently passing through the leaf; said of opposite leaves joined at their bases.

Perforated – a. With holes.

Perianth – n. The floral envelope, of whatever form; the calyx and corolla.

Pericarp – n. The wall of a mature ovary.

Periclinium – n. The involucre of the capitulum in Compositae.

Peridroma – n. The rachis of ferns.

Perigynium – n. The hypogynous setae of sedges; the flask, or utricle of *Carex*; any hypogynous disc.

Perigynous – a. Borne around the ovary, as with calyx, corolla, and stamens borne on the edge of a cup-shaped hypanthium; such cases are said to exhibit perigyny. See also Hypogyny and Epigyny.

Peripheral – a. On or near the margin.

Peripterous – a. Surrounded by a wing or border.

Perisperm – n. The ordinary albumen of a seed restricted to that which is formed outside the embryo sac; the pericarp or even the integuments of the seed.

Persicicolor – a. Peach-colored, rose pink.

Persistent – a. Remaining attached; not falling off.

Personate – a. Said of a bilablate corolla having a prominent palate.

Perspicous – a. Transparent.

Perula – n. The scale of a leafbud; a projection in the flower of orchids, the mentum.

Perulate – a. Scale-bearing, as most buds.

Petal – n. One of the leafy expansions in the floral whorl styled the corolla; of the hop, the scales of the strobile.

Petaliferous – a. Bearing petals.

Petalode – n. An organ simulating a petal.

Petaloid – a. Like a petal, or having a floral envelope resembling petals.

Petiole – n. The stem of a leaf.

Petiole – n. The stem of a leaf.

Petiolule – n. A small petiole; the petiole of a leaflet.

Petricolous – a. Rock-inhabiting.

Petrophilous – a. Preferring rock.

Phaenantherous – a. With stamens exserted.

Phaenocarpous – a. Having a distinct fruit, with no adhesion to surrounding parts.

Phaenogamous – a. Plants sexually propagating by flowers, of which essential organs are stamens and pistils.

Phanerogam – n. A plant with flowers in which stamens and pistils are distinctly developed.

Pharmacognosy – n. The knowledge of the distinctive features of drugs.

Phellem – n. Cork.

Phellophilous – a. Dwelling in rock fields.

Phenological isolation – n. Isolation by time of flowering as either earlier or later than the other species of the genus.

Phenology – n. The science of the relations between climate and periodic biological phenomena, as the flowering and fruiting of plants, the migration of birds, etc.

Phenotype – n. A group of individuals of similar appearance but not necessarily of similar genetic constitution.

Phoeniceous – a. Purple-red.

Phoranthium – n. The receptacle of the capitulum of Compositae.

Photeolic – a. Pertaining to the "sleep" of plants.

Phototropism – n. The act of turning in response to light.

Phototype – n. A photograph of a type specimen, an abbreviation of the word photographotype.

Phragma (pl. phragmata) – n. A spurious dissepiment in fruits.

Phyllary – n. An involucral bract in the Compositae.

Phylloclad – n. A flattened branch assuming the form and function of foliage.

Phyllode – n. Leaflike petiole having no blade, as in some acaclas and other plants.

Phylloid – a. Leaflike.

Phyllome – n. An assemblage of leaves or of inciplent leaves in a bud.

Phyllopodes – pl. n. The dilated sheathing base of a leaf in *Isoetes*.

Phylloptosis – n. The unnatural fall of leaves.

Phyllotaxy – n. The mode in which the leaves are arranged in regard to the axis.

Phylogenetic – a. Pertaining to the ancestral history of the race.

Phylogeny – n. The race history of an animal or plant deduced from development.

Phytogamy – n. Cross-fertilization of flowers.

Phytogenesis – n. The evolution and development of plants.

Phytogeographer – n. An expert on plant distribution.

Phytogeography – n. The science of plant distribution.

Phytography – n. The description and illustration of plants; systematic or taxonomic botany.

Phytological – a. Relating to the study of plants.

Phytologist – n. A botanist.

Pileate – a. With a cap.

Pileorhiza – n. The root-cap, a hood at the extremity of the root.

Piliferous – a. Bearing hairs, or tipped with hairs; hair-pointed.

Piloglandulose – a. Bearing glandular hairs.

Pilose – a. With soft hairs.

Pinna – n. The primary unit of a pinnately compound leaf.

Pinnate – a. Feather-formed, as with the leaflets of a compound leaf placed on either side of a rachis.

Pinnately – a. In a pinnate fashion.

Pinnately veined – n. With the vein pattern simulating a feather.

Pinnatifid – a. Cleft in a pinnate manner.

Pinnatisect – a. Cut down to the midrib in a pinna.

Pinninerved – a. Pinnately veined, running parallel towards the margin.

Pinnule – n. A secondary pinna; the foliaceous unit of a bipinnately compound leaf.

Pip – n. A popular name for the seed of an apple or pear.

Piperaceus – a. Peppery, pepper-like.

Piriform – a. Shaped like a pear.

Pisaceous – a. Pea-green, the color of unripe seeds.

Pisiferous – a. *Pisum*-bearing, pea-bearing.

Pisiform – a. Pea-shaped.

Pistil – n. The female organ of a flower, consisting when complete of an ovary, style, and stigma.

Pistillate – a. Having pistils and no stamens; female.

Pit – n. A small hollow or depression; the endocarp of a drupe containing a kernel or seed.

Pith – n. The spongy center of an exogenous stem.

Pitted – a. Marked with small depressions, punctate.

Place-constant – n. An invariable factor of plant like in a given locality.

Placenta – n. The place in the ovary where ovules are attached.

Placentation – n. The disposition of the placenta.

Plagiodromous – a. Said of tertiary leaf-veins when at right angles to the secondary veins.

Plagiophototropic – a. Assuming an oblique position to the rays of light, as leaflets of *Robinia, Tropaeolum*, etc.

Plaited – a. Plicate, folded like a fan.

Plane – n. Level, even or flat surface.

Plane of symmetry – n. That which divides an object into symmmetrical halves.

Plant – n. A vegetable organism nourished by gases or liquids and not ingesting solid particles of food.

Platanoid – a. *Platanus*-like, like the plane-tree or sycamore.

Pleiomery – n. The state of having more whorls than the normal number.

Pleiopetalous – a. Many-petaled.

Pleiopetaly – n. Doubleness in flowers.

Pleiosepalous – n. Many-sepaled.

Pleiospermous – a. With an unusually large number of seeds.

Pleurogyrate – a. Said of fern sporangia which have a horizontal annulus.

Pleurotribal – a. Said flowers whose stamens are adapted to deposit their pollen upon the sides of insect visitors.

Plicate – a. Folded on the several ribs in the manner of a closed fan, oc-

curing in palmately veined leaves, as in maple and currant.

Plococarpium – n. A fruit composed of follicles arranged around as axis.

Plotophyte – n. A flosting plant, its functional stomata on the upper surface of its leaves.

Plumbeous – a. Lead-colored, greenish-drab.

Plumose – a. Pubescent in a manner simulating a feather or a plume.

Plumule – n. The primary leaf-bud of an embryo.

Plurilocular – a. Many-celled; with many locules.

Poad – n. A meadow plant.

Pod – n. A dehiscent dry pericarp; a rather general uncritical term.

Podocarp – n. A stipitate fruit, that is, one in which the ovary is borne on a gynophore.

Polachena – n. A fruit similar to a cremocarp, but composed of five carpels.

Pollard – n. A tree dwarfed by frequent cutting of its boughs a few feet from the ground of shoots from the place where cut.

Pollarding – n. The cutting back to produce a mop-headed growth.

Pollen – n. The fertilizing dust-like powder produced in the anthers of phanerogams, more or less globular in shape, sometimes spoken of as "microspores" the male gametophyte in seed plants.

Pollen carrier – n. The retinaculum of asclepiads; the gland to which the pollen-masses are attached, either immediately or by caudicles.

Pollen flower – n. A flower which produces pollen but no nectar.

Pollen-mass – n. Pollen grains cohering by a waxy texture of fine threads into a single body.

Pollinate – v. To transfer pollen from the anther to the stigma or female organ.

Pollination – n. The placing of pollen on the stigma or stigmatic surface.

Polliniferous – a. Pollen-bearing.

Pollinium – n. A coherent mass of

pollen, as in orchids and asclepiads.

Polster – n. A cushion plant, a low, compact perennial.

Polyadephous – a. With stamens grouped into several brother-hoods or bundles.

Polyandrous – a. Having an indefinite number of stamens.

Polyanthous – a. Having many flowers, particularly if within the same involucre.

Polycephalous – a. Bearing many heads or capitula.

Polycotyledonous – a. With more than two cotyledons.

Polyembryony – n. The presence of more than one embryo in an ovule.

Polygamodioecious – a. Polygamous but chiefly dioecious.

Polygamodioecious – a. Polygamous but chiefly monoecious.

Polygamous – a. Bearing perfect and unisexual flowers on the same individual.

Polygenesis – n. The production of a new type at more than one place or more than one time.

Polymerous – a. With numerous members to each series or cycle.

Polymorphic – a. With several or various forms, variable as to habit.

Polypetalous – a. With many distinct petals.

Polysepalous – a. With many distinct sepals.

Polyandrous; – with numerous stamens.

Polystemonous – a. Polyandrous; with numerous stamens.

Polystichous – a. With many ranks or rows, as leaves.

Polythalamic – a. Having more than one female flower within the involucre; derived from more than one flower, as a collective fruit.

Polytropism – n. The turning of leaves in order to place themselves vertically and meridionally, the two surfaces facing east and west.

Pomaceous – a. Relating to apples.

Pome – n. A fleshy fruit, the product of a compound pistil with the seeds encased within a papery or cartilaginous cell, as the apple.

Pomeridian – a. Afternoon, as blooming in the afternoon.

Pontohalicolous – a. Dwelling in salt marshes.

Porandrous – a. With anthers which open by pores.

Poricidal – a. Opening by pores.

Porose – a. With small holes or pores.

Porrect – a. Directed outward and forward.

Posterior – a. At or toward the back; opposite the front; toward the axis; away from the subtending bract.

Potamophilous – – a. River-loving.

Praemorse – a. As though bitten off, terminated abruptly.

Precocious – a. Appearing or developing very early, the aments in *Salix* expanding before the leaves.

Prehensile – a. Clasping or grasping, as in tendrils.

Prevernal – a. Early spring-flowering; of early spring.

Prickle – n. A small and weak spinelike body borne irregularly on the bark or epidermis.

Primary – a. First in order of time or development.

Primine – n. The outer integument of an ovule.

Primocane – n. The first year's cane (usually without flowers) of *Rubus* and similar genera.

Prismatic – a. Of the shape of a prism.

Prison flowers – pl. n. Flowers which imprison their insect-visitors until fertilization is effected.

Proanthesis – n. A flowering in advance of the normal period, some flowers appearing in autumn in advance of the ensuing spring.

Process – n. Any projecting appendage.

Procumbent – a. Prostrate, trailing; lying flat upon the ground.

Prohydrotropism – n. The act of turning toward a source of moisture.

Proliferating – a. Producing offshoots.

Propagule – n. See Diaspore.

Prophyllum – n. The bracteole at the base of an individual flower, as in *Juncus;* a numbranous structure between a branch and the main stem in Graminae.

Prostrate – a. Laying flat on the ground.

Protandrous – a. With anthers maturing before the pistils in the same flower.

Protandry – n. A condition in which the anthers mature before the pistil in the same flower, the pollen being dispersed before the pistil is receptive.

Protanthesis – n. The normal first flower of an inflorescence.

Proterandrous – a. With anthers ripening before the pistil in the same flower; protandrous, one kind of dichogamy.

Proterogyny – a. A condition in which the pistil is receptive before the anthers have mature pollen.

Proterotypes – pl. – n. Primary types; all specimens which have served as the basis for descriptions and figures of organisms; further divided into Holotype; cotype (or Syntype), Paratype, Lectotype, and Chirotype.

Prothallium – n. The minute reduced gametophyte of the forns and their allies (Pteridophyta).

Prothallus – n. The gmetophyte stage or generation of Pterido-phytes, a multicellular and usually flattened thallus-like structure on the ground, bearing the sexual organs, as the antheridia and archegonia.

Protogenesis – n. Reproduction by budding.

Protogynous – a. Characterized by protogyny.

Protogyny – n. A condition in which the pistil matures before the anthers.

Protolog – n. The original description of a genus, species, or variety.

Prototype – n. The assumed ancestral form, from which the descendents have become modified.

Pruinose – a. With a waxy powdery secretion on the surface, glaucous.

Pruniform – a. Plum-shaped.

Prurient – a. Causing an itching sensation.

Psammophilous – a. Sand-loving, as the vegetation of dunes.

Pseudobulb – n. The thickened or bulb-formed stems of certain orchids, the part being solid and borne above ground.

Psilicolous – a. Prairie-dwelling.

Psilophilous – a. Prairie-loving.

Psychophilous – a. Pollinated by diurnal lepidoptera.

Psychrocleistogamy – – n. Cleistogamy induced by cold.

Pterocarpous – a. Wing-fruited.

Pterocaulous – a. Wing-stemmed.

Pterospermous – a. With winged seeds.

Pterygopous – a. Having the peduncle winged.

Puberulent – a. Somewhat or minutely pubescent.

Pubescence – n. Hairiness.

Pubescent – a. Covered with short soft hairs, or down.

Pugioniform – – a. Dagger-shaped.

Pulveraceous – a. Covered with a layer of powdery granules.

Pulverulent – a. Powdered, as if dusted over.

Pulvinate – a. Cushion-shaped.

Pulviniform – – a. Having the shape of a cushion or pad.

Pulvinus – n. An enlargement close under the insertion of a leaf; the swollen base of a petiole, as in *Mimosa pudica,* sometimes of the top of the petiole.

Pumpform – a. Applied to Papillionaceous flowers, with concealed anthers, as *Lotus, Coronilla,* and *Ononis.*

Punctate – a. Marked with dots, depressions or translucent glands.

Puncticulate – a. Minutely punctate.

Pungent – a. Ending in a rigid and sharp point.

Puniceous – a. Crimson, reddish-purple.

Purpurescent – a. Becoming or turning purple.

Pustular – a. Having slight blister-like elevations.

Pustulose – a. Blistery, furnished with pustules or irregular raised pimples (not as roughened as papillose).

Putamen – n. The shell of a nut; the hardened endocarp of stone fruit.

Pyramidal – a. Pyramid-shaped.

Pyrene – n. Nutlet, particularly the nutlet in a drupe.

Pyriform – – a. Pear-shaped.

Pyxidate – a. With a lid, as some capsules.

Pyxis – n. A capsule with circumscissile dehiscence, the upper portion acting as a lid.

Q

Quadrangulate – a. Having four angles, which are usually right angles.

Quadrate – a. Nearly square in form.

Quadrifoliate – a. With four leaves or leaflets.

Quilled – a. Said of normally ligulate florets which have become tubular.

Quinary – a. In fives.

Quinate – a. Growing together in fives, as leaflets from the same point.

Quinquncial – a. Arranged in a quincunx; in aestivation, partially imbricated of five parts, two being exterior, two interior, and the fifth one having one margin exterior and the other interior, as in the calyx of the rose.

Quinquecostate – – a. Having five ribs.

Quinquefarious – a. In five ranks.

Quinquefoliate – a. Five-leaved.

Quinquefoliate – a. In five pairs, as of leaflets.

Quinquelocular – a. Five-celled.

Quinquenerved – a. With the midrib dividing into five, that is, the main rib and a pair on each side.

R

Race – n. A variety of such fixity as to be reproduced from seeds; used also in a loose sense for related individuals without regard for rank.

Raceme – n. An indeterminate inflorescence consisting of a central rachis bearing a number of flowers with pedicels of nearly equal length.

Racemiform – a. In the form of a raceme.

Racemose – a. Resembling a raceme; in racemes.

Rachilla – n. A diminutive or secondary rachis or axis; in grasses and sedges, the axis that bears the florets.

Rachis – n. An axis bearing flowers or leaflets; petiole of a fern frond.

Radiant – a. Diverging from a central point.

Radiate – a. Spreading from or arranged around a common center.

Radical – a. Belonging or pertaining to the root.

Radicant – – a. Rooting, usually applied to stems.

Radicicolous – a. With the flower seated immediately upon the crown of the root; dwelling in the root as a parasite.

Radicle – n. The lower portion of the axis of an embryo seedling.

Rain-leaves – pl. – n. Leaves which are adapted to shed the rain from their surfaces, and generally are acuminate-tipped; drip tips.

Ramal – a. Belonging to a branch.

Rameal – a. See Ramal.

Ramentum (pl. ramenta) – n. Used in the plural for the thin chaffy scales of the epidermis, as the scales of many ferns.

Ramose – a. Having many branches.

Ramulose – – a. Having many branches.

Rank – n. A row, especially a vertical

row.

Raphe – n. In a more or less anatropous ovule, a cord or ridge of fibro-vascular tissue connecting the base of the nucellus with the placenta.

Raphis (pl. raphides) – n. A needle-shaped crystal, used in the plural to describe the crystals found in the cells of some plants.

Rapiformis – a. Turnip-shpaed.

Ratoon – n. A shoot from the root of a plant which has been cut down.

Ray – n. One of the radiating branches of an umbel; the marginal, as opposed to the disk, flowers in Compositae or other flower clusters, whenn there is a difference is structure.

Ray flower – n. The primary rays in exogenous stems between the different bundles, passing radially outwards.

Recapitulation theory – n. That every organism in its individual life-history repeats the various stages through which its ancestors have passed in the course of evolution.

Receptacle – n. That expanded portion of the axis which bears the floral organs; torus.

Reclinate – a. Bent down or falling back from the perpendicular.

Recurved – a. Bent or curved downward or backward.

Reduplicate – a. Doubled back; as a term of aestivation, in which the edges are valvate and reflexed.

Reflexed – a. Abruptly curved or bent downward or backward.

Regal – a. Royal.

Region – n. The area occupied by given forms.

Region, austral – n. Southern region.

Region, boreal – n. Northern region.

Region, tropical – n. Region within the tropics.

Regma (pl. regmata) – n. A fruit with elastically opening segments or cooci, as in *Euphorbia*; a form of schizocarp.

Regular – a. Uniform or symmetrical in shape or structure; of a flower, actinomorphic, radially symmetrical.

Relic, relict – n. A species properly belonging to an earlier vegetation type than that in w॰ ॰ it is now found.

Remote – a. Scattered, not close together.

Reniform – a. Kidney-shaped, said of the form of some leaves.

Repent – a. Undulate or wavy, as the margin of some leaves.

Repent – a. Creeping, prostrate, and rooting at the nodes.

Repetum – n. Fruit with the valves connected by threads, persistent after dehiscencee, such as in orchids, *Aristolochia* and some Papaveraceae.

Replum – n. A framelike placenta from which the valves fall away in dehiscence, frequently used so as to include the septum of Cruciferae.

Reprogression – n. A mode of flowering in which the primordial flower at the summit opens first, after which flowering occurs in succesion form the bottom up wards.

Reptant – a. Creeping on the ground and rooting.

Resilient – a. Springing or bending back, as some stamens.

Resin cyst – n. Cell or cavity occluded with resin.

Resiniferous – a. Producing resin.

Rest – n. Dormancy induced in cold climates by lowness of temperature, in hot climates by want of moisture.

Resupinate – a. Upside down, or apparently so.

Reticulate – a. Forming a network.

Reticulation – n. Network, the regular crossing of threads.

Reticulum – n. A membrane of cross-fibers found in palms at the base of the petiole.

Retinaculum – n. The gland to which one or more pollinia are attached in orchids; in asclepiads, a horny elastic body to which the pollen-masses are fixed; in most Acanthaceae, the funiculus, which is curved like a hook and retains the seed until mature.

Retrocurved - - a. Bent back.

Retrofiexed - a. Bent back, reflexed.

Retrorse - a. Directed backweard or downward.

Retuse - a. With a shallow notch at a rounded apex.

Reversion - - n. A change backward, as to an earlier condition.

Revolute - a. Rolled backward, margin rolled toward lower side.

Rhabdocarpous - a. Long-fruited; with fruits shaped like a rod.

Rhachilla - n. See Rachilla.

Rhachis - n. See Rachis.

Rhaphe - n. See Raphe.

Rhaphis (pl. rhaphides) - n. See Raphis.

Rheophilous - a. Creek-loving.

Rheotropism - - n. A turning in response to a current of water.

Rhipidium - n. A fan-shaped cyme, the lateral branches being developed alternately in two opposite directions.

Rhizanthous - a. Root-flowered; flowering from the root or seeming to do so.

Rhizocarp - n. A sporangium such as is produced on rootlike processes or members of the Marsileacae.

Rhizomatose - a. Having the character of a rhizome.

Rhizome - - n. The rootstock or dorsiventral stem having a rootlike appearance, prostrate on or under ground, sending off rootlets, the apex progressively sending up stems or leaves.

Rhizophilous - a. Growing attached to roots.

Rhizophyllous - a. Growing attached to roots.

Rhizophyllous - a. Roots that proceed from leaves.

Rhizotaxia - n. The system of arrangement of roots.

Rhomboidal - a. Approaching a rhomibc outline, quadrangular, with the lateral angles obtuse.

Rhyacophilous - a. Torrent-loving.

Rib - n. A primary vein, especialy the central longitudinal or midrib.

Rictus - n. The mouth or gorge of a bilabiate corollda.

Rigescent - a. Becoming rigid.

Rigid - a. Stiff, inflexible.

Rimose, rimous - a. With chinks or cracks, as in old bark.

Rind - n. The outer bark of a tree, all the tissue outside the cambium; the tough outer layer of some fleshy fruits.

Ringent - a. Gaping, as the mough of an open bilabiate corolla.

Riparious - a. Growing by rivers or streams.

Ripe - a. Mature, characterized by the completion of an organ or organism for its alloted function.

Rivulose - a. Having small sinuate channels; marked with lines like a rivulet.

Rogue - n. A gardener's name for a plant which does not come true from seed; a variation from the type.

Root - n. The descending axis, growing opposite from the stem, without nodes, mostly developing underground, and absorbing moisture from the soil.

Rootlet - n. A very slender root or the branch of a root.

Rootstock - n. Subterranean stem; rhizome.

Rosette - n. A cluster of spreading or radiating basal leaves.

Rostellate - a. The diminutive of rostrate, somewhat beaked.

Rostellum - n. A little beak. a slender extension from the upper edge of the stigma in orchids.

Rostrate - a. With a beak, narrowed into a slender tip or point.

Rostrum - n. Any beak-like extension; the inner segment of the coronal lobes in asclepiads.

Rosula - n. Small rose; a rosette of

leaves, as in houseleek.

Rosulate – a. In the form of a rosette.

Rotate – a. Wheel-shaped, circular, and flat, applied to a gamopetalous coroll with a short tube.

Rotund – a. Rounded in outline, somewhat orbicular, but a little inclined toward the oblong.

Rotundifolious – Round-leaved.

Rubescent – a. Reddish, becoming red.

Rubiginose – a. Rust-colored usually applying to glandular hairs.

Ruderal – n. A plant growing in rubbish or waste places.

Rudiment – n. An imperfectly developed organ or part.

Rudimentary – a. Arrested in an early stage of development.

Rufescent – a. Becoming reddish.

Rufous – a. Reddish-brown.

Rugouse – a. Wrinkled, as a leaf surface with sunken veins.

Rugulose – a. Finely wrinkled.

Ruminate – a. Having a chewed appearance.

Runcinate – a. Saw-toothed or sharply incised, the teeth retrorse.

Runner – n. A stolon, an elongated lateral shoot, rooting at intervals, the intermediate part apt to perish,. and thus new individuals arise.

Rupestral – – a. Growing on walls and rocks.

Rupicolous – a. Growing among the rocks.

Ruptile – a. Dehiscing in an irregular manner.

Rupturing – a. Bursting irregularly.

S

Sabulicolous – a. Growing in sandy places.

Saccate – a. Gag-shaped.

Sacciform – n. A one-seeded indehiscent pericarp, enclosed within a hardened calyx, as the Marvel of Peru; applied to such. fruits as those of *Chenopodium* which burst irregularly.

Sagittate – a. Enlarged at the base into two acute straight lobes, like the barbed head of an arrow.

Saline – a. Of or pertaining to salt.

Salverform – a. With a slender tube and an abruptly expanding limb, as that of the *Phlox*; hypocrateriform, salver-shaped.

Salver-shaped – a. See Salverform.

Samara – n. A winged achene-like fruit.

Samaroid – a. Samara-like.

Sanguine – – a. Blood-colored, crimson.

Sap – n. The juice of a plant; the fluid contents of cells and young vessels, consisting of water and salts absorbed by the roots and distributed through the plant.

Sapid – a. Havifng a pleasant taste.

Saponaceous – a. Soapy, slippery to the touch.

Sapor – n. The property of the taste of a plant, such as bitterness.

Sapromyiophilous – a. Growing in humus.

Saprophyte – – n. A plant deriving all of its nourishment form the bodies of decaying organisms.

Sap wood – n. The new wood in an exogenous tree, so long as it is pervious to the flow of water, the alburnum.

Sarcospore – n. A plant with fleshy disseminules.

Sarment – n. A long, slender runner, or stolon, as in the strawberry.

Sarmentose – a. With long slender runners.

Saurochore – n. A plant disseminated by lizards or snakes.

Sausage-shaped – a. Allantoid.

Sautellus – n. A bulblet such as those of *Lilium tigrinum*.

Saxicolus – a. Dwelling or growing among rocks.

Scabridulous – a. Slightly rough.

Scabrous – a. With short bristly hairs; rough to the touch.

Scalariform – a. Havifng markings suggestive of a ladder.

Scale – n. Any thin scarious body, usually a degenerative leaf, somethimes of epidermal origin.

Scandent – – a. Climbing in any manner.

Scape – n. Leafless peduncle arising from the ground, it may bear scales or bracts but not foliage-leaves and may be one-or many-flowered.

Scaphoid – a. Boat-shaped.

Scapiflorous – a. Having flowers borne on a scape.

Scapiform – a. Bearing or resembling a scape.

Scar – n. A mark left on the stem by the separation of a leaf, or on a seed by its detachment; a cicatrix.

Scarious – a. Thin, dry, and membranous, not green.

Scarlet – a. Vivid red, having some yellow in its composition.

Schistaceous – a. Slate gray.

Schizocarp – n. A pericarp which splits into one-seeded portions, mericarps.

Schizopetalous – a. With cut petals.

Scion – n. A young shoot, a twig used for grafting.

Sciophilous – a. Shade-loving.

Sciophyll – n. A shade leaf.

Scissille – – Separating, easily split.

Sciuroid – a. Curved and bushy like a squlrrl's tail.

Scleranthium – – n. An achene enclosed in an findurated portion of the calyx tube, as in *Mirabilis*.

Scleroid – a. Having a hard texture.

Sclerophyllus – a. Hard-leaved.

Scobiform – a. Having the appearance of sawdust.

Scobina – n. The zigzag rasplike rachilla of the spikelet of some grasses.

Scobinate – a. With a surface that feels rough as though rasped.

Scorpioid – a. Said of a coiled cluster in which the flowers are two-ranked and borne alternately at the right and left.

Scorpioid cyme – n. Cincinnus, the lateral branches developed on opposite sides alternately, as in Boraginaceae.

Scotophilous – a. Dwelling in darkness.

Scrobiculate – a. Marked by minute or shallow depressions, pitted.

Scrotiform – a. Pouch-shaped.

Scurf – n. Small branlike scales on the epidermis.

Scurfy – a. Coverd with small branlike scales.

Scutate – a. Buckler-shaped, like a small shield.

Scutellate – a. Shaped like a small plater.

Scutellum – n. Any of several small shield-shaped parts or orgtans; a conical cap of the endosperm in Cycadeae; the first leaf in a grass embryo attached at the basal node of the mesoctyl and serving as a food storage organ.

Scutum – n. The broad dilated apex of the style in asclepiads.

Seasonal amphichromatism – n The production of two differently colored flowers on the same stock due to season.

Seasonal heterochromatism – n The production of different colors in the flowers of the same inflorescence due to season.

Sebaceous – a. Like lumps of tallow.

Secondary peduncle – n. A Branch of a many-flowerd inflorescence.

Secund – a. Said of parts or organs directed to one side only, usually by torsion.

Seed – n. A mature ovule.

Seed-leaf – n. Cotyledon.

Seedling – n. A plant produced from seed, in distinction to a plant propagated vegetatively; a juvenile plant.

Seed-stalk – n. The funiculus or podosperm.

Segment – n. One of the parts of a leaf, petal, calyx, or perianth that is divided but not truly compound; any of the parts into which an organism naturally separates or is divided; a section.

Segregate - n. That which is kept apart; a segregate is a species separated from a super-species.

Sejugous - a. Having six pairs of leaflets, as some pinnatye leaves.

Selenotropism - n. Movements of plants caused by the light of the moon.

Selfed - a. Fertilized by its own pollen.

Self-pollination - n. Pollination by pollen from the same flower.

Semen (pl. semines) - n. The seed of a flowering plant.

Semester ring- - n. The ring produced in the wood of many tropical trees, in consequence of two periods of growth and rest in a year.

Semiglobose - a. Half-globose; hemispherical.

Semilunate - a. Shaped like a half-moon, crescent-shaped.

Seminiferous - a. Seed-bearing; used for the special portion of the pericarp bearing the seeds.

Seminiferous scale - n. In coniferae, that scale above the bractscale on which the ovules are placed and the seeds borne.

Senescence - n. Aging of protoplasm.

Senescent - a. Growing old or effete.

Sensitive - a. Responsive to stimuli, as the leaves of *Mimosa pudica*.

Sepal - n. One of the separate parts of a calyx.

Sepaloid - - a. With the texture of, or resembling a sepal.

Sepicolous - a. Inhabiting hedges.

Septate - a. Partitioned; divided by partitions.

Septentate - a. Having parts in sevens, as ion a compounmd leaf, with seven leaflets arising from the same point.

Septicidal - a. Said of a capsule that dehisces through the dissepiments or lines of junction.

Septifolious - Seven-leaved.

Septifragal - a. With the valves breaking dfaway from the dissepiments in dehiscence.

Septum - n. Any kind of partition, whether a true of false dissepiment.

Seriate - a. Disposed in series of rows, eigher transverse or longitudinal.

Sericeous - - a. Silky, clothed with closedly appressed, soft, straight pubescence.

Serotinal - a. Produced late in the season, or the year, as in autumn; autumnal.

Serotinous - a. Produced or occuring late in the season.

Serra - - n. The tooth of a serrate leaf.

Serrate - a. With sharp teeth on the margin pointing forward.

Serrulate - a. Serrate with minute teeth.

Sessile - a. Without a stalk of any kind, as a leaf without a petiole.

Seta (pl. setae) - - n. A bristle or bristle-shaped body; the arista or awn of grasses when terminal; a peculiar stalked gland in *Rubus*; used by cyperologists for the bristle within the utricle of certain species of *Carex*; it represents a continuation of the floral axis.

Setacepis - a. Bristle-like, with bristle.

Setiferous - a. Bearing bristles.

Setiform - a. In the shape of a bristle.

Setose - a. Bristly, beset with bristle.

Setulose - a. With minute bristles.

Sex - n. In botany,m male or female functions in plants.

Sexual - a. Pertaining to sex.

Shade leaves - n. Leaves adapted to modified light.

Sheath - n. Any long or more or less tubular structure surrounding an organ or part.

Sheathing - a. Enclosing, as by a sheath.

Shield - n. The staminods of cypripedium; in coniferae, the thick rhomboid extremlty of the conescales.

Shield-shaped - a. In the form of a

shield.

Shoot - - n. A young growing branch or twig; the ascending axis.

Shrub - n. A low, usually several-stemmed, woody plant; a bush.

Sigmoid - a. Said of a leaflet or segment that is curved sidewise in opposing directions; S-shaped.

Siliceous, silicious - a. Composed of or a abounding in silica.

Silicle - n. The short fruit of certain Cruciferae, silicule.

Silicolous - a. Growing in flinty soils.

Silicule - n. A short silique. not much longer than broad, silicle.

Siliqua - n. The peculiar pod of the Cruciferae, with two valves fallin away from a frame (the replum) on which the seeds grow, and across which a false partition is formed.

Silks - pl. - n. The style and sigmas in maize.

Silky - a. Having a covering of soft appresses fine hairs; sericeous.

Silva - n. See sylva.

Silvery - a. With a whitish metallic more of less shiny luster.

Simple - a. Of one piece; not compound.

Simple fruits - n. Those fruits which result form trhe ripening of a single pistil.

Simple inflorescence - n. A flower cluster with one axis, as a spadix, spike, or catkin.

Simple leaf - n. A leaf with one blade, with incomplete segmentaion.

Simple pistil - n. A pistil consisting of one carpel.

Sinistrorse - a. turning to the left; counterclockwis.

Sinuate - a. With a deep wavy margin.

Sinuous - - a. See sinuate.

Sinus - - n. The space between two lobes of a leaf or other expanded organ.

Siotropism - n. Response to shaking, as with *Mimosa*.

Skin - - n. A thin external covering, the cuticle or epidermis.

Skotophilous - a. See Scotophilous.

Sleep - n. The response of plants to the absence of light resulting in changes in position of organs such as leaves.

Sleep movement - n. Positions taken by leaves during the night; nyctitropic movement.

Smooth - a. Without roughness or pubescence.

Snail-plants - pl. - n. Plants which are supposed to be pollinated by snails and slugs; malacophilous plants.

Sobole - n. A shoot, especially from the ground.

Soboliferous - a. Sucker-bearing.

Sole - n. of a carpel, the end far-thest from the apex.

Soleaform - a. Sliper-shaped, almost resembling an hourglass.

Solitary - a. Single, only one from the same place; species of which the individuals are in extrteme isolation.

Sordid - a. Dirty in tint, chiefly applied to pappuses of an impure white.

Sorophore - - n. A gelatinous cushion of the ventral edge of the sporocarp of *Marsilea* and ferns.

Sorose - a. A fleshy multiple fruit, as a mulbery or pinearpple.

Sorus - n. A cluster of sporangia in ferns.

Spadiceous - a. As to color, date brown; having the nature of, or bearing a spadix.

Spadix - n. The thick or fleshy spike of certain plants, as the Araceae, surrounded or subtended jby a spathe.

Spananthus - - a. Having new flowers.

Sparse - a. Scattered.

Spathe - - n. The bract or pair of bracts surrounding or subtending a flower cluster or spadix; it is sometimes colored and flower-like, as in the *Calla*.

Spathella - n. An old name for the glumes of grasses; sometimes the paleae were included.

Spathellula – n. A palea of grass.

Spathe valves – n. The bractlike envelopes beneath the flowerrs in certain monocotyledons, as *Allium* and *Narcissus*.

Spathulate – a. Spatula-shaped.

Speciation – n. The processes wherby new species are formed.

Species (pl. species) – n. Groups of actually or potentially inter-breeding natural populations, which are reproductively isolated from other such.

Specific name – n. The latin appellative appropriated to a given species, usually an adjective, but sometimes a substantive used in apposition.

Specimen – n. A plant, or portion of one, prepared for botanic study.

Speiranthy – n. The state of having a twisted flower.

Spermatophyte – n. A phanerogam, a plant with true seeds.

Spermatozoid – n. A free swimming male gamete.

Sphalerocarpum – n. An accessory fruit, as an achene in a baccate calyx-tube.

Sphenoid – a. Wedge-shaped, cunneate.

Sphingophilous – a. With flowers pollinated by hawkmoths and noctthnal lepidoptera; they usually have a strong sweet smell, and nectar in flower-tubes.

Spicate – a. Like a spike, or disposed in a spike.

Spiciform – a. Spikelike, in the form of a spike.

Spicule – n. A diminutive or secodary spike; a fine, fleshy erct point.

Spiculose – a. With a surface covered with fine points.

Spike – n. An inflorescence consisting of a central rachis bearing a number of sessile flowers.

Spikelet – n. The unit of the inflorescence in grasses, consisting of two glumes and one or more florets; a diminutive spike.

Spikelike – a. Said of a dense panicle in which the pedicels and branches are short and hidden by the spikelets, as in *Phleum*.

Spiladophilous – a. dwelling in clay.

Spindle-shaped – a. fusiform, tapering from the middle toward each end.

Spine – n. A sharp woody out-growth from the stem, usually a modified branch, sometimes a petiole, stipule,or other part.

Spinescent – a. Ending in a spine or sharp point; more or less spiny.

Spinose – a. Spinelike or with spines.

Spinule – n. A little spine or spinelike process.

Spinulose – a. With small spines.

Spiny – a. Beset with spines.

Spiral – a. As though wound around an axis.

Spiral flower – n. A flower with the members arranged in spirals and not in whorls.

Spiricle – – n. A delicate coiled thread in the superficial cells of certain seeds and achenes which uncoils when moistened, as in *Ruellia*

Splint – n. A forester's term for alburnum or sapwood.

Sponsalia – n. Anthesis; the pollination period.

Sporadic – a. Occurring here and there, without continuous range.

Sporangiophore – n. A stalklike structure bearing sporangia.

Sporangiophore – n. A sac endogenously producing spores.

Spore – n. A cell which becomes free and capable of direct development into a new individual.

Sporocarp – – n. A receptacle containing sporangia or spores.

Sporophore – n. A spore-bearing branch or organ.

Sporophyll – – n. A spore-bearing leaf.

Sport – n. A sudden spontaneous deviation or variation of an organism from type, beyoud the usual limits of individual variation.

Spray – n. small branches or branchlets of trees with their leaves.

Spring wood – – n. Wood produced early in the year, characterized by larger ducts and cells than the later growth.

Spumose – a. Frothy.

Spur – n. A short, compact branch with little or no internodal development; a tublar or saclike projection from a blossom, as of a petal or sepal.

Squama (pl. squamae) – n. A scale of any sort, usually the homolog of a leaf.

Squamaceous – a. Scaly.

Squamate – a. Furnished with scales, scalelike leaves, or bracts.

Squamellate – a. With small or secondary scales.

Squamule – n. The hypogynous scale of grasses, the lodicule.

Squamulose – a. With small scales.

Squarrose – a. Spreading or recurved at the tip, said of the tips of some lemmas.

Squarrulose – a. Diminutively squarrose.

Stalk – n. The stem of any organ, as the petiole, peduncle, pedicel, filament, culm or stipe.

Stamen – n. The pollen-bearing organ of the flower, the male organ in the angiosperms.

Stamen, sterile – n. A body belonging to the series of stamens but without pollen.

Staminate – a. Having stamens and no pistil, male.

Stamineal – a. Relating to or consisting of stamens.

Staminode (staminodium, – pl. staminodia) – n. A sterile stamen, or a structure resembling such and borne in the staminal part of the flower; in some flowers, as in Canna. staminodia are petal-like and show.

Standard – n. The upper and broad more of les erect petal of a papilionaceous flower.

Stasad – n. A plant of stagnant water.

Stasimorphy – n. Alternation of form caused by arrested development.

Stasophilous – a. Dwelling in stagnant water.

Staurigamis – n. Cross-fertilization.

Stegium – n. Threadlike appendages sometimes found covering the style of asclepiads.

Stelipilous – a. With stellate hairs.

Stellate – a. Star-shaped or radiating like the points of a star.

Stellate scale – pl. – n. Discs borne bhy their edge or center.

Stelliform – a. Star-shaped.

Stem – n. The main axis of a plant; leaf-bearing and flower-bearing as distinguished from the root-bearing axis.

Stem, subterranean – n. A rhizome, tuber, bulb, or corm.

Stenopetalous – a. Narrow-petaled.

Stenophyllous – a. Narrow-leaved.

Stenotropism – n. The state of having narrow limits of adaptations top varied conditions.

Stereotropism – n. The act of responding to contact stimuli.

Sterile – a. Barren, as a flower destitute of plstil; used for staminate or neuter flowers.

Sternotribal – a. Said of flowers whose anthers are so aranged as to dust their pollen on the under part of the thorax of their insect visitors.

Stigma (pl. stigmas or stigmats) – n. The part of a pistil or style which receives the pollen; a point on the spores of Equisetum.

Stigmatic – a. Pertaining to the stigma.

Stinging hair – n. A hollow hair seated on a gland which secretes an acid substance, as in nettles.

Stipe – n. The "leaf-stalk" of a fern; the suport of a gynoecium or carpel.

Stipitate – a. With a stipe.

Stipular – a. Having stipules or relating to them.

Stipulate – a. Furnished with stipules.

Stipule – n. One of the pair of appendages borne at the base of the leaf in many plants.

Stipulose – a. Having stipule.

Stolon – n. A sucker, runner, or any basal branch which is disposed to root.

Stoloniferous – a. With stolons or runners that take root.

Stoma (pl. stomata) – n. A specialized orifice in the epidermis communicating with intercellular spaces.

Stomate – n. See Stoma.

Stone – n. The hard endocarp of a drupe, a pit.

Stone fruit – n. A drupe such as a plum or peach.

Stool – n. The base of a plant from which offsets or layers are taken; several stems arising from the same root, as in wheat.

Stopple – n. A projection or lid on ao admit passage of the pollen tube.

Stramineous – a. Strawlike or straw-colored.

Strap – n. The ligule of a ray floret in Compositae.

Strap-shaped – a. Ligulate or lorate.

Straw – n. The jointed hollow culm (stem) of grasses.

Streptocarpous – a. With fruits spirally marked; with twisted fruits.

Striate – a. With fine grooves, ridges or lines of color.

Strict – a. Stiffly upright, rigid, erect.

Striga – a. Stiffly upright, rigid, erect.

Striga – n. A small, straight, hairlike scale.

Strigose – a. Beset with sharppointed appressed straight and stiff hairs or bristle; hispid.

Strobilaceous – a. Relating to or resembling a cone.

Strobile – n. An inflorescence made up largely of imbricated scales, as the hop or the pine; a cone.

Strombuliform – a. Spirally twisted like some shell.

Strombus – n. A spirally coiled legume, as in *Medicago*.

Strombus-shaped – a. Shaped like snail shell.

Strophiole – n. An apendage at the hilum of certain seeds; a a caruncle.

Stylar – a. Relating to the style.

Stylar column – n. The column of orchids.

Style – n. The more or less elongated part of the pistil between the ovary anbd the stigma.

Stylopod (-ium) – n. The enlargemnet at the base of styles in Umbelliferae.

Suaveolent – a. Sweet-smelling, fragrant.

Subcordate – a. Almost cordate.

Suber – n. Cork.

Sub-erose – a. Slightly gnawed in apearance.

Suberous – a. Corky in texture.

Subfamily – n. a group of genera within a family.

Submerged – a. Growing under water.

Suborder – n. A group of genera lower than a family.

Subpetiolar – a. Under the petioles, as the buds in *Platanus*.

Subspecies – n. A group of forms ambiguous in rank, and between a variety and a species.

Subtend – v. To stand below and close to, as a bract below a flower or a leaf below a bud.

Subterraneous – – a. Under the ground.

Subtribe – n. A divsion between a tribe and a genus.

Subula – n. A fine, abrupt, sharp point.

Subulate – a. Awl-shaped.

Succulent – a. Juicy or pulpy.

Sucker – n. A shoot of subterranean origin; a hasutorium, some-times restricted to the penetrating organ or papilla.

Suffrutescent – a. Obscurely or

somewhat shrubby.

Suffruticous - a. Pertaining to a low and somewhat woody plant. diminutigely shrubby or fruticose. woody at base.

Sulcate - a. Grooved or furrowed lenghwise.

Sulcus - n. A small groove.

Sulfureous - a. Sulfur-colored.

Sulphur rain - Pine pollen carried in excessive amount by air currents.

Super-axillary - a. Borne above the axil.

Superficiales - pl. - n. Said of leptosporangiate ferns, with sori arising form the surface of the fround.

Superior - a. Growing or placed above; also in a lateral flower on the side next to the axis; the posterior, or upper lip of a corolla is the superior.

Superior ovary - n. An ovary with the floral envelopes inserted below it on the tours; a hypogynous ovary.

Supernatant - a. Floating on the surface.

Superposed - a. Placed vertically over some other part.

Supine - a. Prostrate with face turned upward.

Supra-axillary - a. See Super-axillary.

Surculose - a. Producing suckers.

Surculum - n The rhizpme of a fern.

Surcurrent - a. Having winged expansions from the base of the leaf prolonged up the stem.

Suspended ovule - n. An ovule hanging from the apes of the cell.

Suture - - n. A junction or seam of union; a line of opening or dehisce.

Sword-shaped - a. Ensiform.

Syconium - n. A hollow multiple fruit, as that of a flg.

Sylva - n. An account of the trees of a district, or a discourse on trees.

Sylvestrine - a. Of or pertaining to woods.

Symbiont - n. Organism in symbiosis.

Symbiosis - n. The living together of dissimilar organisms, with benefit to one only, or both; also styled commensalism, consortism, individualism, mutualism, nutricism, prototrophy, and syntropism.

Symbiosis, antagonistic - n. A struggle between two organisms.

Symbiosis, conjunctive - n. An intimate blending of the symbionts so as to form an apparently single body.

Symbiosis, disjunctive - n. Symbiosis with no direct union between the symbionts.

Symbiotic - a. Relating to symbiosis.

Symmetrical - a. Actinomorphic; regular, capable of division by a longitudinal plane into similar halves.

Symmetry - n. A due proportio of the severalk parts of a body to each other.

Symmetry, billateral - n. Symmetrical arrangement capable of equal division in one plane only, as in a pea flower.

Symmetry, radial - n. Symmetrical arrangement capable of equal division in more than one direction through the center. as in the mallow flower.

Sympatric - a. Inhabiting one and the same area.

Sympetalous - - a. With partially or wholly fused petal.

Symphyllode - n. A cone scale of Abietineae.

Symphyllodium - n. The combined ovuliferous scales in the flower of certain Coniferae.

Symphyllodium - n. Coalescence; fusion of like parts.

Synacmy - n. A condition in which stamens and pistils mature to gether;l the opposite of heteracmy.

Synantherous - a. With anthers joined together.

Synanthesis - n. The simultaneous maturing of stamens and pistils.

Syncarp - n. A multiole or fleshy aggregate fruit, as the mulberry or magnolia.

'Syncarpous – a. Composed of two or more united carpels.

Synchronic species – pl. – n Species which belong to the same time level, contemporary.

Syncolliphytum – n. A plant in which the perianth becomes combined with the pericarp.

Synema (pl. synemate) – n. The column of monadelphous stamens, as in Malvaceae.

Synoecious – a Having staminate and pistillate flowers both present in the same head.

Synoym – n. A name with the same meaning as another name in the same language but spelled differently;l in taxonomy, synonyms are two or more scientific names for the the same taxon, one of which is correct and the other incorrect under the International Rules of Nomenclature.

Synoymous – a. Having the same meaning.

Synonymy – n. Discarded names for identical objects.

Syntype – n. One of two or more specimens or elements used by the author when no holotype was designated, or one of two or more specimens simultaneously designated as type.

Synzoochory – n. Dispersion by animals.

Syrtidophilous – a. Dwelling on dry sand bars.

Systellophytum – – n. A plant with a persistent calyx appearing to form part of the fruit.

System – n. A Scheme of classification, as the Natural system.

Systematic botany – n. A study of plants in their mutual relationships and taxonomic arrangement.

Systerophyte – n. A plant which lives on dead matter; a saprophyte.

T

Tabasheer – n. A siliceous concretion occurring in the joints of bamboo.

Tachyspore– n. A plant which quickly disperses its seeds.

Tactile – a. Sensitive to the touch.

Taphrophilous – a. Ditch-dwelling.

Tap-root – – n. The primary descending root, forming a direct continuation from the radicle.

Tartarous – a. With a loose or rough crumbling surface, as some lichens.

Tassel – n. The staminate inflorescence in maize.

Tawny – a. Fulvous, dull brownish-yellow.

Taxis – n. The reaction of free organisms in response to external stimuli by movement.

Taxon (pl. taxa) – n. A general term applied to any taxonomic element, population, or group, irrespective of its classification level.

Taxonomy – n. The systematic classification of organisms.

Tectoparatype – n. A specimen selected to show the microscopic structure of the original type of a species or genus.

Tectum – n. Roof.

Tegule – n. One of the involceral bracts subtending the flower head in Compositae.

Teknospore – n. A spore produced directly from male or female organs of Equisetaceae and many ferns.

Teleolog – n. The doctrine of final causes, or theory of tendency to an end.

Telmatophilous – a. Marsh-loving.

Telmicolous – a. Dwelling in fresh-water marshes.

Telotropism – n. The act of turning to one stimulus to the exclusion of all others.

Temulentous – a. Drunken, nodding in a jerky, irregular manner.

Tendril – n. A rotating or twisting threadlike process, or extension by which the plant grasps an object and clings to it for support; morphologically it may be a modified stem, leaf, leaflet, or stipule.

Tentacle – n. A sensitive glandular hair, as those on the leaf of *Drosera*.

Tepal – n. Used in the plural for sepals and petals of similar form and not readily differentiated.

Teratology – n. The study of malformation and monstrosities.

Terete – a. Circular in transverse section.

Terminal – a. Proceeding from, or belonging to the end or apex.

Ternary – a. In threes, trimerous; the result of the third axial order, as derived from the primary.

Ternate – a. In threes.

Terrestrial – a. Growing in the soil in distinction from growing in water or other habitats.

Terricolous – a. Dwelling in the ground.

Tesselate – a. With the surface marked with sqeare of oblong depressions.

Testa – n. The outer coat of the seed, usually hard and brittle.

Testiculate – a. Shaped like the tubers of orchids and fruit of *Mercurialis*.

Tetrad – n. A group of four objects, as the four pollen-grains formed from one pollen-mother cell.

Tetradymous – a. Having four cells or cases.

Teradynamous – a. Having four long stamens and two short,as in Cruciferae.

Tetragonal – a. Four-angled.

Tetramerous – a. Of four members.

Tetrandrous – a. With four sepals.

Tetrastachyous– a. With four spikes.

Tetrastichous – a. In four vertical ranks.

Thalamus – n. The receptacle of a flower.

Thalloid – a. Resembling or shaped like a thallus.

Thallus – n. A flat leaflike organ; in some Cryptogams, the entir cellular plant body without differentiatiion into stem and foliage.

Thelephorous – a. Covered with nipple-like prominences.

Theoretic diagram – n. A floral diagram of the theoretic components, not necessarily the same as seen on inspection.

Thermocleistogamy – n. Self-pollination taking place with flowers the opening of whose perianth has been inhibited by low temperature.

Thermophilous – a. Dwelling in warm water.

Thermotropism – n. The act of turning in response to heat.

Therophyllous – a. Producing leaves in summer, deciduous-leaved.

Therophyte – n. A plant which completes its development in one season, its seeds remaining latent during the hot season; an annual.

Thigmomorphosis – n. Change in the original structure due to contact, as the adhering discs of *Ampelopsis*.

Thigmotaxis – n. The result of mechanical stimulus.

Thinicolus – a. Dwelling on shifting sand dunes.

Thinophilous – a. Dune-loving.

Thorn – n. A spine, usually an aborted branch, simple or branched.

Throat – n. The Opening or orifice into a gamopetalous corolla or perianth; the place where the limb joins the tube.

Thyrse – n. A compact and more orless compound panicle; more correctly a panicle-like cluster with main axis indeterminate and other parts determinate.

Thyrsoid – a. Resembling a thyrse.

Thyrsula – n. A little cyme which is borne by most Labiates in the axil of the leaves.

Tiller – n. A sucker of branch from the base of the stem.

Tillering – a. Throwing out shoots from the base of the stem.

Timber-line – n. The upper limit of tree vegetation on the mountains or high latitudes.

Tiphed – n. A pond plant.

Tiphophilous – a. Pond-loving.

Tomentose – a. With tomentum; densely woolly or pubescent; with matted soft woollike hairiness.

Tomentulose – a. Somewhat or delicately tomentose.

Tomentum – n. A densely matted pubescence.

Tongue – n. Ligule.

Tooth – n. A small, pointed marginal lobe.

Toothed – a. Dentate.

Topotype – n. A turning towards a place from which a stimulus comes.

Topotype – n. A specimen of a named species from the original locality.

Torose – a. Cylindrical with contractions at intervals, somewhat moniliform.

Torulose – a. The diminutive of torose.

Torus – n. The receptacle of a flower, that portion of an axis on which the parts of the flower are inserted.

Trabecular – a. Like a cross-bar.

Trabeculate – a. Cross-barred.

Trace – n. A strand of vascular tissue connecting the leaf with the stem.

Trachycarpous – a. Rough-fruited.

Trailing – a. Prostrate, but not rooting.

Trap flowers – pl.n. Prison flowers, which confine insect visitiors until pollination has taken place.

Trap-hairs – pl. n. Special hairs which confine insects in certain flowers until pollination is effected.

Trap-prison – n. A flower, such as *Aristolochia*, which confines insect visitors until pollination has taken place.

Traumatic – a. Of or pertaining to a wound.

Traumatism – n. Abnormal growth in consequence of injury.

Traumatropism – n. The sensitiveness of certain plant organs or wounds.

Tree – n. A woody plant that produces one main trunk and a more of less distinct and elevated head.

Triachaenium – n. A fruit similar to a cremocarp, but of three carpels.

Triad – n. A group of three objects.

Triadelphous – a. With stamens in three sets.

Trinadrous – a. Having three stamens.

Triangulate - a. Three-angled.

Tribe – n. A group superior to a genus, but less than an order.

Tricamarous – a. Said of a fruit composed of three loculi.

Tricarinate – a. With three keels or angles.

Tricarpellary - a. Of three carpels.

Trichasium – n. A cymose inflorescence with three branches.

Trichocarpous – a. Hairy-fruited.

Trichocephalous – a. With flowers collected into heads, and surrounded by hairlike appendages.

Tricholoma (pl. tricholmata) – n An edge or border with hairs.

Tricholma – n. Any hairlike out growth of the epidermis, as a hair or bristle.

Trichotomous–a. Three-forked, bran-ching into three divisions.

Tricolor – a. Three-colored.

Tricussate – a. Said of whorls of three leaves each, ternate.

Tridentate – a. Three-toothed.

Tridigitate – a. Thrice digitate, with three fingers.

Tridynamous – a. With three stamens out of six being longer than the rest.

Trifid – a. Divided into three parts.

Trifoliate – a. Having three leaflets.

Trifurcate – a. Having three forks or branches.

Trigamous – a. Bearing three kinds of flowers, trimorphic.

Trigeminous – a. Tergeminate, trijugate, triple.

Trigonous – a. Three-angled.

Trilobate – a. Three-lobed.

Trimerous – a. In threes, three-membered parts.

Trimonoecious – a. With perfect, staminate and pistillate flowers on the one plant.

Trimorphic - a. Occurrring under

three forms, as with long, short, and intermediate styles.

Trimorphism – n. Heterogamy, with long, short, and mid-styled flowers.

Trinervate – a. Three-nerved.

Trioecious – a. With perfect, staminate and pistillate flowers on different individual plants with in the species.

Tripartite – a. Divided into three parts.

Tripinnate – a. Thrice pinnate.

Tripterous – a. Three-winged.

Triquetrous – a. With three salient.

Tristichous – a. In three vertical rows.

Trisulate – a. With three grooves or furrows.

Triternate – a. Three times three; the leaflets or segments of a twice ternate leaf again divided into three parts.

Tropic – a. Reacting to a stimulus by external change in an organism.

Tropism – n. A curvature which results from a response to some stimulus; the disposition to respond by turning or bending.

Trumpet-shaped – a. Tubular with dilated orifices, salver-shaped.

Truncate – a. Ending abruptly, the base or apex nearly or quite straight across.

Tryma (pl. trymata) – n. A drupaceous nut with dehiscent exocarp.

Tuber – n. A short, thickened branch of a subterranean stem, beset with buds or "eyes".

Tubercle – n. A little tuber; any excrescence, as on the roots, ascribed to the action of symbiotic organisms.

Tuberculate – a. Furnished with knoblike excrescences or tubercles.

Tuberiferous – a. Bearing tubers.

Tuberoid – a. Said of a fleshy-thickened root, resembling a tuber, as in many terrestrial orchids.

Tuberous – a. Cespitose, clustered, or clumped.

Tumescent – a. Somewhat tumid, inflated, or stollen.

Tumid – a. Swollen.

Tunic – n. The skin of a seed, the spermoderm; the coat of a bulb; any loose membranous skin not formed from the epidermis.

Tunicated – a. Composed of concentric layers or coats, as the bulb of an onion.

Turbinate – a. Top-shaped.

Turfaceous – a. Pertaining to bogs.

Turgescence – n. The distension of a cell or cellular tissue by water or other liquid.

Turgid – a Swollen from fullness, but not from air.

Turion – n. A scaly, often thick and fleshy, shoot produced from a bud on an underground root-stock, as *Asparagus*.

Tussock – n. A tuft of grass or grasslike plants.

Twig – n. A small shoot or branch of a tree.

Twiner – n. A plant which twines or climbs by winding its stem around a support.

Type – n. A nomenclatural type is that constituent element of a taxon to which the name of the taxon is permanently attached, whether as an accepted name or as a synonym.

Type specimen – n. The original specimen from which a description was written.

Type specimens – pl. n. Ecotypes.

Typical – a. In classification, conforming to the originally described specimen.

Typonym – n. A synonym: a name based on the same type, pecimen, or concept as another and older name.

U

Ubiquitous – a. Occurring everywhere.

Uliginose – a. Growing in swamps.

Umbel – n. An indeterminate inflorescence consisting of several pedicellate flowers having a common point of attachment.

Umbel, compound – n. An umbel with each ray itself bearing an umbel.

Umbel, cymose – n. An apparent umbel, but with the flowers opening centrifugally; a cyme which simulates an umbel.

Umbellent – n. A secondary umbel.

Umbelliferous – a. Bearing umbels.

Umbelliform – a. In the shape of an umbel.

Umbellule – n. An umbellet; a small umbel.

Umbilical cord – n. A vascular strand by which seeds are sometimes attached to the placenta, the funiculus.

Umbilicate – a. Depressed in the center.

Umbilicus – n, The hilum of a seed.

Umbo (pl. umbnes) – n. A boss of protuberance.

Umbonate – a. Bearing an umbo or boss in the center.

Umbonulate – a. Having or ending in a very small boss or nipple.

Umbracticolous – a. Growing in shady places.

Umbraculiferous – a. Bearing an umbrella.

Unarmed – a. Destitute of prickles or other armature; sometimes it means pointless, muticous.

Uncate – a. Hooked, bent, the tip in the form of a hook.

Uncinate – Hooked at the point, with hooks.

Unctuous – a. Having a surface which feels greasy.

Undate – a. Wavy, undulate.

Undulate – a. Furnished with a claw.

Ungulate – a. Clawed.

Unicarpellate – a. With fruit consisting of a single carpel.

Unilateral – a. One-sided, either originating on or, usually, all turned to one side.

Unipetalous – a. Having a corolla of only one petal, the others not being developed (not used for gamopetalous).

Uniseriate – a. In one horizontal row or series.

Unisexual – a. Of one sex; with either stamens or pistil or their representative.

Unitypic – a. Monotypic, of one type.

Unsymmetrical – a. Irregular, asymmetrical.

Urceolate – a. Pitcher-like, hollow and contracted at the mouth like an urn or pitcher.

Urceolus – n. The two confluent bracts of *Carex*, the utricle; any flask-shaped anomalous organ; small pitcher.

Urn – n. The base of a pyxis.

Utricle – n. A small bladdery pericarp, as in *Atriplex*; a membranous sac surropunding the fruit proper in *Carex;* any bladder-shaped appendage.

Utricular – a. Bladder-shaped.

V

Vagina – n. The sheathing petiole which forms a continuous tube, as in sedges or grasses.

Vaginate – a. Sheathed.

Vaginiferous – a. Bearing sheaths.

Vallecula – n. the groooves in the intervals betweeen the ridge in the fruit of Umbelliferae.

Vallecular – a. Of or pertaining to the grooves in the fruit of Umbelliferae.

Valvate – a. Opening by valves, as in most dehiscent fruits and some anthers; parts of a flower bud that meet without over-lapping.

Valve – n. One of the pieces into which a capsule naturally separates at maturity; the segment of a calyx meeting in vernation without overlapping; a partially detached flap of an anther.

Variant – n. A form arising from variation.

Varicose – a. Abnormally enlarged in places, irregularly swollen.

Variety – n. A group of organisms within a species that differs from other members or groups within the species in one or more minor characteristics but not enough to justifiy a new specific epithet.

Vascular – a. With vessels or ducts.

Vasculum – n. A collecting can for botanical specimens.

Vegetable – a. Belonging to or consisting of plants.

Vegetation – n. The sum total of all plants growing on an area.

Vein – n. A strand of vascular tissue in a flat organ such as a leaf.

Velamen (pl. Velamines) – n.   A parch- ment-like sheath or layer of spiral-coated air cells on the roots of some tropical epiphytic orchids and aroids.

Velum – n. The membranous indusium in *Isoetes*.

Velumen – n. Close, short, soft hairs,

Velutinous – a. Velvety, due to a coating of fine soft hairs.

Venation – n. Veining; arrangements or disposition of veins.

Venenose – a. Very poisonous, venomous.

Venomous – a. Poisonous.

Ventral – a. Of or pertaining to the belly; pertaining to or designating that surface of a carpel, petal, etc which faces toward the center of the flower.

Ventural suture – n.   The ventral seam or line of dehiscence in a carpel.

Ventricose – a. Swelling or inflated on one side, as the corolla of some Labiates and Scrophu-lariaceae.

Venuloso-hinoideous – a.   Said of veins which form the midrib and are parallel and cross-veined.

Vermicualr – a. Worm-shaped.

Vermiform – a. Worm-shaped.

Vernal – a. Pertaining to spring.

Vernalization – n. The process of shortening the vegetative period of plants by seed treatment.

Vernation – n. The disposition or arrangement of leaves in the bud.

Vernicose – a. Shiny as though varnished.

Verrucose – a. Covered with wart-like elevations.

Versatile – a. Hung or attached near the middle and usually moving freely, as an anther attached crosswiswe on the apex of a filament and capable of turning.

Versicolor – a. Variously colored, as of one color blending into another, or changing in color.

Verticil – n. A whorl, or circular arrangement of similar parts round an axis.

Verticillaster – n. A false whorl, composed of a pair of opposed cymes, as in Labiatae.

Verticillastrate – a. Bearing or arranged in clusters resembling whorls.

Verticillate – a. Whorled with two or more leaves at a node, cyclical.

Vesicle – n. A small bladder or cavity.

Vesicular – a. Composed or covered with little bladders or blisters.

Vespertine – a. Appearing or expanding in the evening.

Vestige – a. The remaining trace of an organ which was fully developed in some ancestral form.

Vestigial – a. Rudimentary.

Vexillum – n. The standard or large posterior petal of a Papilionaceous flower.

Viatical – a. Growing by roadsides or paths.

Villose – a. With long, silky, straight hairs.

Vimineous – a. Bearing long and flexible twigs.

Vinaceous – a. Wine-colored, purplish-red.

Vine – n. The plant which bears grapes, *Vitis vinifera;* applied to any trailing or climbing stem, or runner.

Vinicolor – a. The color of wine, dark or purple-red.

Viniferous – a. Wine-bearing.

Violaceous – a. Violet-colored.

Virescent – a. Turning green.

Virgate – a. Wand-shaped, twiggy.

Viridescent – a. Becoming green.

Viscid – a. Sticky from a tenacious coating or secretion.

Vitreous – a. Transparent, hyaline, formerly used for the light green of glass.

Vitta – n. An aromatic oil tube of the pericarp of most Umbelliferae.

Viviparous – a. Germinating or sprouting from seed or bud while attached to the parent plant.

Volute – a. Rolled up in any way.

W

Wart – n. A hard or firm excrescence.

Web – n. A network of interlacing threads or fibers.

Wedge-shaped – a. Cuneate.

Weed – n. A plant detrimental to man's interest, displeasing to the eye, or of no apparent value.

Weedy – a. With the attributes of a weed.

Weel – n. An arrangement of hairs which keeps out unhidden insect guests from flowers.

Whorl – n. Cyclic arrangement of appendages at a node.

Wind-pollinated – a. With the pollen conveyed the the agency of air; anemophilous.

Wing – n. Any membranous expansion attached to an organ; the lateral petal of a Papilonaceous corolla.

Witch's broom – n. A disease shown by tufts of shoots, due to attack by fungi or mites.

Wood – n. The lignified portion of plants, included within the cambium, but exclusive of the pith.

Woolly – a. Lanate, tomentose, clothed with long and tortuous or matted hairs.

Wrinkled – a. Rugose, creased.

X

Xanthic – a. Tending toward yellow.

Xanthophyll – n. A yellow substance insoluble in water and associates with chlorophyll.

Xanthorrhiza – Yellow-root.

Xenia – n. The direct influence of foreign pollen on the parts of the mother plant.

Xenochroma – n. The effect of foreign pollen producing a change in the color of the fr

Xenogamy – n. Cross- fertilization.

Xeriobole – n. A plant whose seeds are scattered by dehiscence due to dryness.

Xerochase – n. A fruit that opens in dry air and closes in humid air.

Xerochastic – a. Said of plants whose fruits burst by desiccation thereby scattering their seeds or spores.

Xerophilous – a. Growing in arid places.

Xerophyte – n. A plant which can subsist with a small amount of moisture, a desert plant.

Xerotropism – n. The tendency of plants or parts thereof to alter their position to protect them-selves from desiccation.

Xylem – n. The wood elements of a vascular bundle.

Xylem rays -pl. n. – A radial plate of xylem between two medullary rays.

Z

Zenotropism – n Negative geotropism.

Zigzag – a. Having short bends or angles form side to side.

Zonate – a. Marked circularly as the leaves of *Pelargonium zonale*; zoned, banded.

Zoned – a. Colored in rings or circles.

Zoochore – n. A plant distributed by animals.

Zoophilous – a. Pollinated by the agency of animals.

Zoospore – n. A free-moving spore of the lower Cryptogams; an asexual reproductive cell with cilia.

Zygomorphic – a. Capable of division by only one plane of symmetry.

Subject Classification

Aestivation

Aestivation – n. The manner in which the floral parts are arranged in the bud before expansion.

Convolute – a. With petals rolled up in such a way that the outer part of each covers the inner part of the one in front o. it, while in turn its inner part is covered by the one behind it. In cross section, the petals re-semble curved spokes in a wheel.

Corrugate, or crumpled – a. Characterized by the irregular crumpling of otherwise plane petals due to rapid growth in a confined space.

Imbricate – a. With the outer parts overlapping the inner parts, as the shingles on a roof; they break joints.

Induplicate – a. Valvate with the margins of each part projecting inward.

Involute – a. Valvate with the margins of each part rolled inward.

Plicate – a. With the parts folded ingthwise.

Quinquncial – a. In aestivation, partially imbricated of five parts, two being exterior, two interior, and a fifth one having one margin exterior and the other interior, as in the calyx of the rose.

Reduplicate – a. Valvate with the margins projecting outward.

Valvate – a. With the parts meeting by their abrupt edges without overlapping or turning.

Agents of Pollination

Animal-pollinated:
  Zoochore – n. A plant distributed by animals (usually applied to plant dispersal).

Zoophilous – a. Animal-loving; flowers pollinated by animals.

Bat-pollinated:
  Chiropterophilous – a. Said of flowers pollinated by birds.
  Ornithophilous – a. Said of flowers pollinated by bats.

Bird-pollinated:
  Ornithogamous – a. Said of flowers pollinated by birds.
  Ornithophilous – a. Said of flowers pollinated by birds.

Insect-pollinated:
Cantharophilcus – a. Said of flowers pollinated by beetles.

Dipterid – n. Fly-floweer, Flowers visited by dipterous flies.

Entomogamous – a. Pollinated by insects.

Entomogamy – n. The pollination of flowers by insects.

Entomphilous – a. Said of flowers dependent upon insects for pollination.

Hover-fly flowers – n. ' Flowers adapted for pollination by *Syrphidae*.

Humble - bee – (bumble - bee) flowers – n. Flowers especially adapted to pollination by the species of *Bombus*.

Insect-pollination – n. The transfer of pollen from the anther to the stigma by insects.

Lepidopterid – n. Flower adapted for pollination by moths and butterflies.

Moth-flowers – n. Flowers adap-ted for pollination by moths; they are usually white night-blommers.

Muscarian flowers – n. Flowers that attract flies by a putrid stench.

Necrocolcopterophilus – a. Said of flowers pollinated by carrion beetles.

Psychophilous – a. Said of flowersx pollinated by diurnal lepidoptera.

Man-pollinated:
  Anthropochorous – a. Distributed by man (usually applied to plant dispersal).
Self-pollinated:
  Autogamous – a. Characterized by self-fertilization.
  Autogamy – n. The fertilization of a flower by its own pollen.
  Cleistogamous – a. With small closed self-pollinated flowers.
  Close-pollinated – a. Pollinated by its own pollen.
  Self-pollination -n. Pollination by its own pollen.
Snail-pollinated:
  Malacophilopus – a. Said of flowers pollinated by snails and slugs.
Water-pollinated:
  Hydrocarpic – a. Said of aquatic plants which are pollinated above the water but withdraw the fertilized flowers below the surface for development, as in *Vallisnaria*.
  Hydrochore – n. A plant distributed by water (usually aplied to plant dispersal).
  Hydrophilous – a. Water-loving; said of flowers polinated by water.
Wind-polinated:
  Anemochorous – a. Wind-distributed (usually applied to plant dispersal).
  Anemogamous–a.Wind-pollinated.
  Anemophilous – a. Wind-pollinated.

Corolla

Achilary – a.Without a lip, as in some orchids.
Achlamydeous – a. Without a perianth, as in willows.
Actinomorphic – a. With radial symmetry, regular.
Acyclic – a. With the parts arranged spirally, not in whorls.
Aestivation – n. The manner in which the parts of a flower are folded up in the bud.
Ambigenous – a. Said of a perianth whose exterior is caltycine and the interior corolline, as in *Nymphaea*.
Amphichromy – n. A display of two different colors when in flower.
Apetalous – a. Without petal.
Banner – n. The standard of a Papilionaceous flower.
Bilabiate – a. Two-lipped.
Calcarate – a. Spurred.
Campanulate – a. Bell-shaped.
Cardiopetalous – a. With heart-shaped petals.
Carina – n. A keel; used either for the two combined lower petals of a Papilionaceous corolla or for a salient longitudinal projection on the center of the lower face of an organ, as on the lemmas of many grasses.
Cement-disk – n. The retinaculum of orchids.
Chasmogamous – a. Said of a flower whose opening precedes pollination; see *Cleistogamous*.
Chasmogamy – n. The opening of the perianth at flowering time, the opposite of cleistogamy.
Chloranthy – n. The reversion of petals to green leaves.
Choripetalous – a. Polypetalous, with separate petals.
Claw – n. The long narrow petiole-like base of the petals or sepals in some flowers.
Clip – n. The seizing mechanism in the flowers of asclepiads.
Corolla – n. The inner floral envelope composed of separate or connate petals.
Corolline – a. Seated on a corolla; corolla-like; petaloid; belonging to a corolla.
Corona – n. Crown. coronet; any appendage or intrusion that stands between the corolla and stamens, or on the corolla, as the cup of a daffodil, or that is an outgrowh of the staminal part or circle, as in the milkweed.
Cyclic – a. In whorls, not spirals.
Deflorate – a. Past the flowering state.

Dialypetalous – a. With separate petals, ploypetalous.

Dichlamydeous – a. With a double perianth, calyx and corolla.

Disk flowers – n. The tubular flowers in the center of heads of *Compositae*, as distinguished from the ray flowers.

Epipetalous – a. Borne upon the petals; placed before the petals.

Euephemerous – a. Said of flowers which open and close within 24 hours.

Faucal – a. Pertaining to the throat of a gamopetalous corolla.

Faux – (pl. **fauces**) – n. Usually used in the plural to designate the throat in a gamopetalous corolla.

Floral – a. Pertaining to flowers.

Floral diagram – n. A drawing to show the relative position and number of the constituent parts of a flower.

Floral envelope – n. The perianth leaves; the calyx and corolla.

Florepleno – a. With full or double flowers.

Floret – n. A small flower, usually one of a cluster; in grasses, the flower with the two subtending bracts.

Floscule – n. A little flower, a floret, a floret.

Flower – n. A modified plant structure concerned with the production of seeds in angio-sperms.

Funnelform – a. With tube gradualy widening upward and passing insensibley into a limb, as in many flowers of *Convolvulus;* infundibuliform.

Galea – n. A petal shaped like a helmet, placed next to the axis, as in *Aconitum.*

Galeate – a. Hollow and vaulted as in many labiate corollas.

Gamopetalous – a. With petals united, corolla in one place.

Gorge – n. The throat of a flower.

Haplochlamydeous – a. Monochlamydeous, having a single perianth.

Hemeranthous – a. Day-flowering.

Herkogamy – n. Applied to hermaphrodite flowers in which some structural peculiarity prevents self-pollination.

Hypochil – n. The (often fleshy or otherwise modified) basal portion of the labellum or lip in *Orchidaceae.*

Hypocrateriform – a. Salverform.

Infundibuliform – a. Funnelform.

Keel – n. The united lower petals of a Papilionaceous flower.

Labellum – n. The third petal of orchids, usually enlarged and by torsion of the ovary becoming anterior from its normal posterior position; a lip.

Labiate – a. Lipped; of or pertaining to the Labiatae.

Labium – n. The lower lip of a labiate flower.

Lapeanthium – n. A petal that contain a nectary.

Ligule – n. A strap-shaped body such as the limb of the ray florets in *Compositae;* the lobe of the outer corona in *Stapelia.*

Limb – n. A border, the expanded part of a gamopetalous corolla, as distinct from the tube or throat; the lamina of a petal.

Lip – n. One of the two divisions of a bilabiate corolla or calyx, that is, a gamopetalous corolla cleft into an upper and lower portion; the labellum of orchids.

Mitra – n. The galea of a corolla.

Monochlamydeae – pl.n. A large division of Phanerogams which have only one set of floral envelopes.

Monochlamydeous – a. With only one set of floral envelopes.

Monopetalous – a. Capable of being dissected equally in one plane only; zygomorphic; bilaterally symmetrical.

Palate – n. In personate corollas, a rounded projection or prominence of the lower lip closing the throat or very nearly so.

Peloria, pelory – n. Reversion, on the part of the individual, to the production of regular flowers, when the species typically has asymmetrical or bilaterally symmetrical flowers.

Perianth – n. The floral envelope, calyx and corolla.

Personate – a. Said of a bilabiate corolla having a prominent palate.

Petal – n. One of the leafy expansions in the floral whorl styled the corolla.

Petaliferous – a. Petal-bearing.

Petalode – n. An organ simulating a petal.

Petaloid – a. Like a petal, or having a floral envelope resembling petals.

Pitfall flowers – n. Transitional flowers, such as *Asarum*, which detain small *Diptera*.

Pleiopetalous – a. Many-petaled.

Pleiopetaly – n. Doubleness in flowers.

Polypetalous – a. With several distinct petals.

Protanthesis – n. The normal first flower of an inflorescence.

Quincuncial – a. Arranged in a quincunx; in aestivation, partially imbricated of five parts, two being exterior, two interior, and the fifth one having one margin exterior and the other interior, as in the calyx of the rose.

Rictus – n. The mouth or gorge of a bilabiate corolla.

Ringent – a. Gaping, as the mouth of an open bilabiate corolla.

Rotate – a. Wheel-shaped, circular and flat, applied to a gamopetalous corolla with a short tube.

Salverform – a. With a slender tube and an abruptly expanding limb, as that of the Phlox; hypocra-teriform.

Schizopetalous – a. With cut petals.

Seasonal amphichromatism – n. The production of two differently colored flowers on the same stock due to season.

Seasonal heterochromation – n. Different colors in the flowers of the same inflorescenc due to season.

Standard – n. The upper and broad more of less erect petal of a Papilionaceous flower.

Stenopetalous – a. Narrow-petaled.

Symmetrical – a. Actinomorphic, regular, capable of division by a longitudinal plane into similar halves.

Throat – n. The opening or orifice into a gamopetalous corolla or perianth; the place where the limb joins the corolla tube.

Trap flowers – n. Prison flowers which confine insect visitors until pollination has taken place.

Trumpet-shaped – a. Tubular with a dilated orifice, salverform, hypocra-teriform.

Unilabiate – a. One-lipped.

Vexillum – n. The standard or large posterior petal or a Papilionaceous flower.

Dehiscence

Assumentum (pl. assumenta) – n. One of the two valves of a siliqua.

Circumscissile – a. Opening or dehiscing by a horizontal line around the fruit or anther.

Dehisce – v. To open spontaneously when mature, as seed capsules.

Dhiscence – n. The method or process of opening of a seed-pod or anther.

Dehiscent – a. Said of that which dehisces, as the opening of a fruit or anther along lines of suture.

Fissile – a. Tending to split or easily split.

Hydrochastic – a. Said of plants in which the bursting of the fruit and the bursting of the fruit and the dispersion of the seeds are caused by absorption of water.

Loculicidal – a. With dehiscence on the back between the partitions into the cavity.

Operculate – a. Opening by a lid.

Porocidal – a. Opening by pores.

Ruptile – a. Dehiscing in an irregular manner.

Septicidal – a. With dehiscence along

lines of union of the carpels.

Septifragal – a. With the valves breaking away from the dissepiments in dehiscence.

Suture – n. A junction or seam of union; a line of opening or dehiscence.

Ventral suture – n. The ventral seam or line of dehiscence in a carpel.

Xerochase – n. A fruit that opens in dry air and closes in humid air.

Xerochastic – a. Said of plants whose fruits burst by desiccation, thereby scattering their seeds or spores.

Direction

Amphigean – a. Native around the world.

Austral – a. Southern.

Boreal – a. Northern.

Deflexed – a. Bent or turned abruptly downward.

Dextrorse – a. Turning to the right, clockwise.

Eutropic – a. Twining with the sun, clockwise, dextrorse.

Geonasty – n. The act of curving toward the ground.

Geotropic – a. Turning toward the earth.

Hesperal -a. Of the West.

Homalotropous – a. Said of organs which grow in a horizontal direction.

Hyperboreal – a. Of the far North.

Impressed – a. Bent inward, hollowed or furrowed as if by pressure.

Meridional – a. Southern (in the Northern Hemisphere).

Occidental – a. Western.

Oriental – a. Eastern

Parallel – a. Extended in the same direction, but equally distant at every point.

Porrect – a. Directed outward and forward.

Septentrional – a. Nothern.

Sinistrorse – a. Turning to the left or counterclockwise.

Zigzag – a. Having short bends or angles from side to side.

Dispersal

Aelophilous – a. Disseminated by wind.

Anemochore – n. An organism that is disseminated by the wind.

Anemophilous – a. Distributed by wind.

Anthropophilous – a. Plants which follow man.

Blastochore – n. A plant distributed by offshoots or buds.

Bolochore – n. A Plant distributed by propulsion.

Bradyspore – n. A plant which disperses its seeds slowly.

Brotochore – n. A plant dispersed by man.

Centrospore – n. A plant with spiny disseminules.

Clitochore – n. A plant which is distributed by falling or sliding.

Disseminule – n. A seed, fruit or spore modified for dispersal.

Edobole – n. A plant whose seeds are scattered by propulsion are scattered by propulsion through turgescence.

Glacospore – n. A plant with viscid disseminules.

Hydrochore – n. A plant distributed by water.

Hydrophilous – a. Water-loving; distributed by water.

Migration – n. Any movement by which the range of a species is extended (strictly speaking, it means moving under its own power).

Migrule – n. The unit of migration, as seed, fruit, runner, bulb, etc.

Ornithophilous – a. Bird-longing; distributed by birds.

Sarcospore – n. A plant with fleshy disseminules.

Saurochore – n. A plant disseminated by lizards or snakes,

Synzoochory – n. Dispersion by animals.

Xeriobole – n. A plant whose seeds

are scattered by dehiscence due to dryness.

Zoophilous – a. Distributed by animals.

Fruits (carpography)

Acarpic – a. Without fruit.

Acarpotropic – a. Not throwing off its fruits.

Achene – n. A small, hard, dry, indehiscent one-seeded fruit in which the ovary wall is free from the seed.

Achenodium – n. A double achene, as the cremocarp of *Umbelliferae.*

Aggregate fruit – n. A cluster of ripened ovaries traceable to separate pistils of the same flower and inserted on a common receptacle.

Akene – n. See Achene.

Amphicarpous – a. Producing two kinds of fruit.

Amphore – n. The lower part of a pyxis, as in henbane.

Anthocarpous – a. Said of fruits with accessories, sometimes pseudocarps, as in the strawberry and pineapple.

Apogamous – a. Developed without fertilization.

Apyrenous – a. Said of fruit which is seedless.

Article – n. The portion of a fruit (especially in Leguminosae) separated from others by a constriction or joint, as in *Desmodium.*

Assumentum – (pl. **assumenta**) – n. One of the two valves of a siliqua.

Atrocarpous – a. Black-fruited.

Autocarp – n. A fruit obtained by self-fertilization.

Baccate – a. Berry-like, pulpy or fleshy.

Balausta – a. The fruit of a pomegranate with a firm rind, berried within, and crowned with the lobes of an adnate calyx.

Beak – n. A long prominent and substantial projection; applied particularly to the prolongation of fruits and carpels.

Berry – n. A mature, fleshy, few-to many- seeded. ovary of a single pistil.

Biferous – a. Producing two crops of fruit in one season.

Bilocular – a. Two-celled, with two compartments.

Bivalvular – a. With two valves.

Brachycarpous – a. Short-fruited.

Bradycarpic – a. Fruiting after winter, in the second season after flowering.

Bur – n. Any fruit with a rough or prickly envelope, whether a pericarp, a persistent calyx, or an involucre, as of the sandbur and burdock.

Calyptra – n. A hood or lid; particularly, the hood or cap of the capsule of moss or lid in the fruit of *Eucalyptus.*

Capsella – n. Seed vessel; a small capsule.

capsular – a. Pertaining to a capsule; formed like a capsule.

Capsule – n. A simple fruit, the product of a compound pistil splitting along two or more lines of suture.

Carpel – n. A simple pistil; one unit of a compound pistil; the cone scale in conifers.

Carpography – n. The description of fruits.

Carpophopre – n. A portion of receptacle prologed between the carpels as in *Umbelliferae.*.

Caryopsis – n. A small, dry, indehiscent fruit in which the seed coat is adherent to the ovary wall.

Censer-action – n. The action of capsules which like censers (incense-burners), partially open by valves, the seeds being gradually shaken out by wind, as in *Papaver* and *Stramonium.*

Circumscissile – a. Opening or dehiscing along a horizontal line around the fruit or anther, the valve usually coming off like a lid.

Cochlea – n. A closely coiled legume.

Cochleate – a. Spiral, like a snail

shell.

Coelospermous – a. Hollow-seeded; said of the seedlike carpels of *Umbelliferae*, with the ventral face incurved at the top and bottom.

Coenocarpium – n. The collective fruit of a entire inflorescence, as a fig or pineapple.

Columella – n. A persistent central axis around which the carpels of some fruits are arranged, as in *Geranium;* the receptacle bearing the sporangia of *Trichomanes* and other ferns.

Commissure – n. The place of joining or meeting, as the face by which one carpel joins another.

Cone – n. a. Fruit of the pine family *pinaceae* and of *Cycads;* strobilas.

Conelet – n. A little cone, applied to a cone of the first year.

Conocarpium – n. An aggregate fruit consisting of many fruits on a conical receptacle, as the strawberry.

Conoid – a. Conelike.

Creatospore – n. A plant with nut fruits.

Cremocarp – n. A dry, seedlike fruit, composed of two one-seeded carpels invested by an epigynous calyx, separating when mature into mericarps.

Cupule – n. The cup of such fruits as the acorn; an involucre composed of bracts adherent at least by their base.

Cyamium – n. A kind of follicle resembling a legume.

Cynarrhodion – n. A Fruit like that of the rose, fleshy, hollow, fand enclosing achenes.

Cypsela – n. An achene invested by an adnate calyx, as the fruit of *Compositae*.

Dasycarpous – a. Thick-fruited.

Dialycarpic – a. Having a fruit composed of distinct carpels.

Dicarpellary – a. Composed of two carpels.

Didymous – a. Found in pairs, as the fruits of *Umbelliferae;* divided into two lobes.

Dischisma – (pl. **dischismate**) – n. The fruit of *Platystemon*, which divides into longitudinal carpels, each of which again divides transversely.

Dissepiment – n. A partition in an ovary or pericarp caused by the adhesion of the sides or the carpellary leaves.

Dissilient – a. Bursting asunder.

Drupaceous – a. Resembling a drupe, possessing its character, or producing similar fruit.

Drupe – n. A fleshy, one-seeded indehiscent fruit with the seed enclosed in a stony endocarp.

Drupelet – n. One drupe of a fruit made up of aggregate drupes, as in raspberry.

Endocarp – n. The inner layer of the pericarp.

Eriocarpous – a. Woolly-fruited.

Exocarp – n. The outer layer of the pericarp.

Flask – n. The utricle of *Carex*.

Follicle – n. A single carpellate dry fruit dehiscing along one line of suture.

Fructiferous – n. The act of fruiting.

Fruit – n. A mature ovary or ovaries with or without closely related parts.

Fruit dots – n. The sori of ferns.

Galbulus – n. The fruit of *Taxodium;* a modified cone, the apex of each carpellary scale being enlarged and somewhat fleshy.

Gourd – n. A fleshy, one-celled, many-seeded fruit with parietal placentation.

Gynobase – n. An enlargement or prolongation of the receptacle bearing the ovary.

Hemicarp – n. A half-carpel, a mericarp.

Hesperidium – n. A berry with a tough, leathery rind, as the orange.

Heterocarpous – n. Producing more than one kind of fruit.

Hip – n. The fruit of the rose; technically, a cynarrhodion.

Hygrochastic – a. Said of plants in which the bursting of the fruit and the dispersion of seeds are caused by absorption of water.

Indehiscent – a. Not opening by valves or along regular lines.

Jugum – n. A ridge one the fruits of *Umbelliferae.*

Key fruit – n. The samara of the ash.

Lasiocarpous – a. Pubescent-fruited.

Legume – n. A dry fruit of a simple pistil usually dehiscing along two lines of suture; the fruit of *Leguminosae.*

Locule – n. A compartment or cell of a pistil or anther.

Loment – n. A flattened legume which is constricted between the seeds, falling apart at the constrictions when mature into one-seeded joints.

Lomentaceous – a. Bearing or resembling loments.

Mace – n. The aril of the nutmeg.

Malicorium – n. The rind of a pomegranate.

Mast – n. The fruit of such trees as beech, oak, hickory etc.

Megasporocarp – n. The developed megasporangium in *Azolla,* finally containing a single perfect megaspore.

Mericarp – n. One of the achene-like carpels or a closed half-fruit as *Umbelliferae..*

Monolocular – a. One-celled, unilocular, applied to ovaries.

Multiple fruit – n. A cluster of ripened ovaries traceable to the pistils of separate flowers, as in mulberry.

Multiseptate – a. With many parti-·tions.

Naucum – n. The fleshy part of a drupe; seed with a very large hilum.

Naucus – pl. n. Certain Cruciferous fruits which have no valves.

Nuciferous – a. Bearing or producing nuts.

Nut – n. An indehiscent, usually one-celled, one-seeded fruit (though usually traceable to a compound ovary) with a bony, woody, leathery, or papery wall and in general, partially or wholly enclosed in an involucre or husk.

Nutlet – n. A small nut.

Nux – n. A nut.

Operculum – n. A lid or cover which separates by a transverse line of division, as in a pyxis.

Paracarpous – a. Said of ovaries whose carpels are joined together by their margins only.

parietal – a. Borne on, or belonging to a wall.

Parthenocarpy – n. The production of fruit without true fertilization.

Parthenogenesis – n. A form of apogamy in which the oosphere develops into a normal product of fertilization without a preceding sexual act.

Pentacamarous – a. With five locules.

Pentacarpellary – a. With five carpels.

Pepo – n. A ground type of fruit, a one-celled, many-seeded, inferior fruit with parietal placentas and a pulpy interior.

Pericarp – n. The wall of a mature ovary, consisting of an exocarp, a mesocarp, and an endocarp.

Phaenocarpous – a. Having a distinct fruit,with no adhesion to surrounding parts.

Phragma – (pl. **phragmata**) – n. A spurious dissepiment in fruits.

Pit – n., The endocarp of a drupe with the enclosed seed.

Placenta – n. The place in an ovary where the ovules are attached.

Placentation – n. The disposition of the placenta.

Plococarpium – n. A fruit composed of follicles arranged around an axis.

Plurilocular – a. With many cells or locules.

Pod – n. A dry dehiscent fruit from an unilocular ovary with marginal placentation.

Podocarp -n. A stipitate fruit, that is, with the ovary borne on a gynophore.

Polachena – n. A fruit similar to a cremocaryp but composed of five carpels.

Pome – n. A fleshy fruit, the product of a compound pistil with the seeds encased within a cartilagionous wall, as in the apple.

Pomiferous – a. Pome-bearing.

Porocidal – a. Openning by pores.

Putamen – (pl. **putamines**) – n. The shell of a nut; the hardened endocarp of stone fruits.

Pyrene – n. Nutlet, particularly the nutlet in a drupe.

Pyxis – n. A capsule with circumscissile dehiscence, the upper portion acting as a lid.

Quinquelocular – a. Five-celled.

Regma (pl. regmata) – n. A fruit with elastically opening segments or cocci, as in *Euphorbia;* a form of schizocarp.

Repletum – n. A fruit with the valves connected by threads, persistent after dehiscence such as orchids, *Aristolochia,* and some *Papaveraceae.*

Replum – n. A framelike placenta from which the valves fall away in dehiscence, frequently used so as to include the septum of *Cruciferae* in the term.

Rhizocarp – n. A sporangium such as is produced on rootlike processes of members of the *Marsileaceae.*

Ripe – a. Mature, characterized by the completion of an organ or organism for its allotted function.

Samara– n. Winged, achene-like fruit.

Schizocarp – n. A pericarp which splits into one-seeded portions or mericarps.

Schleranthium – n. An achene enclosed in an indurated portion of the calyx tube, as in *Mirabilis.*

Septifragal – a. With the valves breaking away from the dissepiments in dehiscence.

Silicule – n.A Short siliqua, not much longer than than broad.

Sillicle – n. The short fruit of certain *Cruciferae.*

Silique – n. The peculiar pod of the *Cruciferae,* two valves falling away from a frame, the replum, on which the seeds grow and across which a false partition is formed.

Simple fruit – n. A fruit which results from the ripening of a single pistil.

Sorose – n. A fleshy multiple fruit, as a mulberry or a pineapple.

Sorus – n. A cluster of sporangia in ferns.

Sphalerocarpum – n. An accessory fruit, as an achene in a baccate calyx-tube.

Strobile – n. A fruit made up largely of imbricated scales, as in the hop and the pine; a cone.

Strombuliform – a. Said of fruit that is spirally twisted.

Syconium – n. A multiple hollow fruit, as that of a fig.

Syncarp – n. A multiple or fleshy aggregate fruit, as the mulberry and *Magnolia.*

Syncarpous – a. Composed of two or more united carpels.

Syncolliphytum – n. A plant in which the perianth becomes combined with the pericarp.

Systellophytum – n. A fruit in which a persistent calyx appears to form a part of the fruit.

Trachycarpous – a. Rough-fruited.

Triachaenium – n. A fruit similar to a cremocarp but of three carpels.

Tricarpellary – a. Of three carpels.

Trichocarpous – a. Hairy-fruited.

Tryma – (pl. **trymata**) – n. A drupace-ous nut with dehiscent exocarp.

Unicarpellate – a. With fruit consisting of a single carpel.

Urn – n. The base of a pyxis.

Utricle– n. A small bladdery pericarp, as in *Atriplex;* a membranous sac surrounding the fruit proper in

Carex; any bladder-shaped appendage.

Vallecula – n. The grooves in the intervals between the ridges in the fruit of *Umbelliferae.*

Valve – n. A segment into which a capsule naturally separates at maturity.

Xerochastic – a. Said of plants in which the bursting of the fruit and dispersion seeds is caused by desiccation.

Habitats

(Terms concerned with the various kinds of habitats).

Alpine
Acrophilous – a. Dwelling in the alpine region.
Alpestrine – Nearly alpine, subalpine.
Coryphad – n. An alpine meadow plant.
Subalpine – a. Below alpine, almost alpine.

Bank
Ochthad – n. Bank plant.
Ochthophilous – a. Bank-loving.

Bog
Turfaceus – a. Pertaining to bogs.
Turfophilous–a. Bog-loving, found in bogs.

Clay
Argillaceous – a. Clayey, pertaining to clay, or clay-colored.
Spiladophilous – n. Dwelling in clay.

Cold
Coryphad – n. An alpine meadow plant.
Crymophilous – a. Long polar regions; inhabiting polar regions.
Frigid – a. Cold, of cold regions.

Dark
Scotophilous– a. Darkness-loving, dwelling in darkness.
Skotophilous–a. See Scotophilous.

Ditch
Taphrophilous – a. Dith-loving. growing in ditches.

Dry
Cheradophilous – a. Loving dry habitats; dwelling in dry places.
Cherswad – n. A plant growing in dry places.
Eremophilous – a. Desert-loving, dwelling in deserts.
Xerophilous – a. Loving dry places, dwelling in dry places.

Dung
Fimetariours – a. Growing on or among dung.
Fimicolous–a. Inhabiting manure.

Earth
Epigeous – a. Above the soil; growing above the soil.
Geophilous – a. Earth-loving; said of plants which fruit under-ground.
Hypogeous – a. Below the soil; growing or remaining below the soil.
Terricolous – a. Dwelling on the ground.

Field
Agrophilus–a. Loving grain fields.
Campestrine – a. Of or pertaining to fields.
Hemerophilous – a. Loving cultivation readily cultivated.
Nomad – n. A pasture plant.
Nomophilous – a. Pasture-loving, inhabiting pastures.
Poad – n. A meadow plant.

Forests
Alsad – n. A grove plant.
Ancophilous – a. Loving mountain glens or valleys.
Dendrocoluous – a. Dwelling on trees.
Dendrophilous – a. Dweiling on or among trees, loving trees.
Helohylophilous – a. Loving wet forests.
Hylacolous – a. Tree-dwelling.
Hylocolous – a. Inhabiting forests.
Hylodophilous – a. Loving dry wooods; dwelling in dry woods.
Neorose– a. Growing in the woods.
Nemus (pl. nemores) – n. Woods.
Orgadophilous – a. Loving open woodland; dwelling in open foerests.
Stenophyllophilous – a. Loving deciduous forests.
Sylvatic – a. Growing among the

trees.
Sylvestrine– a. Growing in woods.

Gravel
Chalicad – n. A gravel-slide plant.
Chalicodophiluus – a. Loving gravel- slides; inhabiting gravel-slides.
Glareose -a. Frequenting gravel.

Hedges
Sepicolous – a. Inhabiting hedges.

Humus
Sapromyiophiluus – a. Humus-loving; inhabiting humus.

Lakes or ponds
Lacustrine – a. Belonging to, or inhabing lakes and ponds.
Lentic – a. Pertaining to, or living in quiet or still water.
Limnophilous – a. Dwelling in lakes.
Tiphad – n. A pond plant.
Tiphophilous -a. Pond-loving; inhabiting ponds

Limestone
Calcareous – a. Or pertaining to limestone.
Calcicolous – n. Inhabiting limestone soils.
Gypsophilous – a. Limestone-loving; inhabiting gypsum soils.

Loam
Melangeophilous– n. Loam-loving; inhabiting loam soils.

Marshes
Banados – pl. n. Shallow swamps (Paraguay).
Helad – n. A marsh plant.
Limnodophilous – a. Dwelling in marshes.
Paludose – a. Growing in marshy places.
Palustrine – a. Of or growing in marshes.
Pontohalicolus – a. Dwelling in salt marshes.
Stasad – n. A plant of stagnant water.
Telmatophilous – a. Loving wet meadow.
Telmicolous – a. Dwelling in fresh water marshes.

Mountains
Montane – a. Pertaining to mountains, as plants which grow on them.
Orophilous – a. Mountain-loving; inhabiting mountains.

Mud
Limicolous – a. Inhabiting muddy places, as on the margins of pools.
Limose – a. Of marshes.
Luticolus – a. Mud- or mire-loving; inhabiting muddy places.

Prairie
Graminicolus – a. Grass-inhabiting.
Psilicolous – a. Prairie-dwelling.
Psilophilous – a. Prairie-loving.

Rain
Ombrophilous– a. Rain-loving, inhabiting places of frequent rains.

Salt
Drimophilous – a. Salt-loving.
Halophilous– a. Salt-loving; growing in salty soil.
Halophyte – n. A plant which grows in saline soil.
Saline – a. Of or pertaining to salt.

Sand
Amathicolous – a. Growing on sandy plains.
Amathophilous – a. Growing in sandy plains or in sandy hills.
Ammochthad – n. A sand-bank plant.
Ammophilous – a. Sand-loving; inhabiting sand.
Arenarious – a. Of sand or sandy places.
Arenicolous – a. Inhabiting sandy places.
Cheradad – n. A sand-bar plant.
Cheradophilous – a. Loving dry habitats; dwelling in dry places.
Enaulophilous – a. Loving sand draws.
Psammophilous – a. Sand-loving; inhabiting sand.
Saabulicolous – a. Growing in sandy places.
Thinicolous – a. Dwelling on shifting sand dunes.
Thinophilous – a. Dune-loving; inhabiting dunes.

Sea
Agad – n. A beach plant.

Aigialophilous – a. Beach dwelling.

Aigicolous – a. Inhabiting a stony strand or beach.

Littoral -n. Belonging to or growing on the seashore.

Marine – a., Growing within the influence of the sea, or immersed in it.

Maritime -a. Belonging to the sea. or confined to the sea coast.

Shade

Sciophilous – a. Shade-loving; inhabiting shady places.

Umbracticolous – a. Inhabiting shady places.

Snow

Chionad – n. A snow plant.

Chionic – a. Of snow fields.

Niveous – a. Growing in or near the snow, pertaining to snow.

Springs

Crenad – n. A plant of springs.

Crenophilous – a. Loving springs.

Stone

Chasmophilous – a. A cranny-loving plant.

Lapidose – a,. Growing among stones.

Petricolous – a. Rock-inhabiting.

Petrophilous – a. Stone-loving, dwelling among stones.

Phellophilous – a. Loving rock fields.

Rupestral – a. Pertaining to rocks.

Silicolous – a. Growing in flinty soils.

Streams

Crenicolous – a. Dwelling in spring-fed brooks.

Fluvial – a. Said of plant growing in streams.

Namatad – n. A plant growing in or near a brook.

Potomophilous – a. River-loving; dwelling in or near rivers.

Rheophilous – a. Creek-loving; dwelling in torrents.

Riparious – a. Growing by rivers or streams.

Sun

Heliad – n. A heliophyte or sun-loving plant.

Thickets

Aithalophilous – a. Dwelling in evergreen thickets.

Capoe – n. A palm thicket (Brazil).

Driodad – n. A plant of a dry thicket.

Lochmocolous – a. Inhabiting thickets.

Lochmodophilous – a. Inhabiting thickets.

Lochmodophilous – a. Loving dry thickets; growing in dry thickets.

Lochmophilous – a. Thicklet-loving; found in dry thickets.

Walls

Rupestral – a. Growing on walls and rocks.

Waste places

Chledocolus – a. Dwelling in waste places.

Chledophilous – a. Loving waste places.

Water

Emersed – a. Raised above and and out of the water, emerged.

Hydrophilous – a. Loving wet places of water. pollinated by water.

Hydrophyte – n. A water plant.

Natant – a. Floating.

Inflorescence

Ament – n. A catkin; a more of less flexible, usually pendulous spike bearing apetalous unisexual flowers.

Amentiferous – a. Bearing aments.

Amentum – n. Catkin.

Anthela – n. The panicle of *Juning* the main axis.

Anthelate – a. With elongate flower-bearing branches, as in some *Junci*.

Anthemy, anthemia – n. A flower-cluster of any kind.

Capitate – n. With a head.

Catkin – n. See Ament.

Centrifugal – a. In inflorescences, blooming from the top downward.

Centripetal – a. In inflorescences,

blooming from the outside in-ward, or from the base upward.

Cincinnus -n. A one-branched scorpioid cyme.

Compound inflorescence – n. An inflorescence composed of secondary inflorescences.

Corymb – n. A short, broad, more or less flat-topped, indeterminate flower-cluster, the outer flowers opening first.

Corymbiform – a. In the shape of a corymb.

Cyanthum – n. The ultimate inflorescence of *Euphorbia* consisting of a cuplike involucre bearing the flowers from its base.

Cyme – n. A broad, more or less flat-topped, determinate flower-cluster, with central flowers blooming first.

Cymose – a. Cymelike.

Cymule – n. A small cyme.

Definite – a. Determinate, terminating in a flower bud.

Definite inflorescence – n. A determinate inflorescence, terminating in a flower bud, blooming from the inside outward or from the top downward.

Dichasium – a. Said of an inflorescence in which the terminal flower blooms slightly in advance of its nearest associates.

Dichasium – n. A cyme with two lateral axes.

Dicymose – a. Doubly cymose.

Diffuse – a. Loosely branching or spreading; of open growth.

Drepanium– n. A sickle-shaped cyme.

Ecblastesis – n. The appearance of buds within a flower, proliferation of an inflorescence.

Fascicle – n. A condensed or close cluster.

Fasciculate – a. In condensed or close clusters.

Fasciculate – a. In condensed or close clusters.

Glome – n. A rounded head of flowers.

Glomerate – a. In a dense or compact cluster of clusters.

Glomerule – n. A cluster of heads in a common involucre.

Head – n. A dense spherical or flat-topped inflorescence of sessile flowers clustered on a common receptacle.

Indefinite – a. In an inflorescence, indeterminate.

Indefinite inflorescence – n. An inflorescence that is indeterminate, blooming from the outside in-ward.

Inflorescence – n. Mode of flower-bearing; technically less correct but much more common in the sense of a flower-cluster.

Intercalary inflorescence – n. An inflorescence in which the main axis continues to grow vegetatively after giving rise to the flowers.

Julaceous – a. Bearing catkins, amentaceous.

Mixed inflorescence – n. One in which partial inflorescences develop differently from the main axis, as centrifugal and centripetal together.

Monochasium – n. A one-branched cyme, either pure or resulting from th reduction of cymes.

Nucamentum – n. An amentum, or catkin.

Panicle – n. A compound or branched raceme.

Paniculate – a. Having a panicle type of inflorescence.

Phoranthium – n. The receptacle of the head of *Compositae*.

Polythalamic – a. Having more than one female flower within the involucre; derived from more than one flower, as a collective fruit.

Raceme – n. An indeterminate inflore-scence consisting of a central axis bearing a number of flowers with pedicels or nearly equal length.

Racemose – a. Resembling a raceme: in racemes.

Racemiform – a. In the form of a raceme.

Rhipidium – n. A fan-shaped cyme, the lateral branches being developed alternately in two opposite directions.

Scape – n. A leafless peduncle arising from the ground; it may bear scales or bracts but not foliage-leaves and may be one-or many-flowered.

Scapiform – a. Resembling a scape.

Scapose – a. Bearing or resembling a cape.

Scorpioid – a. Said of a coiled cluster in which the flowers are two-ranked and borne alternately at the right and left.

Scorpioid cyme – n. Cincinnus, the lateral branches developed on opposite sides alternately, as in *Boraginaceae*.

Simple inflorescence – n. A flower-cluster with one axis, as a spadix, spike, or catkin.

Spadix – n. The thick or fleshy spike of certain plants, as the *Araceae*, surrounded or subtended by a spathe.

Spathe – n. The bracts or leaf surrounding or subtending a flower-cluster of spadix; it is sometimes colored and flower-like, as in the *Calla*.

Spiciform – a. Spikelike.

Spike – n. An inflorescence consisting of a central rachis bearing a number of sessile flowers.

Tassel – n. The staminate inflorescence in maize.

Thyrse – n. Compact and more of less compound panicle; more correctly a panicle-like cluster with main axis indeterminate and other parts determinate.

Thyrsoid – a. Resembling a thyrse.

Thyrsula – n. A little cyme which is borne by most Labiates in the axil of the leaves.

Trichasium – n. A cymose inflorescence with three branches.

Umbel – n. An indeterminate inflorescence consisting of several pedicellate flowers having a common point of attachment.

Umbel, compound – n. An umbel in which each ray bears a small umbel.

Umbel, cymose – n. An apparent umbel, but with the flowers opening centrifugally; a cyme which simulates an umbel.

Umbellate – a. With o, ertaining to umbels.

Umbellet – n. A secondary umbel.

Umbelliferous – a. With umbels.

Umbelliform – a. In the shape of an umbel.

Umbellule – n. An umbellet; a small umbel.

Verticillaster – n. A false whorl, composed of a pair of opposed cymes, as in Labiates.

Leaves (Forms)

Aciculate – a. Slender, needleshaped.

Awl-shaped – a. Narrow and gradually tapering to a sharp point.

Cochlear – a. Said of a form of imbricate aestivation with one piece exterior.

Cordate – a. Heart-shaped, with the notch basal.

Cuneate – a. Wedge-shaped, with the broad end apical.

Dltoid – a. Triangular; delta shaped.

Drepaniform – a. Sickle-shaped.

Elliptical – a. Oblong with rounded ends.

Ensiform– a. Sword-shaped, gladiate.

Falcate – a. Sickle-or scythe shaped.

Gladiate – a. Sword-shaped, ensiform.

Halberd-shaped – a. Sagittate with the basal lobes turned outward, hastate.

Hastate – a. Arrow-shaped with the basal lobes turned outward, halberd-shaped.

Heart-shaped – a. Cordate, broadly ovate with two rounded lobes at the base.

Lanceolate – a. Lance-shaped, much longer than wide and tapering upward.

Linear – a. Long and narrow with margins parallel or nearly so.

Lyrate – a. Lyre-shaped.

Needle – n. The stiff linear leaf of a *Pinaceae*.

Nephroid – a. Kindney- shaped, reniform.

Obcordate – a. Heart-shaped with the notch apical.

Oblanceolate – a. Inverted lanceolate.

Oblong – a. Longer than broad, with the margins nearly parallel.

Oboovate – a. Oval, but broader toward the apex.

Orbiculate – a. Round or circular.

Oval – a. Elliptical with the width greater than half the lenght.

Ovate – a. Oval, but broader to ward the base.

Pandurate – a. Fiddle-shaped.

Panduriform – a. Fiddle-shaped.

Peltate – a. Shield-shaped with the petiole attached to the under side.

Reniform – a. Kidney-shaped nephroid.

Rhomboidal – a. Rhombic-shaped.

Sagittate – a. Arrow-shaped.

Scale – n. Any thin scarious body, usually a degenerate leaf, sometimes of epidermal origin; sometimes used meaning glume.

Scale-leaves – n. Modified leaves on underground stems; small flat leaves as those on *Cupressus* and *Selaginella*.

Spatulate – a. Spatula-shaped.

Subulate – a. Awl-shaped.

Wedge -shaped – a. Cuneate.

Apexes

Acuminate – a. Tapering to a prolonged point.

Acute – a. Distinctly and sharply pointed, but not drawn out.

Apiculate – a. With a minute pointed tip.

Aristate– a. Awned, bearing an arista.

Cuspidate – a. Tipped with a sharp rigid point.

Emarginate – a. With a shallow notch at the apex.

Mucronate – a. With a mucro; bristle-tipped.

Obcordate – a. Heart-Shaped with the notch at the apex.

Obtuse – a. blunt or rounded at the end.

Retuse – a. An obtuse tip with a slight depression in the middle.

Truncate – a. As though cut off by a straight transverse line.

Bases

Acuminate – a. With prolonged tapering to the petiole.

Acute – a. Distinctly and sharply pointed but not drawn out.

Amplexicaul – a. Clasping the stem.

Auriculate – a. With an auricle or a claw.

Clasping – a. With the base clasping the stem.

Connate-perfoliate – a. Having opposite leaves joined at the bases.

Cordate – a. Heart-shaped with the notch at the base.

Cuneate – a Wedge-shaped.

Decurrent -a. Said of a leaf which extends down a stem below the point of insertion.

Hastarte – a. Similar to sagittate but with the lobes pointing outward.

Oblique – a. With one side of the base being larger than the other.

Obtuse – a. Blunt or rounded at the end.

Ocrea – n. A legging-shaped or tubular structure formed by the union of two stipules.

Peltate – a. With the petiole attached to the under side rather than the margin.

Perfoliate – a. With the stem apparently passing through the leaf.

Sagittate – a. With basal lobes pointing downward, like the base of an arrow.

Surcurrent - a. With winged expansions from the base of a leaf prolonged up the stem.

Truncate – a. As if cut off by a straight transverse line, blunt.

Modification

Bracts	Scales
Fronds	Sepals
Petals	Spines
Phyllode	Stamens
Pistils	Tendrils

Complexity

Simple – a. With one blade with incomplete or no segmentation.

Compound – a. With two or more blades called leaflets.

Pinnately – adv. With the leaflets arranged on opposite sides along a common rachis.

Odd pinnate – a. With a terminal leaflet.

Abrupt pinnate – a. Without a terminal leaflet.

Palmately – adv. With the leaflets radiating in all directions from the apex of the petiole.

Decompound – a. More than once compound.

Margins

Cleft – a. Divided into lobes separated by narrow or acute sinuses which extend more than half way to the midrib.

Crenate – a. With rounded or blunt teeth.

Crenulate – a. Finely crenate.

Crisp – a. Curly or wavy, as the leaves of *Rumex crispus*.

Dentate – a. With sharp teeth pointing otuward.

Denticulate – a. Minutely or finely dentate.

Dissected – a. Deeply divided or cut into many segments.

Divided – a. With lobing or segmentation extending to the base or midrib.

Doubly serrate – a. With small serrations on larger ones.

Entire – a. With an even margin; not interrupted by toothing, lobing, or other division.

Fimbriate – a. With the margin bordered with long slender processes.

Glandulose-serrate – a. With serrations tipped or bordered with glands.

Incised – a. Cut sharply and irregularly, more or less deeply.

Inflexed – a. Turned in at the margins.

Intramarginal – a. Within and near the margins.

Lacerated – a. Torn or irregularly cleft.

Laciniate – a. Cut into lobes separated by deep, narrow, irregular incisions.

Lobed – a. With lobes extending to near the middle.

Multifid – a. Cleft into many lobes or segments.

Palmate – a. Lobed or divided, so that the sinuses point ot the apex of the petiole.

Palmatifid – a. Lobed or divided, so that the sinuses point nearly to the apex of the petiole.

Parted – a. Divided by sinuses which extend nearly to the midrib.

Partitioned – a. Having the deepest division into which a leaf can be cut without becoming compound.

Pectinate – a. With narrow segments set close together like the teeth of a comb.

Peltate – a. With the petiole attached to the under side instead of the margin.

Peltified – a. Said of a peltate leaf that is cut into segments.

Peripheral – a. On or near the margins.

Pinnafifid – a. Cleft almost to the midrib.

Repand – a. Undulate or wavy.

Revolute – a. With margins rolled toward the lower side.

Runcinate–a. Saw-toothed or sharply incised with retrorse teeth.

Serrate – a. With sharp teeth pointing forward.

Serrulate - Serrate   with minute

teeth.

Sinuate–a. With a deep wavy margin.

Sinuous – a. Wavy.

Undulate – a. Wavy, repand.

Venation

Anadromous – a. Said of ferns, in which the first set of nerves in each segment of the frond is given off on the upper side of the midrib toward the apex, as in *Aspidium* and *Asplenium.*

Anametadromous – a. Said of the venation of ferns in which the weaker pinnule are anadromous and the stronger are catadromous.

Anastomosing – a. Characterized by the union of one vein with another, the connection forming a reticulation.

Arcuate – a. Moderately curved, bent like a bow, said of leaf venation of *Cornus, Caenothus,* etc.

Argyroneurous – a. With silver-colored veins.

Basinerved– a. Veined from the base.

Campylodromous – a. Said of venation in which the secondary veins curve toward the margins but do not form loops.

Catadromous – a. Said of the venation of ferns in which the first set of nerves in each segment of the frond is given off on the basal side of the midrib, as in *Osmunda.*.

Convergent – a. Said of veins which run from the base to the apex of a leaf in a curved manner.

Costa – n. A rib, as a midrib.

Craspedodromous – a. With the lateral veins running from midrib to margin without dividing.

Diadromous – a. With fan-shaped venation, as in *Gingko biloba.*

Dictyodromous – a. With reticulate venation.

Feather-veined – a.With secondary veins proceeding from the midrib; pinninerved.

Hinoideous – a. With veins proceeding from the midrib parallel and unbranched.

Hyphodromous – a. With the veins sunken in the leaf and not readily visible.

Infossous – a. With the veins sunken but leaving a visible channel.

Intercostal – a. Between the ribs or veins.

Internerves – n. The space between the nerves.

Leuconeurous – a. White-nerved.

Marmorate – a. With veins of color, as maribled, or mottled.

Midrib – n. The main rib or central vein of a leaf or leaflike structure.

Multicostate – a. Many-ribbed.

Nervation – n. Venation, the manner in which foliar nerves or veins are arranged.

Nerve – n. A simple or unbranched vein or slender rib.

Net-veined – a. Reticulated; net-veined with any system or irregularly anastomosing veins.

Palmately veined – a. With veins arranged in a palmate manner.

Parallelodromous – a. With parallel veins, as in lilies.

Parallel-veined – a. With the lateral veins straight, as in *Alnus;* with the entire system straight as in grasses.

Penniveined – a. Veined in a pinnate manner.

Pinninerved – a. Pinnately veined, the veins running parallel towards the margin.

Plagiodromous – a. Said of tertiary leaf-veins when at right angles to the secondary veins.

Quinquenerved – a. With the midrib dividing into five, that is, the main rib and a pair on each side.

Radiately veined – a. With veins radiating from a centrally attached petiole.

Reticulate – a. Forming a network.

Rib – n. A primary vein, especially the central longitudinal or midrib.

Trinervate – a. Three-nerved.

Vein – n. A strand of vascular tissues in a flat organ, such as a leaf.

Venation – n. Veining; the arrangement or disposition of veins.

Onyms

Antonym – n. A word of opposite meaning; a counterterm.

Basonym – n. The original epithet, retained when transferred to a new position.

Homonym – n. A name having the same spelling as another name in the same language but different in meaning; in taxonomy, homonyms are two or more names having the same spelling but applied to two or more taxa of the same rank based upon different types A later homonym is illegitimate.

Hyponym – n. A name to be rejected for want of an identified type.

Metonym – n. A name that is rejected because there is an older valid name based on another member of the same group.

Synonym – n. A name with the same meaning as another name in the same language but spelled differently; in taxonomy, synonyms are two or more scientific names for the same taxon, one of which is correct and the others incorrect under the International Rules of Nomenclature.

Synonymous – a. Having the smae meaning.

Synonymy – n. Discarded names for identical objects.

Typonym – n. A name rejected because an older name was based on the same type.

Ovules and Seeds

Albuminous – a. Having albumin or an endosperm.

Amphitropous – a. Half-inverted and straight, with the hilum lateral,

Anatropous – a. With the ovule reversed, with the micropyle close to the side of the hilum and the chalaza at the opposite end.

Angiospermous – a. With the seeds borne within a pericarp.

Angiosperm – n. A plant having its seeds enclosed in an ovary.

Aril – n. An appendage or an outer covering of a seed, growing out from the hilum or funciculus; sometimes it appears as a pulpy covering.

Campylotropous – a. Said of an ovule or seed which is curved in its formation so as to bring the micropyle or true apex down near hilum.

Caruncle – n. An excrescence or appendage at or about the hilum of the seed.

Chalaza – n. That part of the ovule or seed where the nucellus joins the integuments; it is the base of the nucellus and is always opposite the upper end of the cotyledons.

Cotyledon – n. A seed leaf. a primary leaf in the embryo.

Dicotyledons – pl. n. Plants of the class denoted by their possession of two cotyledons.

Dispermous – a. Two-seeded.

Embryo – n. The rudimentary plant formed in the seed.

Endosperm – n. The albumen of a seed in Angiosperms; in Gymnosperms, the prothallium within the embryo sac.

Exalbuminous – a. Without albumin, used only of seed in which the embryo occupies the entire cavity within the testa.

Foramen – n. An aperture, especially that in the outer integuments of the ovule; microphyle.

Funicle – n. See Funiculus.

Funiculus – n. A stem or thread-like structure that connects the ovule or seed to the placenta.

Hilum – n. The scar or mark on a seed indicating the point of attachment.

Hylum – n. See Hilum.

Hypocotyl – n. The axis jof an embryo below the attachment of the cotyledons.

Incumbent cotyledons – pl. n. Cotyledons so arranged that the back of one lies against the radicle.

Integument - n. The envelope of an ovule; the seed coat.

Kernel - n. The nucellus of an ovule, or of a seed, that is, the whole body within the seed coats.

Melanospermous - a. With black or dark-colored seeds, or spores.

Mesocotyl - n. An interpolated node in the seedling of grasses, so that the sheath and cotyledon are separated by it.

Micropyle - n. The aperture in the integument of a seed formerly the foramen; it marks the position of the radicle.

Monocotyledon - n. A plant having but one cotyledon or seed lobe.

Monocotyledonous - a. With only one cotyledon.

Monospermous - a. One-seeded.

Nucleus - n. A kernel of an ovule, which by fertilization becomes a seed; a dense protoplasmic structure near the center of living cells.

Olibospermous - a. Few-seeded.

Oncospore - n. A seed with hooks which aid in dispersal.

Oospore - n. The fertilized egg in the archegonium of Cryptogams from which the new plant develops directly.

Orthotropous - a. Said of an ovule or seed with a straight axis, chalaza at the other end.

Ovuliferous - a. Bearing ovules.

Parthenogenesis - n. The production of seeds without fertilization.

Perisperm - n. The ordinary albumen of a seed, restricted to that which is formed outside the embryo sac; the integuments of the seed.

Pip - n. A popular name for the seed of an apple or pear.

Pleiospermous - a. With an unusually large number of seeds.

Plumule - n. The primary leaf-bud of an embryo.

Primine - n. The outer integument of an ovule.

Pterospermous - a. With winged seeds.

Raphe - n. An adnate cord or ridge or fibro-vascular tissue which in a more or less anatropous ovule, connects the hilum with the chalaza.

Rhaphe - n. See Raphe.

Seed - n. A mature ovule.

Seed leaf - n. A cotyledon.

Semen - n. The seed of flowering plants.

Seminiferous - a. Seed-bearing; used for the special portion of the pericarp bearing the seeds.

Strophiole - n. An appendage at the hilum of certain seeds; a caruncle.

Testa - n. The outer coat of the seed, usually hard and brittle.

Umbilical cord - n. A vascular strand by which seeds are attached to the placenta; the funiculus.

Umbilicus - n. The hilum of the seed.

Pistils

Adynamogyny - n. Loss of function the female organ of a flower.

Agamogynaecism - n. In Compositae, the state of having female and neuter floweers in the same individual.

Apocarpous - a. With separate carpels, not syncarpous.

Apocarpy - n. The condition of having the carpels separate.

Bursicule - n. The pouchlike expansion of the stigma into which the caudicle of some orchids is inserted.

Carpel - n. A simple pistil; one unit of a compound pistil; in conifers, the cone scale of the female cone.

Carpellate - a. Possessing carpels.

Compound pistil - n. With two or more carpels coalesced into one body.

Dodecagynous - a. Possessing twelve distinct pistils or carpels.

Gynoecium - n. The pistil or pistils of a flower, the female portion as a whole.

Gynophore - n. The stipe or stalk of an ovary prolonged within the calyx.

Gynosporangium – n. The receptacle in which gynospores are developed.

Gynospore – n. One of the larger reproductive bodies (female) in the Isoetaceae.

Heterodistyly – n. Dimorphism, the presence of two kinds of plants within a species, one with long, the other with short styles.

Hexagynous – a. With six pistils.

Homostyly – n. The same relation of length between all styles and anthers of the same species.

Hypogynium – n. The perianth-like structure subtending the ovary in *Scleria* and some other Cyperaceae..

Hypogynous – a. Free from, but inserted beneath, the pistil or gynoecium.

Inferior – a. Beneath, lower, as an inferior ovary, one that is below the calyx-leaves.

Inferior ovary – n. An ovary with the perianth located on top.

Macrostylous – a. Long-styled.

Mesocarp – n. The middle layer of a pericarp.

Metandry – n. A condition in which the female flowers mature before the male; protogyny.

Monogynous – a. With one carpel.

Monostylous – a. With a. single style.

Octagynia – n. A Linnean order of plants with eight-styled flowers.

Octagynous – a. With eight-styled flowers.

Ovary – n. That part of the pistil which contains the ovules.

Pentagynous – a. With five pistils or styles.

Perigynium – n. The hypogynous setae of sedges; the flask or utricle of *Carex*; any hypogynousdisc.

Perigynous – a. Borne around the ovary and not beneath it, as with calyx, corolla, and stamens borne on the edge of a cupshaped hypanthium.

Pistil – n. The female organ of a flower, consisting when complete of an ovary, style, and stigma.

Pistillate – a. With pistils and no stamens; female.

Protogynous – a. With pistil maturing before stamens in the same flower.

Pterogynous – a. With a winged ovary.

Scutum – n. The broad dilated apes of the style in Asclepiads.

Sterile – a. Barren, as a flower without a pistil.

Stigma – n. The part of a pistil that receives the pollen; a point on the spores of *Equisetum*.

Stigmatic–a.Pertaining to the stigma.

Stylar – a. Relating to the style.

Style – n. The more or less elongated part of the pistil between the ovary and the stigma.

Stylopod – n. The enlargement at the base of the styles in Umbelliferae.

Superior ovary – n. An ovary with all of the floral envelopes inserted below on the torus.

Positions

Accumbent – a. Lying against another organ.

Accumbent cotyledons – pl. n. Cotyledons having their edges against the radicle.

Apressed – a. Lying flat against an organ.

Ascending – a. Sloping upward; produced somewhat obliquely or indirectly upward.

Assurgent – a. Ascending, rising.

Cernusous – a. Drooping; inclining somewhat from the perpendicular, nodding.

Coarctate – a. Crowded together.

Decumbent – a. Reclining or lying on the ground, but with the ends ascending.

Descending – a. Tending gradually downward, as the branches of some trees and as the roots.

Dextrorse – a. Turning to the right, or clockwise.

Erect – a. Upright, perpendicular.

Extrorse – a. Turned or faced outward or away from the axis, as an anther turned away from the center of the flower.

Incumbent – a. Resting or leaning upon.

Introrse – a. Turned or faced inwad or toward the axis, as an anther turned toward the center of the flower.

Inverted – a. Turned over; end-for-end; top side down.

Juxtapositionn – n. The relative position in which organs are placed; a placing or being placed side by side.

Nodding – a. Curved somewhat from the vertical, drooping.

Procumbent – a. Lying upon the ground, prostrate, trailing.

Prostrate – a. Lying flat, procumbent.

Reclinate – a. Bent down or falling back from the perpendicular.

Recurved – a. Bent or curved downward or backward or backward.

Reflexed – a. Abruptly curved or bent downward or backward.

Repent – a. Prostrate and rooting.

Reptant – a. Creeping on the ground and rooting.

Resupinate – a. Upside down, or apparently so.

Retrocurved–a. Bent or curved back.

Retroflexed – a. Bent back, reflexed.

Retrorse – a. Directed backwards or downwards.

Scandent – a. Climbing in any manner.

Sinistrorse – a. Turning to the left, or counterclockwise.

Strict – a. Stiff, upright, rigid.

Subterraneous – a. In or under the soil.

Supine – a. Prostrate with face turned up.

Trailing–a. Prostrate, but not rooting.

Sepals

Asepalous – a. Without sepals.

Caducous – a. Falling off early or prematurely, as the sepals in some plants.

Calyanthemy – n. Petalody of the calyx; the formation of colored petal-like structures in place of a normal calyx.

Calyculate – a. Calyx-like, bearing a part resembling a calyx, particularly, with bracts against or underneath the calyx resembling a supplementary or outer calyx.

Calyx – n. The outermost circle of floral envelopes.

Disepalous – a. With two sepals.

Epicalyx – n. A series of bracts close to and resembling the calyx.

Gamosepalous – a. With the sepals united.

Monosepalous – a. With one sepal.

Octosepalous – a. With eight sepals.

Pentasepalous a. With five sepals.

Pleiosepalous - a. Many-sepaled.

Polysepalous – a. With many distinct sepals.

Sepal – n. One of the separate parts of a calyx.

Sex Distribution

Agamohermaphrodite – a. With hermaphrodite and neuter flowers on the same plant.

Agamospermy – n. Seed production without fertilization.

Allautogamia – n. The state of having two methods of pollination, one usual, and the other facultative.

Allogamous – a. Reproducing by cross-fertilization.

Allogamy – n. The pollination of a flower with pollen another flower, see Geitonogamy and Xenogamy.

Amixia – n. Cross-sterility.

Androgynous–a.Hermaphrodite, with both male and female flowers in the same inflorescence, occasionally used meaning monoecious.

Autoicous – a. See Monoecious.

Autophilous – a. Self-pollinated.

Cenanthy – n. The suppresion of the stamens and pistil leaving the perianth empty.

Column – a. A combination of stamens and styles into a solid central body, as in orchids.

Cleistogene – n. A plant which bears cleistogamous flowers.

Cleistogeny – n. The state of bearing cleistogamous flowers.

Cleistogenous – a. Cleistogamous.

Dichogamous – a. Hermaphrodite with one sex maturing earlier than the other, the stamens and pistil not synchronizing.

Dichogamy – n. A condition in perfect lowers in which the sexes do not mature simultaneously.

Diclinism – n. The separation of anthers and stigma in space, as dichogamy does in time.

Diclinous – a. Unisexual, having the stamens in one flower and the pistil in another.

Diecious – a. See Dioecious.

Dimorphic – a. Occurring under two forms, as with long and short styles.

Dioecious – a. Unisexual, with staminate flowers on one plant and the pistillate on another.

Disanthic – a. With fertilization by pollen from another plant.

Epigymolus – a. Borne on the ovary, the ovary inferior and not perigynous.

Exogynous – a. With the style longer than the corolla and projecting beyound it.

Flos. (pl. flowres) – n. A perfect flower with some protecting envelope.

Frustraneous – a. Said of Compositae with disk flowers hermaphrodite, and those of the ray neuter or imperfect.

Geitonogamy – n. Pollination by pollen from another flower on the same plant.

Gnesiogamy – n. Fertilization between different individuals of the same species.

Gynandrous – a. With the stamens adnate to the pistil, as in orchids.

Gynecandrous – a. With staminate and pistillate fowers in the same spike, the pistillate at the apex, used chiefly in the Cyperaceae.

Gynodioecious – a. Dioecious with some flowers hermaphrodite, others pistillate only, on separate plants.

Gynomonoecious – a. Monoecious with female and hermaphrodite flowers on the same plants.

Gynostemium – n. The compound structure resulting form the union of the stamens and pistil in Orchidaceae.

Hermaphrodite – a. Bearing two kinds of flowers.

Homocephalic – a. Delpino's term in reference to homogamy in which the pollen fertilizes another flower in the same inflorescence.

Heterocephalous – a. With staminate and pistillate flowers on separate heads on the same plant.

Homoclinous – a. Delpino's term in reference to that kind of homogamy in which a complete flower is fertilized by its own pollen.

Homogamous – a. Bearing one kind of flower.

Homogony – n. The condition in which the pistils and stamens of all flowers are of uniform relative length.

Homotropic – a. Fertilized by pollen from the same flower.

Imperfect – a. Said of flowers lacking one of the essential organs.

Misogamy – n. Reproductive isolation.

Monoclinous – a. Having both stamens and pistils in the same flower, applied to the heads of Compositae that have only hermaphrodite flowers.

Monoecious – a. Having unisexual flowers with both sexes on the same plant.

Motion-dichogamy – n. A condition in which the sexual organs very in lenth or position during flowering.

Monoecious – a. Having unisexual

flowers with both sexes on the same plant.

Motion-dichogamy – n. A condition in which the sexual organs vary in length or position during flowering.

Neuter – a. Sexless, as a flower that has neither stamens nor pistils.

Neutral – a. Without stamens or pistils, sexless.

Nyctigamous – a. Said of flowers which open at night and close by day, marrying at night.

Panmixy – n. Free and more of less unlimited cross-fertilizaion.

Perfect – a. Said of flowers having both sex organs present and functioning.

Phaenogamous – n. Said of plants sexually propagating by flowers, the essential organs of which are stamens and pistil.

Phenological isolation – n. Isolation by a time of flowering, as either earlier of later than the other species of the genus.

Phytogamy – n. Cross-fertilizaion, but chiefly dioecious.

Polygamomonoecious – a. Polygamous, but chiefly monoecious.

Polygamous – a. Bearing perfect and unisexual fower on the same individual.

Psychrocleistogamy – n. Cleistogamy induced by low temperature.

Staurigmia – n. Cross-fertilizatio.

Superior – a. Growing or placed above, hypogynous.

Synacmy – n. A condition in which stamens and pistil mature simultaneously, the opposite of heteracmy.

Synathesis – n. The simultaneous maturation of stamens and pistil, synacmy.

Synoecious – a. With staminate and pistillate flowers both present in the same head.

Thermocleistogamy – n. Self-pollination taking place within flowers the opening of whose perianth has been inhibited by low temperature.

Trigamous – a. Bearing three forms of flowers, trimorphic.

Trimorphic – a. Occurring in thre forms, as with long, short, and intermediate styles.

Unisexual – a. Of one sex, stamens or pistil only.

Xenogamy – n. Cross-fertilization between sexual elements borne by different individuals.

Stamens

Agamandroecism – n. The condition in Compositae of having male and neuter flowers in the same individuaal.

Agynic – a. Said of stamens which are free from the ovary, destitute of pistils.

Androecium – n. The stamens of a flower (a. collective term).

Androphore – n. A support or column on which the stamens are raised.

Anther – a. That portion of the stamen which bears the pollen.

Antheriferous – a. Anther-bearing.

Antheroid – a. Anther-like.

Bicruris – a. Two-legged, as the pollen masses of asclepiads.

Clinandrium – n. The anther bed in orchids, that part of the column in which the anther is concealed.

Cryptandrous – With hidden anthers, clesistogamous, the stamens remaining enclosed in the flower.

Decandrous – a. With ten stamens.

Diadelphous – a. With stamens formed in two groups by the union of their filaments.

Diandrous – a. With two stamens.

Didynamous – a. Four-stamened with stamens in pairs, two long, two short, as in many Labiates.

Dodecandrous – a. Normally possesing twelve stamens, occasionally extended to more than twelve.

Emasculation – n. The removal of the anthers from a bud or flower.

Enneandrous – a. With nine stamens.

Exserted – a. Sticking out, projecting beyond, as stamens from perianth, not included.

Filament - n. The part of a stamen that supports the anther.

Gynostegium - n. A sheath or covering of the gynoecium, of whatever nature.

Heterandrous - a. With two sets of stamens, as flowers with two kinds of stamens.

Hexandrous- a. With twenty or more stamens.

Homoeandrous - a. With only one kind of stamen.

Icosandrous- a. With twenty or more stamens.

Incumbent anther - n. An anther attached to the inner face of its. filament.

Isadelphous - a. Equal brother-hood, the number of stamens in each group being equal.

Massule - n. A group of cohering pollen-grains produced by one primary mother cell in orchids, also styled pollen-mass.

Melantherous - a. With black anthers.

Monadelphous - a. With stamens united by their filaments into a tube or column.

Monandrous - a. With one stamen.

Nototribal - a. With stamens arranged so as to deposit pollen on the backs of their insect visitors.

Octostemonous - a. With eight fertile stamens.

Oligandrous- n. An abortive stamen, a staminode.

Pentadelphous - a. With five fraternities or bundles of stamens.

Pentandrous - a. With five stamens.

Phaenantherous - a. With stamens exserted.

Pleurotribal - a. Said of flowers whose stamens are adapted to deposit their pollen upon the sides of insect visitors.

Polyadelphous - a. With stamens disposed into several brother-hoods or groups.

Polyandrous - a. With an indefinite number of stamens.

Plystemonous - a. Polyandrous, with numerous stamens.

Porandrous - a. With anthers opening by pores.

Protandrous - a. With the anthers maturing bbefore the pistil in the same flower, one kind of dichogamy.

Protandry - n. A condition in which the stamens mature before the pistil in the same flower.

Proterandrous- a. With the stamens maturing before the pistil in the same flower; protandrous, one kind of dichogamy.

Proterogyny - n. A condition in which the pistil matures before the anthers.

Psilostemon - a. Smooth-stamened; naked-stamened.

Resilient - a. Springing or bending back, as some stamens.

Sporophyll - n. A spore-bearing leaf.

Stamen - n. The pollen-bearing organ of the flower, the male organ.

Stamen, sterile - n. A body belonging to the series of stamens but without pollen.

Staminate - a. With stamens but no pistil; male.

Stamineal - a. Relating to or consisting of stamens.

Staminode - n. A sterile stamen, or a structure reesmbling such and borne in the staminal part of the flower; in some flowers (as in *Canna*) staminodia are petal-like and showy.

Sternothibal - a. Said of flowers whose anthers are so arranged as to dust their pollen on the under part of the thorax of their insect visitors.

Sulphur rain - n. Pine pollen carried in excessive amounts by air currents.

Synantherous - a. With anthers joined to form a tube.

Synema (pl. synemata) - n. the column of monadelphous stamens, as in Malvaceae..

Triadelphous - a. With stamens in

three sets.

Triandrous – a. With three stamens.

Tridynamous – a. With three stamens out of six being longer than the rest.

Stems

Acaulescent – a. Stemless or apparently so.

Armed – a. Possessing any kind of strong and sharp defense, as thorns, spines, prickles, or barbs.

Articulate – a. Jointed; with nodes or joints, or places where separation may naturally take place.

Axil – n. The upper angle formed between the axis and any organ that arises from it.

Axillary – a. Situated in the axil.

Axis – n. The main or central line of development of any plant or organ; the main stem.

Bast – n. The fibrous constituent of the bark of many species.

Bole – n. The main trunk or a tree with a distinct stem.

Brachiate–a.Spreading with branches suggesting arms.

Branch – n. A lateral division of the stem or axis of growth.

Branchlet – n. The ultimate division of a branch.

Bud – n. An embryonic axis with its appendages.

Bulb – n. A modified bud usually underground; **imbricated**– with scaly modified leaves, as in the **lily; tunicated** – With modified leaves forming concentric layers around the bud, as the onion.

Bulbiferous – a. Bearing bulbs.

Bulbil – n. A bulb arising from the mother bulb.

Bulbet – n. A little bulb produced in the leaf axil, inflorescences, or other unusual places.

Bulbose – a. Having bulbs or the structure of a bulb.

Caespitose – a. Growing in tufts.

Caudex – n. The woody base of a perennial plant.

Caulescent – a. More of less stemmed or steam-bearing; having an evident stem above ground.

Cauliculous – a. With a small stem.

Cauline – a. Pertaining or belonging to the stem.

Cladophyll – a. A branch assuming the form and function of a leaf; a cladode.

Corm – n. A solid bulblike structure, usually subterranean, as the "bulb" of *Gladiolus*.

Cormel – n. A corm arising from a mother corm.

Crown – n. Corona; the base of a tufted, herbaceous, perennial grass; the hard ring or zone at the summit of the lemma of some species of *Stipa*. the part of a stem at the surface of the ground; a part of a rhizome with a large bud, used in propagation.

D. B. H. – n. Diameter Breast High

Deliquescent – a. Dissolving or melting away; said of a stem that loses itself by repeating branching; opposed to excurrent.

Digonous – a. Two-angled, as the stems of some Cacti.

Excurrent-a. With the stem remaining central and other parts being regularly disposed around it; running through to the apex.

Gemma – n. A bud or a body analogous to a bud capable of producing a new plant.

Haplocaulous – a. Having a simple unbranched stem.

Infra-axillary – a. Below the axil, sub-axillary.

Internode – n. The part of a stem between two successive nodes.

Melanoxylon – n. Black wood.

Nodal – a. Pertaining to the node.

Node – n. That point on a stem which normally bears a leaf or leaves.

Nodiferous – a. Bearing nodes.

Nodose – a. With nodes.

Ramal – a. Pertaining to a a branch.

Ramal – a. Pertaining to a branch.

Rameal – a. See Ramal.

Rhizome– n. A dorsiventral, root-like, underground stem which produces roots and shoots; root-stock.

Sapwood – n. The new wood in an exogenous tree, so long as it is pervious to water; the alburnum.

Sautellus – n. A bulblet, such as those of *Lilium tigrinum.*

Scapose – a. Bearing or resembling a scape.

Semester ring – n. The ring produced in the wood of many tropical trees in consequence of two periods of growth and rest in a year.

Stem – n. The main axis of a plant leaf-bearing as distinguished from the root bearing axis.

Stems, subterranean – pl.n. Rhizomes, tubers, bulbs, and corms.

Stipitate – a. With stipe.

Sub-axillary – a. Borne below the axil.

Supra-axillary – a. Borne above the axil, super-axillary.

Triquetrous – a. With three salient angles and three concave faces.

Tuber – n. A short thickened branch of a subterraean stem, beset with buds or "eyes".

Tubercle – n. A little tuber.

Surfaces and Vestures

Shiny surfaces

Glittering – a. With luster as from a polished surface which is not uniform.

Illustrous – a. Bright, brilliant, lustrous.

Lucid – a. Shiny. bright.

Lustrous – a. Glossy, shiny.

Micaceous – a. Glittering, sparkling, mica-like.

Nacreous – a. With pearly luster.

Nitid – a. Smooth and clear, lustrous, glitterig.

Vernicose – a. Shiny, as though varnished.

Smooth surfaces (without hairs, spines, bristles, or scales)

Regular

Alepidote – a. Destitute of scurf of scales.

Bloom – n. The white, waxy, or pruinose covering on many fruits, leaves, and stems.

Glabrate – a. Nearly glabrous, or becoming glabrous with maturity or age.

Glabrous – a. Smooth, devoid of pubescence or hair of whatsoever form.

Glaucous – a. Covered with a "bloom" or a whitish substance that rubs off, as of a plum or cabbage leaf.

Laevigate – a. Smooth, as if polished.

Naked – a. Wanting its usual covering, as without pubescence.

Pruinose – a. Having a waxy powdery secretion on the surface, a "bloom".

Unctuous – a. Having a surface which feels greasy.

Vernicose – a. Shiny, as though varnished.

Viscid – a. Sticky from a tenacious coating or secretion.

Irregular

Alveolate – a. Honey-combed.

Areola – n. A small area marked out on a surface.

Arcolate – a. With areola.

Bullate – a. With surface blistered or puckered, as the leaf of a Savoy cabbage.

Canaliculate – a. Channeled longitudinally.

Colliculose – a. Covered with little round elevations or hillocks.

Corrugate – a. Wrinkled.

Faveolate – a. Honey-combed, alveolate.

Fluted – a. Regularly marked by alternating ridges and groovelike depressions.

Foveolate – a. Marked with small pitting.

Furrowed – a. With longitudinal channels or grooves, sulcate, striate on a large scale.

Lacuno-rugose – a. Marked with irregular wrinkles, as the stone of a peach.

Mammiform - a. Breast-shaped, conical with rounded apex.

Mammilla - n. A nipple or teat.

Mammillate - a. Having teat-shaped processes.

Mammnose-a. Having teat-shaped processes.

Nodulose - a. With little knobs or knots.

Papilla (pl. papillae) - n. A minute nipple shaped projection.

Papillary - a. Resembling papillae.

Papilose - a. Bearing papillae.

Pitted - a. Marked with small depressions, punctate.

Pustulose - a. Blistery, furnished with pustules or irregular raised pimples (not as roughened as papillose).

Rugose - a. Wrinkled, as leaf surface with sunken veins.

Rugulose - a. Somewhat wrinkled.

Scrobiculate - a. Marked by minute or shallow depressions, pitted.

Striate - a. With fine grooves, ridges, or lines of color.

Sulcate - a. Grooved or furrowed lengthwise.

Tesselate - a. Having the surface marked with square or oblong depressions.

Thelephorous - a.Covered with nipple-like prominences.

Verrucose - a. Covered with wartlike elevations.

Wrinkled - a. Rugose, irregularly creased.

Granular and scaly surfaces

Farinose - a. Covered with a mealiness.

Fornix (pl. fornices) - n. A small arched scale.

Granular - a. Cowered with very small grains; minutely or finely mealy.

Grumose - a. Grumby.

Lepis (pl. lepides) - n. A scale, usually attached by its center.

Lepidote - a. Beset with small scurfy scales.

Pulveraceous - a. Covered with a layer of powdery granules.

Ramentum (pl. ramenta) ·· n. Used in the plural for the thin chaffy scales of the epidermis, as the scales of many ferns.

Scobinate - a. Rough as though rasped.

Scurf - n. Small branlike scales on the epidermis.

Scurfy - a. Covered with small scales.

Squamaceous - a. Scaly.

Squamose - a. Furnished with scales.

Squamose - a. Squamate, full of scales.

Stellate scales - pl. n. Trichomes, discs borne by their edge or center.

Tartareous - a. With a loose or rough crumbling surface, as some lichens.

Hairy surfaces

Straight hairs

Canescent - a. Gray-pubescent or becoming so.

Cilium (pl. cilia) - n. Used generally in the plural to desginate marginal hairs.

Ciliate - a. Said of a margin fringed with hairs.

Comose -a. Bearing a tuft or tufts of hair.

Crinus - n. A stiff hair on any part.

Down - n. Soft pubescence; the pappus of such plants as thistles.

Glochidiate - a. Pubescent with barbed bristles.

Hair - a. An outgrowth of the epidermis consisting of one to several cells.

Hirsute - a. With stiff or bristly hairs.

Hirsutulous - a. Slightly hirsute.

Hirtellous - a. Softly or minutely hirsute or hairy.

Hispid - a. Beset with rough hairs or bristles.

Hispidulous - a. Somewhat or minutely hispid.

Hosary - a. Covered with a close white or whitish pubescence.

Multiciliate - a. With many cilia.

Piliferous – a. Bearing hairs, or tipped with hairs; hairpointed.

Pilose – a. With soft hairs.

Plumose – a. Pubescent in a manner simulating a feather or a plume.

Puberulent – a. Somewhate or minutely pubescent.

Pubescence – n. The hairiness or plants.

Pubescent – a. Covered with short soft hairs; down.

Scabrous – a. With short bristly hairs. rough to the touch.

Sericeous – a. Silky, clothed with closely appressed, soft, straight pubescence.

Silky – a. Said of a condition produced by a covering of soft, appressed, fine hairs,; sericeous.

Strigose – a. Beset wioth sharppointed, appressed, straight, and stiff hairs or bristles. hispid.

Trichome – n. Any hairlike outgrowth of the epidermis, as a hair or bristle,

Tufted – ja. Cespitose, comose, having a small cluster of hairs.

Velutinous – a. Velvety, due to a coating of fine, soft hairs.

Velumen – n. Close, short, soft hair.

Villose – a. Covered with long, silky, straight hairs.

Interwoven hairs

Arachnoid – a. Cobwebby, composed of slender entangled hairs; spider-like.

Eriophorous – a. Wool-bearing, densely cottony.

Felted – a. Matted with intertwined hairs.

Floccus – a. Bearing flocci.

Flocculose – a. Like wool.

Gossypine – a. Cottony, flocculent; like the hairs on the seeds of *Gossypium.*

Holosericeous – a. Woolly-silky.

Indument – n. Hairy or pubescent with rather heavy covering.

Lanate – a. Clothed with woolly and interwoven hairs.

Lanose – a. Woolly.

Lanuginose – a. Woolly or cottony.

Manicate – a. Said of pubescence so dense and interwoven that it may be stripped off.

Panniform – a. Having the appearance or texture of felt or woollen cloth.

Pannose – a. Having the appearance or texture of felt or woollen of very close texture.

Tomentose – a. With tomentum; densely woolly or pubescent; with matted soft woollike hairiness.

Tomentulose – a. Somewhat or delicately tomentose.

Tomentum – n. A densely matted pubescence.

Velutinous – a. Velvety, due to a coating of fine soft hairs.

Web – n. A cluster of slender, soft hairs.

Woolly – a. Lanate, tomentose, clothed with long and tortuous or matted hairs.

Hooked hairs

Aduncate – a. Bent or crooked, as a hook.

Aduncous – a. Hooked.

Uncinate – a. Hooked, bent at the tip like a hook.

Branched hairs

Candelabra bairs – pl.n. Stellate hairs in two or more tiers.

Stellipilous – a. With stellate hairs.

Stellate – a. Starry, often said of hairs that have radiating branches from base of separate hairs similary aggregated.

Spiny or prickly surfaces

Acantha – n. Thorn.

Aculeate – a. With prickles.

Aculeolate – a. Beset with small prickles.

Asperate – a. Rough.

Asperous – a. Rough or harsh to the touch.

Barbed – a. Furnished with retrorse projections.

Barbellate – a. Finely barbed.

Barbulate – a. Finely bearded.

Bearded – a. Bearing or furnished with long or stiff hairs.

Bristle – a. A stiff hair.

Bristly – a. Bearing stiff strong hairs.

Echinate – a. Armed with prickles.

Exasperate – a. Rough, with hard, projecting points.

Glochid – n. A barbed hair or bristle.

Glochideous – a. Pubescent with barbed bristles.

Muricate – a. Rough with short, hard points.

Muriculate – a. Very finely muricate.

Pungent – a. Ending in a rigid and sharp point.

Scabridulous – a. Slightly rough.

Scabrous – a. With short bristly hairs, rough to the touch.

Setigerrous -a. Bristly,l bristle-bearing.

Setose – a. Bristly, beset with bristles.

Setulose – a. With minute bristles.

Spiculose – a. With a surface covered with fine points.

Spinescent – a. Ending in a spine or sharp point.

Spinopse – a. Spinelike; with spines or thorns.

Spinulose – a. With small spines or spinules.

Spiny – a. Beset with spines.

Miscellaneous surfaces

Cilium (pl. cilia)– n. Used generally in the plural to desingnate marginal hairs.

Ciliolate – a. Minutely ciliate.

Derma (pl. dermate) – n. Surface of an organ, bark rind, or skin.

Fimbria – n. A fringe.

Fimbriate – a. Fringed. the hairs longer or coarser as compared with ciliate.

Fimbrillate – a. With a minute fringe.

Malpighiaceous hairs – n. Hairs which are straight and appressed but but attached, by the middle,frequent in the family Malpighiaceae.

Pelta – a. A bract attached by its middle as in peppers.

Pilogl: ndulose – a. Bearifng glandular I airs.

Process – n. Any projecting appendage.

Punctate – a. Dotted with depressions or punctures.

Stinging hair – n. A hollow hair seated on a gland which secretes an acid substance, as in nettles.

Tentacle – n. A sensitive glandular hair, as those on the leaf of *Drosera*.

Trap-hairs – n. Special hairs which confine insects in certain flowers until pollination is effected.

Tricholoma – n. With the edge or border furnished with hairs.

Texture

Callose– a. Hard and thick in texture.

Cartilagineous – a. Like cartilage or gristle.

Cereous – a. Waxy.

Chaffy – a. With small membranous scales.

Chartaceous – a. Having the texture of writing paper.

Coriaceous – a. Leather-like.

Corneous – a. Horny, with a horny texture.

Crystalline – a. Resembling a crystal.

Fibrous – a. Having much woody fiber. as the rind of a coconut.

Flaccid–a. Withered and limp, flabby.

Feshy – a. Succulent.

Fragile – a. Weak, easily broken.

Frutescent – a. Becoming shrubby.

Fruticose – a. Shrubby or shrub-like in the sense of being woody.

Gelatinous – a. Jelly-like.

Glutinous – a. Sticky.

Granular – a. Composed of grains; divided into little knots or tubercles.

Herbaceous – a. With the texture, colors, and properties of a herb, not woody.

Hyaline – a. Thin and translucent or transparent.

Indurated – a. Hardened.

Ligneous – a. Woody

Membranaceous – a. The same as membranous.

Membranous – a. Thin, more or less flexible, and translucent.

Mucilaginous – a. Slimy, composed of mucilage.

Oleaginous – a. Oily and succulent.

Oleiferous – a. Oil-bearing.

Oleraceous – a. Herbaceous.

Osseous – a. Bony.

Ossifled – a. Made hard as bone, as the stones of drupes, such as the peach or plum.

Paleaceous – a. Chaffy.

Papery – a. Having the texture of paper, chartaceous.

Papyraceous – a. Papery, white as paper.

Besinous – a. Like, or pertaining to resin.

Sap – n. The juice of a plant; the fluid contents of cells and young vessels consisting of water and salts absorbed by the roots and distributed through the plant.

Saponaceous – a. Soapy, slippery to the touch.

Scieroid – a. Having a hard texture.

Sebaceous – a. Like lumps of tallow.

Spumose – a. Frothy.

Suberous – a. Corky in texture.

Succulent – a. Juicy or pulpy.

Suffruticose – a. Shrubby at the base.

Viscid – a. Sticky from a tenacious coating or secretion.

Time and Season

Aestival – a. See Estiva.

Allochronic species – pl. n. Species which do not belong to the same time level, as opposed to contemporary or synchronic.

Allogenous flors – pl. n. Relic plants of an earlier prevailing flora and envioronment; epiblotic plants.

Annotinous – a. A year old, or in yearly growths.

Annual – a. Of one year's duration; completing its life cycle in one year.

Asynbgamic – a. Unable to cross by reason of differences in time of flowering.

Autumnal – a. Of or pertaining to autumn, flowering in autumn.

Biennial – a. Of two seasons' duration from seed to maturity and death.

Crepuscular – a. Of or pertaining to twilight.

Diplobiont – n. A plant flowering or fruiting twice each season.

Diurnal – a. Occurring in the daytime , sometimes used for ephemeral.

Efflorescence – n. The season of flowering; anthesis.

Ephemer – n. A flower which closes after a short term of expansion.

Ephemeral – a. Persisting for one day only, as flowers of spiderwort.

Estival – a. Of or pertaining to summer.

Frutescence – n. The time of maturity of fruit.

Hemeranthous – a. Day-flowering.

Hibernaculum – n. The winter resting part of a plant, as a bud or underground stem.

Hiberanal – a. Hibernating, relating to winter.

Hibernation – n. Passing the winter in a dormant state.

Hiemal – a. Relating to winter.

Horary – a. Lasting an hour or two, as the petals or *Cistus*.

Hyemal – a. See Hiemal.

Matutinal – a. Pertaining to morning; plants flowering early, as *Ipomoea purpurea*.

Menstrual – a. Lasting for a month or so.

Nocturnal – a. Occurring at night, or lasting one night only.

Nox (pl. noctes) – n. Night.

Nyctanthous – a. Said of night-flowering plants.

Nyctigamous – a. With flowers which open at night and close by day; pollinated at night.

Nyctitropic – a. Turning in response

to darkness.

Nyctitropism – n. The act of assuming the sleep position.

Paraheliotropism – n. Diurnal sleep, the movement of leaves of avoid the effect of intense sun-light.

Perennation – n. A lasting or perennial state.

Perennial – a. Said of a plant which lasts several years, not perishing normally after one flowering and fruiting.

Photeolic – a. Appearing or developing very early as aments in *Salix* expanding before the leaves.

Prevernal – a. Of early spring.

Serotinal – a. Produced late in the season, or the year, as in autumn.

Serotinous – a. Produced or ocurring late in the season.

Sleep – n. The response of plants, with changes in position of organs such as leaves due to the absence of light.

Sleep movement – n. Positions taken by leaves during the night, nyctitropic movement.

Synchronic species – pl. n. Species which belong to the same time level, contemporary.

Therophyllous – a. Producing leaves in summer, deciduous-leaved.

Therophyte – n. A plant which completes its development in one season, its seeds remaining latent during the hot season; an annual.

Trimestris – a. Of three months, as lasting that time or maturing in it.

Vernal – a. Pertaining to spring.

Vespertine – a. Appearing or expanding in the evening.

Tropisms

Tropism is the innote tendency of on organism to react in a definite manner to an external stimulus.

Anemotropism – n. The tropic response of organisme to wind and air currents.

Aphaptotropism – n. The state of not being influenced by touching stems or other surfaces.

Apheliotropism – n. The act of turning away from the sun; negative phototropism.

Aphercotropism – n. The act of turing away from an obstruction.

Aphototropism – n. The act of turning away from the light.

Autonyctitropism – n. Regularly assuming the position usual during the night.

Autoorthotropism – n. The tendency of an organ to grow in a straight line forward.

Barotropism – n. The response of an organism to changes in barometric pressure.

Chemotropism – n. Curvature in response to chemical stimuli.

Cryotropism – n. Movements induced by cold or frost.

Diaphototropism – n. The act of turning at right angles to incident light, as the leaves of some plants.

Diatropism – n. The act of organs placing themselves crosswise to an operating stimulus.

Dromotropism – n. The tropic movement of climbing plants which results in their spiral growth.

Edaphotropism – n. Tropic responses to the soil.

Epigeotropism – n. Tropism resulting in growth on the surface of the soil.

Geodiatropism – n. A function which places an organ at right angles to the force of gravity.

Heliotropism – n. The act of turning in response to the sun.

Homalotropism – n. The act of turning to a horizontal position.

Hydrotropism – n. The act of turning in response to the influence of water.

Nyctitropism – n. The tendency of certain plant organs, as leaves, to assume special "sleeping" positions or to make curvatures under the influence of darkness.

Ombrotropism – n. Tropic responses

of organisms to the stimulus of rain.

Orthoosmotropism – n. The act of assuming an erect position due to osmotic action.

Orthotropism – n. Growth in a vertical position.

Paraheliotropism – n. Diurnal sleep; movements of leaves to avoid the effects of intense sun-light.

Paraphototropism – n. The act of turning at right angies to the incident light.

Phototropism – n. the act of turning in response to light.

Polytropism – n. The act of leaves placing themselves vertically and meridionally, the two surfaces facing east and west respectively.

Prohydrotropism – n. The act of turning toward a source of moisture.

Rheotropism – n. The act of turning in response to a current of water.

Selenotropism – n. Movements of plants caused by the light of the moon.

Siotropism – n. Response to shaking, as with *Mimosa*.

Stenotropism – n. A condition with narrfow limits of adaptations to varied conditions.

Stereotropism – n. Response of contact stimuli.

Telotropism – n. The act of turning to one stimulus to the exclusion of all others.

Thermotropism – n. The act of turning in response to heat.

Thigmotropism – n. The act of turning in response to a mechanical stimulus.

Topotropism – n. The act of turning toward the place from which a stimulus comes.

Traumatropism – n. The sensitiveness of certain plant organs to wounds.

Zenotropism–n.Negative geotropism.

Type Terminology

Agrotype – n. An agricultural race.

Biotype – n. A group of individuals all of one genotype.

Chirotype – n. The ๏ecimen of which a manuscript . ๅe is based.

Clastotype – n. A fragment form the original

Cotype – n. An additional or associate type specimen from which a taxon is described.

Generitype – n. The type species of a genus.

Genotype – n. The type of a a genus, the species upon which the genus, was established.

Holotype – n. The one specimen or other element used by the author of the name, or designated by him, as the nomenclatural type (i.e., the element to which the name of the taxon is permanently attached).

Icotype – n. A type serving for identification, but not previously used in literature.

Isotype – n. A. specimen believed to be a duplicate of the holotype.

Lectotype – n. A specimen or other element selected from the original material to serve as the nomenclatural type, when the holotype was not designated at the time of publication, or when the holotype is missing.

Logotype – n. A type determined historically from two or more original species.

Merotype – n. A specimen collected from the original type in cultivation by means of vegetative reproduction.

Metatype – n. A specimen from the original locality, recognized as authentic by the describer him-self.

Mimotype – n. Forms distinctly resembling each other, fulfilling similar functions, and thus representing each other in different floras.

Monotypic – a. Having only one ex-
ponent, as a genus with but one
species.

Necrotype – n. A form that for-merly
existed but is now extinct.

Neotype – n. A specimen selected to
serve as the nomenclatural type of
a taxon in a situation when all
material on which the taxon was
based is missing.

Paratype – n. A specimen cited with
the original description other than
the holotype.

Phenotype – n. A group of individuals of similar appearance but
not necessarily of similar genetic
constitution.

Phototype – n. A photograph of a
type specimen.

Proterotypes – pl. n. Primary types.
all specimens which have served as
a basis for descriptions, figures of
organisms; further divided into
holotype, cotype (or syntype),
paratype, lectotype, and chirotype.

Prototype – n. The assumed ancestral form from which the descendents have become modified.

Spermotype – n. A specimen cut from
a seedling grown from the original
type.

Syntype – n. One of two or more
specimens or elements used by an
author when no holotype is designated, or one of two or more
specimens simultaneously designated as type.

Tectoparatype – n. A specimen
selected to show the microscopic
structure of the original type of a
species or genus.

Topotype – n. A specimen of a named
species from the type locality.

Type specimen – n. The original
specimen from which a description
is written.

Typical – a. In classification, conforming to the originally described
specimen.

Unitypic – a. See Monotypic.

Vernation

Vernation deals with the disposition of
foliage leaves in the bud. It does not
treat of the insertion of the leaves on
the axis as this comes under
phyllotaxy. The disposition of floral
parts in the bud is treated under
Aestivation.

Circinate – a. Coiled from the top
downward, as the leaves of *Frosera*
and the fronds of true ferns.

Complicate – a. Folded upon itself.

Conduplicate – a. Folded lengthwise,
or doubled up flat on the midrib,
the upper face of the leaf always
within, as in *Magnolia*.

Convolute – a. Rolled up from one
margin, one margin on the inside
and the other on the outside.

Equitant – a. Folded over as if
astride; used for conduplicate
which unfold each other in two
ranks, as in *Iris*.

Involute – a. With both margins
rolled towards the midrib on the
upper surface, as the leaves of
water lily and violets.

Plicate – a. Folded on the several ribs
in the manner of a closed fan. It
occurs in palmately veined leaves,
as in maple and currant.

Reclinate (inflexed) – a. With the
upper part bent on the lower, or
the blade on the petiole, as in
Liriodendron.

Revolute – a. With both margins
rolled toward the midrib on the
lower face, as the leaves of *Azalea.*

Some Specific Epithets with their Meanings

The specific epithet is the second element in a scientific name. It may be a noun (in the nominative or the genitive), or an adjective. When adjectival in form, and not used as a substantive, it agrees in gender with the generic name. The epithetrs are listed in the masculine gender, followed by the feminine and neuter endings. Present participles, which have the same endings, are indicated below.

A

abbreviatus, -a, -um – abbreviated, shortened.

abietinus, -a, -um – like the abies; or firlike.

abruptus, -a, -um – abrupt.

absinthoides, -es, -es – absintha-like.

abyssinicus, -a. -um – Abyssinisan.

acantha (noun) – thorn.

acanthocomus, – a, -um – spinhaired or crowned.

acaulis, – is, -e – stemless.

accumulatus, -a, -um – accumulated.

acephalus, -a, -um – headless.

acerbus, -a, --um – harash or sour.

acerifolius, -a, -um – maple-leaved.

aceroides, -es, -es – maple-like.

acetosum, -a, -um – acetic, sour.

acicularis, -is, -e – needle-like.

acidissimus, -a, -um – very sour.

acidus, -a, -um – acid, sour.

acinaceus, -a,-um – scimitar -or saber-shaped.

acris, -is, -e – acid, sharp.

acrostichoides, -es, -es – acrostichum-like.

acrotrichis, -is, e – hairy-tipped.

aculeatus, -a, -um – with prickles.

acuminatus, -a, um – tapering at the end.

acutifolius, – a, -um – with sharp leaves.

acutus, -a, -um – acute, sharp-pointed.

adenophorus,-a, -um–gland-bearing.

adenophyllus, -a, -um – glandular-leaved

adiantifolius, -a, -um – adiantum-leaved.

admirabilis, -is, -e – admirable, noteworthy.

adnatus, -a, um – adnate, joined to.

adpressus, -a, -um – pressed against.

adscendens(pres. part) – ascending.

adsurgens (pres. part.) – ascending.

aduncus, -a, -um – hooked.

adventivus, -a, -um – adventituous.

advenus, -a, -um – newly arrived, adventive.

aerius, -a, um – aerial.

aestivalis, -is,, -e – summer.

aestivus, -a, -um- summer.

affinis, -is, -e – related (to another species).

agavoides, -es, -es – agave-like.

agglomeratus, -a, -um – heaped up, crowded together.

aggregatus, -a, -um – aggregate, clustered.

agrestis, -is, - e -growing wild.

agrostoides, -es, -es – similar to *Agrostis*.

airoides, -es, -es – like *aira*.

alaris, -is, -e – wing-shaped.

alatus, -a, -um – winged.

albescens (pres. part) – whitish, becoming white.

albicans (pres. part.) – whitening.

albicaulis, -is, -e – white -stemmed.

albidus, -a, -um – white.

albiflorus, -a, -um – with white flowers.

albipilosus, -a, -um – with white shaggy hair.

albispinus, - -a, -um – white-spined.

albus, -a, -um – white.

alcicornis, -is, – e – elk-horned.

alexandrinus, -a,- um – of or pertaining to Alexandria (Egypt).

alliaceus, a, -um – of the alliums, garlic-like, usually connoting odor.

almus, -a, – um – bountiful.

alnifolius, -a, -um – alder- leaved.

alopecuroides, -es, -es – foxtail-like.

alpestris, -is, -e – nearly alpine.

alpicolus, -a, -um – living in high mountains.

alpinus, -a, -um – alpine.

alternifolius, – a, -um – with alternate leaves.

altifolius, -a, -um – tall-leaved.

altissimus, -a, -um – very tall.

altus, -a, -um – high.

alveatus, – a, -um – honey-combed.

alveolaris, – is, -e – honey-combed.

amabilis, – is, -e – lovely.

amarus, -a, -um – bitter.

amazonicus, -a, -um – of the amazon River region.

ambigens (pres. part.) – ambiguous.

ambiguus – doubtful.

amblyodus, -a, -um – blunt toothed.

amentum (noun) – catkin.

amethystinus, -a, -um -amethystine, violet-colored.

amoenus, – a, – um – pleasant, charming.

amphibius, -a. – um – amphibious, growing on land or in water.

amplexicaulis, -is, -e–stem-clasping.

amplexifolius, -a. -um– leaf-clasping.

amurensis, -is, -e – of the Amur River region.

amygdaloides, -es, -es – almond-like.

anacanthus, -a, -um– without spines.

anceps, -ceps, -ceps – two-headed or two-edged.

aneurus, -a, -um – nerveless.

anfractuosus, -a, – um -twisted.,

angelicus, -a, um – twisted.

angelicus, -a, -um – English.

angularis, -is -e – angular.

angyustifolisus, -a, -um – with narrow leaves.

angustus, -a, -um – narrow.

annotinus, -a, -um – year-old.

annularis, – is, -e – ring -shaped.

annuus, – a, -um – annual, yearly.

anomalus, -a, -um – unusual, out of the ordinary.

anserinus, -a, -um – of a goose.

antarcticus, -a, -um – of the Antarctic regions.

apertus,-a,-um – uncovered, bare, open.

apetalus, -a, um – without petals.

aphyllus, -a, -um – without leaves.

apiculatus, -a, -um – tipped with a point.

apodus, – a, -um – footless.

appendiculatus, a, -um – apendaged.

applatanatus, -a, – um – flattened.

apterus, -a, -um – wingless.

aquaticus, -a, – um – aquatic.

aquatilis, -is, -e – aquatic.

aquifolius, -a, -um – with pointed leaves.

aquila (noun) – eagle.

arabicus, -a, -um – Arabian.

arachnifer, -era, – erum – spider-webby. web-bearing.

arachnoides, -es, -es – spider-like, cobwebby.

arborescens (pres. part.) – like a tree.

arboreus, -a, -um – treelike , pertaining to a tree.

articus, – a, -um – arctic, northers.

arcuatus, -a. um – bowlike.

arenarius, -a, -um – of sand or sandy places.

arenicolus, -a, -um – sand-in – habiting.

areolatus, -a, – um – areolate, pitted.

argentatus, -a, -um – silvery, silvered.

argenteus, -a. -um silvery.

argillaceus, -a, -um – clay, growing in clay, or clary-colored.

argophyllus, -a, -um – with silvery leaves.

argutus, -a, -um – sharp-toothed,.

argyphyllus, -a, – um – silver-leaved.

argyreus, a, -um – silvery.

argyrocomus, -a, -um – silver-haired.

argyroneurus, -a, -um – with silver-colored nerves or veins.

aridus, -a, -um – arid, dry.

arietinus, -a, -um – like a ram's head.

aristatus, - a, -um – with an awn or beard.

armatus, -a, -um – armed (as with thorns).

armillaris, -is, -e – like a bracelet.

aromaticus, -a, -um – aromatic.

articulatus, -a, -um – with joints.

arundinaceus, -a, -um – reedlike.

arvensis, -si, -e – field -growing.

ascendens (pres. part.) – ascending.

asper, -era, -erum – rough.

asperifolius, - a,- um – with rogh leaves.

asplenioides, -es, -es – asplenium-like.

assimilis, - is, -e – similar or like.

asteroides, -es, -es – aster-like.

ater, atra, - atrum – aster-like.

ater, - atra, -atrum – coal-black.

atlanticus, a, -um – Atlantic, growing in Atlantic regions.

atratus, -a, -aum – blackened.

atriplicifolius, -a, -um – atriplex-leaved.

atrocaeruleus, - a, - um – dark cerulean or blue.

atrococcineus, - a, -um dark scarlet.

atropuniceus, -a, - um- dark reddish-purple.

atropurpureus, -a, um- dark purple.

atrorubens (pares, part.) – becoming dark red.

atrosanguineus, -a, -um–dark blood-red.

atrovirens (pres. part.) – becoming dark green.

attenuatus, -a, um – attenuated, produced to a point.

augustissimus, - a, - um – dark blood-red.

atrovirens (pres. part.) – becoming dark green.

attenuatus, -a, -um – attenuated,

produced to a point.

augustissimus, -a, -um – very not-able.

augustus, -a, -um – august, not-able, majestic.

aurantiacus, -a, -um – orange red.

aurantifolius, -a, -um – golden-leaved.

auratus, -a, -um – gold-colored.

aureolus, -a, -um – golden.

aureus, -a, -um – golden -yellow.

auricomus, -a, -um – golden-haired.

auriculatus, -a, -um – eared.

australis, -is, -e – southern.

austriacus, -a, -um -Austrian.

austrinus, -a, -um – southern.

autumnalis, -is, -e – autumnal, ·of autumn.

axillaris, -is, -e – axillary, pertaining to the axils.

azureus, -a, -um – sky blue.

B

baccatus,-a, -um – berry-like.

baccifer, -era, -erum – berry-bearing.

balsamifer, -era, -erum – balsam-bearing.

barbatus, -a -um – bearded, barbed.

barbinervis, -is, -e – with nerves bearded.

barbinervis, - is, -e – with nerves bearded.

barbinodis, - is, -e – barbed or bearded at nodes.

barbulatus, -a, -um – small-bearded.

basilaris, -is, -e – pertaining to the base or bottom.

basirameus, -a, -um – with branches low; toward bottom.

bellidifolius, - a, -um – beautiful-leaved.

bellus, -a, -um – nice, handsome.

benedictus, -a, - um – blessed.

benghalensis, -is, -e – of Bengal (India.)

berolensis, -is, -e – of Berlin.

betaceus, -a, -um – of the beet, beet-like.

betulifolius, - a, -um – birch-leaved.

betulinus, -a, -um – birchlike.

bicarinatus, -a, -um – twice-leaved.
bicolor, -or, -or – two-colored.
bicornis, -is -e – two-horned.
bidens, -ens, ens – with two teeth.
bidentatus, -a, -um – two-toothed.
biennis, -is, -e – biennial.
bifidus, -a, -um – two-forked.
biflorus, -a, -um – two-flowered.
bifolius, -a, -um – with two leaves.
bifurcatus, -a, -um – twice-forked.
bigibbus, -a, -um – with two swellings oir projections.
bijugus, -a, -um – yoked, two to gether.
bilobus, -a, -um – with two lobes.
bipinnatifidus, -a, -um – twice pinnately cut.
biserratus, -a, -um – twice-toothed.
bistortus, -a, -um – twice -twisted.
bisulcatus, -a, -um – two -grooved.
blandus, -a, -um – bland, mild, pleasant.
blephariglottis (noun) – fringed tongue.
blepharophyllus, -a, -um – fringed-leaved.
boliviensis, -is, -e, – of Bolivia.
borealis, -is, e – northern.
botryoides -es, -es – cluster-like, like the grape.
brachiatus, -a, -um – branched at righ angles.
brachycarpus, -a, -um– short-stamened.
brachycarpus, -a, um– short-fruited.
brachyphyllus, -a, -um– short-leaved.
brachystachys, -yx, -ys -short-spiked.
bracteatus, -a, -um – bearing bracts.
brasiliensis, -is, -e – of Brazil.
brevicaulis -is, -e – short-stemmed.
brevipes, -pes, -pes – short-lipped.
brevis, -is, -e – short.
breviscapus, -a, -um – short-scaped.
brevisetus, -a, -um – short-bristled.
brunescens (pres. part.) – becoming brown.
brunneus, -a, -um – deep brown.
bufonius, -a, -um – pertaining to the toad.
bulbifer, -era, -erum – with bulbs.
bulbosus, -a, -um – bulbous.
bullatus, -a, -um – blistered, puckered.

C

cachemivicus, -a, -um – of Kashmere (Asia).
caerulescens (pres. part.) – becoming bluish.
caeruleus, -a, -um – blue.
caesius, -a, -um – bluish-gray.
caespitosus, -a, -um – growing in tufts.
caffer, caffra (noun) – Kafir (S. Africa).
calathinus, -a, -um – basket-like.
calcaratus, -a, -um – spurred.
calceiformis, -is, -e – shoe-shaped.
calceous (noun) – small shoe.
calceus (noun) – shoe.
callianthemus, -a, -um – with beautiful flowers.
callicarpus, -a, -um – with beautiful fruit.
callistachyus, -a, -um – with beautiful spikes.
callizonus, -a, -um – beautiful zoned.
callosus, -a, -um – callous, hardened.
calocomus, -a, um – beautiful-haired.
calvus, -a, -um – bald, hairless.
calycinus, -a, -um – calyx-like.
calycosus, -a, -um – calyx-like.
campanula, (noun) – small bell.
campanularius, -a, -um – bell-shaped.
campestris, -is, -e- growing in fields or plains.
camphoratus, -a, -um – pertaining to camphor.
camphylacanthus, -a, um – with crooked spines.
campschaticus, -a, -um – of kamchatka.
camptchaticus, -a, -um – of kamchatka.
canaliculatus, -a, um – channeled, grooved.
canariensis, -is, -e – of the Canary

Islands.

cancellatus, -a, -um – latticed.

candicans (pres. part.) – white-hairy or white-woolly.

candidus, -a, -um – white, shining.

canescens pres. part.) – becoming grayish, becoming ray.

caninus, -a, -um – pertaining to a dog.

cannabinus, -a, um -like *Cannabis* or hemp.

cantabricus, -a, -um – Cantabrian, of Cantabris (N. Spain).

cantonensis, -is, -e – of Canton (in S. China).

canus, -a, -um – ash-colored, hoary.

capensis, -is, -e – of the Cape (of Good Hope).

capillaris, -is, -e, – hairlike.

capillus (noun) -hair.

capitatus, -a, -um – with a head.

capitellus (noun) – little head.

capreolatus, -a, -um – winding, twining.

capsularis, -is, -e – having capsules.

cardinalis, -is, -e – of cardinal -red color.

cardiopetalus, -a, -um – with heart-shaped petals.

carinatus, -a, -um – keeled.

carminatus, -a, -um – carmine.

carmineus, -a, -um – carmine.

carmosinus, -a, -um – crimson.

carneus, -a, --um – flesh-colored.

carnosus, -a, -um – fleshy.

carolonanus, -a, um – pertaining to the Carolinas.

carolinus, -a, - um – of the Carpathian region.

cashmerianus, -a, -um – of Kashmere (Asia).

catharticus, -a, -um – cathartic.

caucasicus, -a, - um – belonging to the Caucasus.

caudatus, -a, - um – with a tail.

caulescens (pres. part) – becoming a stem.

caulialatus, -a, -um – wing-stemmed.

caulinus, -a, -um – belonging to the stem.

cayennensis, -is, -e – of Cayenne (French Guiana).

cellularis, -is, -e – cellular.

centifolius, -a, um – hundred-leaved, many-leaved.

cepa (noun) – onion.

cephalatus, -a, -um – bearing heads.

cephalonicus, -a, -um – of Cephalonia (one of Ionia Islands).

cephalotes, -es, -es – headlike.

cerasifer, -es, -es – arum – bearing like a cherry.

ceratocaluis, -is, -e – horn-stalked.

cerealis, -is, -e – pertaining to Ceres or agriculture.

cereus, -a, um – wax-colored.

cerifolius, -, -um -wax – leaved.

cernuus, -a, -um – drooping, nodding.

chalcedonicus, -a, -um – of Chalcedon (on the Bosphorus).

chalepensis -is, -e – of Aleppo (S.W. Asia).

chamaedrifolius, -a, -um – with leaves like a dwarf oak.

chartaceus, -a, -um, – chartaceous papery.

cheilanthus, -a, us, – lip-flowered.

chilensis, -is, -e – belongin to Chile (S. America).

chiloensis, -is, -e – of Chile (S. America).

chionanthus, -a, -um-snowflowered.

chloranthus, -a, -um- green-flowered.

chrysanthus, -a, -um – golden-flowered.

chrysocarpus, -a, -um – golden-fruited.

chrysocomus, -a, -um – golden-haired.

chrysolepis, -is, -is – golden-scaled.

chrysomallus, -a -um – with golden wool.

chrysophyllus, -a, -um – golden-leaved.

chrysostomus, -a, um – golden-mouthed.

cicutarius,-a, -um– of or like Cicuta.

cilianensis, -is, -e – hairy.

ciliatifolius, -a, um – hairy leaves.

ciliatus, -a, -um – ciliate.

cilicicus, -a, -um – of Cilicia (S. E. Asia Minor).

ciliosus, -a, -um – ciliate, fringed.

cinctus, -a, -um – girdled.

cinereus, -a, -um – ash gray.

cinnabarinus, -a,– um–cinnabar-red.

cinnamomeus, -a, -um – c i n n a m o n brow.

circinalis, -is, -e – coiled.

circularis, -is, – e -round.

cirrhosus, -a, -um – tendriled.

citratus, -a, – um– citrus-like.

citreus, -a, -um – lemon-colored.

citrinus, -a, -um – lemon yellow.

clandestinus, -a, um – concealed.

clausus, -a, -um – shut, closed.

clavatus, -a, um – club-shaped.

claviformus, -a, -um – club-shaped.

coarctatus, – a , - u m – a p r e s s e d , crowded together.

cocciger, -era, -erum–berry-bearing.

coccineus, -a, -um – scarlet.

cochlearis, is, -e, – spoonlike.

coelestinus, -a, -um – sky blue.

coelestis, -is, -e – sky blue.

coerulescens (pres. part.) – becoming blue.

coeruleus, -a, -um – blue.

colchieus, -a, -um – Colchis (eastern Black sea region).

collinus, -a, -um – pertaining to a hill.

colonus, -a. -um – cultivated.

columellaris, -is, -e – pertaining to a small pillar or pedicel.

columnarius, -a, – um – columnar.

comans, -ans, -ans – with hair, or hairlike.

comatus, -a, -um – with hair.

communis, -is -e – growing in common, gregarious.

commutatus, -a, -um – changing.

comosus, -a, -um – long-haired.

compactus, -a, -um– compact, dense.

complanatus, -a, – um – flattened.

complexus, -a,-um – encircled, embraced.

compressus, -a, -um – compressed.

conchaefolius, -a,-um – shell-leaved.

concinnus, -a, -um– neat, well made, elegant.

concolor, -or, -or – of the same coion.

confertus, -a, -um – crowded.

confinis, -is, -e – bordered, bound.

conglomeratus, -a, -um– conglom - eerate, crowded together.

conoidus, -a, -um – conelike.

conspessus, -a, -um – scatered.

constrictus, -a, -um – constricted.

contortus, -a, -um– contorted, twisted.

contractus, -a, -um – narrowed.

convallis, -is, -e – valley.

convolvulus, -a, -um – winding, twining.

copallinus, -a, -um – like copal or resin.

cordatus, -a, – um – heart-shaped.

cordiformis, -is, -e – shaped like a heart.

coriaceus, -a, -um – heart-shaped.

cordiformis, – is, -e – shaped like a heart.

coriaceus, -a, -um – like leather.

corniculatus, -a,, -um – with horns, horned.

corniger, -era, -erum– hornbearing.

cornutus, -a, -um – horny.

corollatus, -a, -um – corollate, like a corolla.

coronarius, -a, -um – suitable for wreaths.

coronatus, -a, -um – crowned.

corrugatus, -a, um – corrugated, wrinkled.

corymbosum, -a, -um – corymb-like.

costatus, -a, -um – costate, ribbed.

crassifolius, -a, -um – thick-leaved.

crassipes, -es, -es – thick-footed.

crassus, – a, -um – thick.

crebrus, -a, – um – close, frequent, repeated.

crenatiflorus,-a,-um – crenate-flowered.

crenatus, -a, – um – crenellated, scalloped.

crenulatus, -a, -um – crenulate, finely scalloped.

creticus, -a, – um – of Crete (island of

Crete).

crinitus. -a. -um – provided with long hair.

crispifolius, -a, -um – with leaves crisped or curled.

crispus, -a, -um – crisp, curied.

cristagalli (noun) – cockscomb.

cristatus, -a, -um – crested.

crocatus, -a, -um – saffron yellow.

croceus, -a, -um – saffron-colored, yellow.

cruciatus, -a,-um– cruciate, crosslike.

cruciformis, – is, -e – cross-shaped.

cruentus, -a, -um – blood red, bloody.

crusgalli (noun) – cockspur.

crustatus, -a, -um – encrusted.

cryptandrus, -a, -um – with hidden stamens.

crystallinus, -a, -um – crystalline.

ctenoides, -es, -es – comblike.

cucullatus, -a, -um – hooded.

cultriformis, -is, -e – shaped like a broad knife blade.

cuneatus, -a, -um – wedge-shaped.

cuneifolius, -a, -um – with wedge-shaped leaves.

cuneiformis, – is, -e – wedge-shaped.

cupreatus, -a, -um – copery.

cupreus, -a, -um – copper-colored.

curtipedicellatus, -a, -um – short-pediceled.

curtipendulus, -a, -um – with short pendula.

curvatus, -a, -um – curved.

cuspidatus, -a, -um – sharp-pointed.

cyananthus, -a, -um – blue-flowered.

cyaneus, -a, – um – cornflower blue.

cyanophyllus, -a, -um – blue-leaved.

cyanthus, -a, -um – cylindrical.

cymousus, -a, -um – bearing cymes.

D

dactylifer,-era,-erum–finger-bearing.

dactyloides, -es, -es – finger-shaped.

dactylon (noun) – finger.

dahuricus, -a, -um – of Dahuria or Dauri (trans-Baikal, Siberia).

damascenus, -a, -um – of Damascus.

dasycarpus, -a, -um – thick-fruited.

dasyphyllus, -a, -um – thick-leaved.

dasystachys, -ys, -ys – thick-spiked.

dauricus, -a, -um – of Dahuria or Dauria (Trans-Baikal, Siberia).

davuricus, -a, -um – of Dahuria or Dauria (Trans-Baikal, Siberia).

dealbatus, -a, -um – whitened, white -washed.

debilis, -is, -e – weak, frail.

decapetalus, -a, -um – ten-petaled.

deciduus, -a, -um – deciduous.

decipiens (pres. part.) – deceptive.

decorus, -a, -um – elegant, comely.

decumbens (pres. part) – decurrent, running down the stem.

deflexus, -a, -um – bent down-ward.

deliciosu, -a, -um – delicious.

deltoides, – es, -es – triangular.

demersus, -a, -um – under the water.

demissus, -a, -um – low, weak.

dendroideus, a, – um – treelike.

densus, -a, -um – dense.

dentatus, -a, -um – toothed.

denticulatus, a, – um – with teeth.

denudatus, -a, – um – nude, naked.

depauperatus, -a, – um – stunted, dwarfed.

depressus, -a, -um – depressed.

deustus, -a, -um – burned.

diacanthus, -a, -um – two -spined.

diandrus, -a, um – two -stamened.

dichotomiflorus, -a, -um – flowering in twos.

dichotomus, -a, -um- forked in pairs.

dicoccus, -a, -um – with two berries.

diffissus, -a, -um – split.

diffusus, -a, -um– loosely branching.

digitatus, -a, -um – finger-shaped.

digitus, (noun) – finger.

dilitatus, -a, -um – dilated, expanded.

dimidiatus, -a, -um – halved.

dioicus, -a, – um – dioecious.

dipsacus, -a, -um – of the teasel or Dipsacus.

dipterocarpus, -a, – um – with two winged carpel or fruit.

discolor, -or, -or – of different colors.

dissectus, -a, – um – dissected, deeply cut.

dissitiflorus, -a, -um – remotely or loosely flowered.

distachyus, -a, -um – two-spiked.

distichus, -a, -um – t w o - r a n k e d ; leaves or flowers in rank on opposite sides of stem.

diurnus, -a, – um – of the day, as day-flowering.

divaricatus, -a,-um–spreading, widely divergent.

divergenus (pres. part)– wide-spreading.

diversiflorus, -a, -um – d i v e r s e l y flowered, flowers varible.

diversifolius, -a, -um – with variable leaves.

divisus, -a, um – divided.

dodecus, -a, -um – twelve.

dolabratus, -a, -um–mattock-shaped.

domesticus, -a, -um – indigenous, at home.

donax, -ax, -ax – reed or cane.

dracocephalus, -a, um–dragon-head.

dracunculoides, -es, – es – tarragon -or dragon-like.

drepanophyllus, -a, -um – with sickle-shaped leaves.

dubius -a, -um – doubtful.

dulcis, -is, -e – sweet.

dumosus, -a, -um – bushy.

duracinus, -a, -um – hard-berried.

duriusculus, -a, – u m – s o m e w h a t hard or rough.

E

ebracteatus, -a, -um– without bracts.

eburneus, -a, -um – ivory white

echinatus, -a, -um – prickly, like a headgehog.

edulis, -is, – e – edible.

effusus, -a, um– very loose-spreading.

elasticus, -a, -um – elastic.

elatior, -ior, -ius – taller.

elatus, -a, -um – tall, high.

elegans, -ans, -ans – elegant.

elegantissimus, -a, -um – most

elegant.

elevatus, -a, -um – elevated.

ellipsoidalis, -is, -e – elliptic.

elliptieus, -a, – um – elliptical.

elongatus, -a, -um– lengthened, long.

emarcidus, -da, -um – faded.

emarginatus, -a, -um – with a shallow notch at apex.

emersus, -a, -um – emersed.

enodis, -is, -e – without nodes.

ensatus, -a, -um – sword-shaped.

ensiformis, -is, -e – sword-shaped.

ensifolius, -a, -um– with sword-shaped leaves.

ephemericus, -a, -um – ephemeral, lasting but a day.

epigaeus, -a, -um – above the soil.

equus (noun) – horse.

erectus, -a, -um – upright.

eriacanthus, -a, -um – woolly-spined.

eriantherus, -a, -um – woolly-an-thered.

erianthus, -a, -um – woolly flowered.

ericoides, -es, -es – like Erica, heath-like.

eriocarpus, -a, -um – woolly-fruited.

eriopodus, -a, -um – having hairs at base.

erosus, -a, -um – erose, as if gnawed.

erubescens (pres. part) – blushing.

erythrocarpus, -a, -um – red-footed, or red-stalked.

erythrosepalus, -a, um– red-seplaed.

erythrosorus, -a, -um– with red sori.

esculentus, -a, -um – edible.

europaeus, -a, -um – European.

evertus, -a,-um–expelled, turned out.

exaltatus, – a, -um–exalted, very tall.

excelsior, -ior, -ius – taller.

excelsus, -a, -um – tall, high.

excisu, – a, -um – excised, cut away.

excolor, -or, -or – colorless.

exiguus, -a, -um – little, small, poor.

eximius, -a, -um – distinguished, out of the ordinary, excellent.

exoticus, -a, -um– exotic, from

another couintry.

expansus, -a, -um – expanded.

extensus, -a, -um – spread out of stretched out.

F

fabarius, -a, -um – pertaining to beans.

falcatus, -a, -um – sickle-shaped.

fallax, -ax, -ax – deceptive.

farinosus, -a, um–farinaceous, mealy.

fasciatus, -a, -um– fasciae, coalesced.

fasciculatus, -a, -um – fascicled, clustered.

fastigiatus, -a, -um – fastigiate, branches erect and close together.

fastuosus, -a, -um – pround.

fatuus, -a, um – foolish, simple, insipid.

fecundus, -a, -um – fertile.

femininus, -a, -um – feminine.

fenestralis, -is, -e – lattice-shaped.

fenestrellatus, -a. -um – with small window-like openings.

ferox, -ox, -ox–ferocious, very thorny.

ferrugineus, -a, -um – rusty, of the color of iron rust.

fertilis, -is, -e – fertile, fruitful.

fibrilla (noun) – small fiber.

fibrosus, a, -um – fibrus, with fibers,

ficifolius, -a, -um – with leaves like a fig.

flcoides, -es, -es – figlike.

filamentosus, -a, – um – filamentour.

filicifolius, -a, um – fern-leaved.

filifer, -era, -erum – with filaments or threads.

filifolius, -a, -um – thread-leaved.

filiformis, -a, -um – long, very slender, thread like.

fimbriatus, -a, -um – fringed, fimbriate.

firmus, -a, -um – strong, firm.

fissilis, -is -e – cleft or split.

fistulosus, -a, -um – hollow-cylindrical.

flabellatus, -a, -um – with fanlike parts.

flabelliformis, -is, -e – fan-shaped.

flaccidus, -a, -um – flaccid, soft.

flagellaris, -is, -e – whiplike.

flagelliformis, -s, -e, – whipshaped.

flammeus, -a, -um – fire red, flame-colored.

flavens (pres. part.) – becoming yellowish.

flavescens (pres. part) – becoming yellow.

flavicomus, -a, -um – yellow-wooled or yellow haird.

flavispinus, -a, -um – yellow-spined.

flavissimus, -a, -um – deep yellow, very yellow.

flavus, -a, -um – yellow.

flexibilis, -is, -e – flexible.

flexicaulis, -is, -e, – pliant-stemmed.

flexilis, -is, -e – pliant.

flexuous, -a, -um – alternately bent in different directions.

floralis, -is, -e – floral.

florentinus, -a, -um – of Florence.

florepleno – with full or double flowers.

floribundus, -a, -um – free-flowering, blooming profusely.

floridanus, -l, -um – flowering, or from Florida.

floridus, -a, -um – flowering, full of flowers.

fluminensis, -is, -e – of a river.

fluitans (pres. part.) – floating.

fluvialis, -is, -e – fluvial.

foeminus, -a, -um – feminine.

foetidissimus, -a, -um – very fetid.

foetidus, -a, -um – having an of fensive odor.

foliatus, -a, -um – with leaves.

foliosus, -a, -um – -leafy.

follicularis, -is, -e – bearing follicles.

fontinalis, -is, -e – pertaining ot a spring of water.

formosissimus, -a, – um – most of very beautiful.

formosus, -a, -um – beautiful, handsome.

foverlatua, – a, -um – pitted, with small depressions.

fragilis, -is,- e – fragile, brittle.

fragrans, -ans, -ans – f r a g r a n t ,

odorous.

fragrantissimus, -a, -um – very fragrant.

frigidus, -a, um – cold, of cold regions.

frondosus, -a, -um – leafy.

fructigenus, -a, -um – fruitful.

frumentaceus, – a, – um – pertaining to grain (or corn).

frutescens (pres. part) – becoming shrubby.

fruticosus, -a, -um – shrubby.

fucatus, -a, -um – painted, dyed.

fulgens, -ens,-ens–shining, glistening.

fulgidus, -a, -um – shining, glistening.

fuliginosus, -a, -um – sooty, black-colored.

fulvidus, -a, -um – slightly tawnny.

fulvus, -a, -um – brownish-yellow.

funebris, -is, -e, – funeral.

funiculatus, -a, -um – of a slender rope or cord.

furcatus, -a, -um – forked.

fuscus, -a, -um – brown, dusky.

fusiformis, -is, -e – spindle -shaped.

G

galeatus, -a, -um – helmeted.

galericulatus, – a,-um – helment-like.

gallicus, -a, -um – of Gaul or France; also pertaining to a cock or rooster.

gandavensis, -is, -e, – belonging to Ghent, Belgium.

geminatus, -a, -um–double or paired.

geminiflorus, -a, -um– twin-flowered.

geminispinus, -a, -um – twin-spined.

gemmatus, -a, -um – bearing buds.

gemmifer, -era, -erum -bud-bearing.

generalis, -is, -e – general, prevailing.

geniculatus, -a, -um – bent more or less like a knee.

germanicus, -a, -um – German.

gibberosus, -a, -um – humped, hunch -backed.

gibbiflorus, -a, -um – swollen on one side.

gibbus, -a, -um – of Gibralter.

giganteus, -a, -um– gigantic, very large.

giganthes, -es, – es -giant-flowered.

glabellus, -a, -um – smoothish.

glaberrimus, -a, -um – Very smooth, smoothest.

glabratus, -a, -um – somewhat glabrous.

glabrescens (pres. part) – becoming smooth.

glabriflorus, -a, -um – smooth-flowered.

gladiatus, -a, – um – swordlike.

gladius (noun) – sword.

glandulifer, -era, -erum – gland-bearing.

glandulosus, -a, -um – swordlike.

gladius (noun) – sword.

gra, -erum – gland-bearing.

glandulosus, -a, um – glandular.

glaucescens (pres. part) – becoming glaucous.

glaucophyllus, -a, -um – glaucous - leaved.

glaucus, -a, -um – glacous, with a bloom.

globosus, -a, -um – -like a globe.

globulus, -a, -um – like a small globe.

glomeratus, -a -um – with dense or comopact cluster or clusters.

gloriosus, -a, -um – glorious, superb.

glumaceus, -a, -um – with glumes or glumelike structures.

glutinosus, -a, -um – sticky.

gnaphalodes, -es, -es – like Gnaphalium.

gomphocephalus, -a, -um – club-headed.

goniatus, -a, -um – angled, cornered.

gosypinus, -a, -um -gossypium – like , cotton-like.

gracilis, -is, -e – slender.

gracillimus, -a, -um – very slender.

graecizans (pres. part.) -becoming widespread.

graecus, -a, -um – of Greece, Greek.

gramineus, -a, – um -grassy, grass-like.

graminifolius, -a, -um – with grass-like leaves.

grandidentatus, -a, -um – large-toothed.

grandiflorus, -a, -um – with large flowers.

grandifolius, -a, -um – with large leaves.

grandipunctatus, -a, -um– with large spots.

grandis, -is, -e – large, big.

granulatus, -a, -um – granular, covered with small grains.

granulosus, -a, um – granulate.

ghratiosus, -a, um – agreeable.

gratissimus, -a, -um – very pleasing or agreeable.

gratissimus, -a, -um – very pleasing or agreeable.

graveolens (pres. part.) – strong scented.

griseus, -a, – um – gray.

grumosus, -a, um – crummy.

gumminfer, -era, -erum -gum - bearing.

guttatus, -a, -um – with drops.

gymnocarpus, -a, -um – naked-fruited.

H

haemanthus, -a, -um– with blood-red flowers.

halepensis, -is, – of Aleppo (S.W. Asia).

halophilous, -a, -um – salt-loving.

hamatus, hamousus, -a, -um - hooked.

hastatus, -a, -um -spear-shaped.

bebecarpus, -a, -um – pubescent-fruited.

hederaceus, -a, -um -of - or like the ivy (Hedera).

helianthoides, -es, -es – helianthus-like.

helodoxa (noun) – marsh beauty.

helvinus, -a, -um – yellowish.

helvolus, -a, -um – pale yellow.

herbaceus, -a, -um - herbaceous, not woody.

hesperius, -a, -um -of the west, or evening.

heteranthus, -a, -um- various-flowered.

heterocarpus, - -a, -um -various-fruited.

heterolepis, -is, -is - with hetero geneous scales.

heterophyllus, -a, -um – with leaves of more than opne shape.

hexagonus, -a, -um -six-angled.

hexapetalus, -a, -um – six-petaled.

hians (pres. part.) – open, gaping.

hibernicus, -a, -um - of Ireland.

hibernus, -fa, -um – belonging to winter, wintry.

hiemalis, -is, -e, - pertaining to winter.

himalicus, -a, -um – Himalayan.

hircinus, -a, -um – with a goat's s odor.

hirsutisimus, -a, -um – very hairy.

hirsutus, -a, -um– dbristly or prickly.

hirtellus, -a, -um – somewhat hairy.

hirtiflorus, -a, -um– -hairy-flowered.

hirtipes, -es, -es - hairy-footed.

hirtus, -a, -um – hairy.

hispanicus, -a, -um– Spanish, of Spain.

hispidulus,-a,-um– somewhat bristly.

hispidus, -a, -um – with stiff hairs.

hollandicus, -a, -um – of Holland.

holosericeus, -a, -um – woolly-silky.

homolepis, -is, -is – with one kind of scales.

horizontalis, -is, -e – horizontal.

horidus, -a, -um – horrid.

hortensis, -is, -e – pertaining to a garden.

hortulanus, -a, -um – belonging to a garden, or gardens.

humifusus, -a, -um – sprawling on the ground.

humilis, -is, -e - dwarf, low growing.

hybridus, -a, -um – hybrid.

hyemalis, -is, -e – of witner.

hymenanthus, -a, -um – membran-aceous-flowered.

hymenodes, -es, -es– membrane-like.

hymenosepalus, -a, -um – with membranous sepals.

hyperboreus, -a, – um of the far

North.

hypnoides, -es, -es – moss -like.

hypocrateriformis, -is, -e – salver-shaped.

hypogaeus, -a, – um–under the earth or soil.

hypoglaucus, -a, -um – glaucous beneath.

hypoleucus, -a, -um – Whitish, pale beneath.

hystrix, -ix, ix – bristly, procupine-like.

I

ibericus, -a, -um – of Iberia (the Spanish peninsula).

igneus, -a, -um – fire red.

ignis (noun) – fire.

ilicifolius, -a, -um – ilex-leaved, holly-leaved.

illustris, -is, -e – bright, brilliant, lustrous.

imberbis, -is, -e – without beards or spines.

imbricatus, -a, -um – overlapping like shingles.

immaculatus, -a, -um – immaculate, spotless.

immersus, -a, -um – under water.

impatiens, -ens, -ens – impatient.

imperator, -or, -or – commanding, imperious.

imperialis, -is, -e – imperial.

incanus, -a, -um – gray, hosary.

incarnatus, -a, -um – flesh-colored.

incertus, -a, -um – uncertain, doubt-ful.

incisifolius, -a, – um – cut-leaved.

incisus, -a, -um – incised.

incrassatus, -a, -um – thickened.

indentatus, -a, -um – indented.

indicus, -a, -um – Indian, of India or the East Indies.

indivisus, -a, -um – undivided.

induratus, -a, -um – hardened.

inequalis, -is, -e – unequal.

inermis, -is, -e – unarmed.

infectorius, -a, -um–used for dyeing, pertaining to dyes.

inferus, -a, -um – inferior.

inflatus, -a, -um – inflated.

infundibuliformis, -is, -e – funnel-shaped.

ingens, -ens, – - ens -enormous.

inodorus, -a, -um – without odor, scentless.

inquinans, -ans, -ans – polluting, discoloring.

insectus, -a, -um – incised.

insignis, -is, -e – striking.

insititius, -a, -um – grafted.

insularis, -is, -e – pertaining to an is-land.

integerrimus, -a, -um -very entire.

integrifolius, -a, -um with entire.

intermedius, -a, -um - intermediate.

interruptis, -is, -e – not continuous, irregular.

introrsus, -a, -um – intermediate.

interruptis, -is, -e – not continuous, irregualr.

introrsus, -a, -um – introrse, turned inward.

intumescens, -ens, -ens – tumid, swollen, enlarged, distended.

involucratus, -a, -um – with an in-volucre.

involutus, -a, -um – rolled inward.

ioensis, -is, -e – of Iowa (state in U.S.A.).

ionanthus, -a, -um – violet-flowered.

ischaemum (noun) – a blood stypotic, an astringent.

isophyllus, -a, -um – equal-leaved.

italicus, -a, -um – Italian.

ixocarpus, -a, -um– sticky, or glutin-ous-fruited.

J

javanicus, -a, -um – javan, of java.

jubatus, -a, -um – crested, with a mane.

jucundus, -a, -um – delightful, pleas-ing.

jugosus, -a, -um – joined, klyoked.

junceus, -a, -um – rushlike.

juniperinus, -a, -um – like juniper.

K

kamtschaticus, a, -um – of Kamchatka.

kermesinus, -a, -um – crimson.

kewensis, -is, -e – belonging to Kew (Royal Botanic Gardens, Kew, England).

L

labiatus, -a, -um – with a lip.

labiopsus, -a, -um – lipped.

labrosus, -a, -um – large-lipped.

laceratus, -a, -um – lacerate.

laciniatus, -a, -um – cut into narrow pointed.

laciniosus, -a, -um – laciniose, laciniate, torn.

laciniosus, -a, -um – laciniose, laciniate, torn.

lactatus, -a, -um – milky.

lacteus, -a, -um – milky -white.

lactiflorus, a, -um – with milkcolored flowers.

lacustris, -is, -e – growing in lakes.

ladanifer, -era, -erum – ladanuim-bearing.

laetiflorus, -a, -um – with bright, or pleasing flowers.

laetivirens, -ens, -erum – ladanum-earing.

laetiflorus, -a, -um – with brigh, or pleasing flowers.

laetivirens, -ens, -ens – bright green.

laetus, -a, -um – bright, vivid.

laevigatus, -a, -um – smooth.

laevipes, -pes, -pes – smooth-footed.

laevis, -is, -e – smooth.

lagemarius, -a, -um – of a bottle or flask.

lanatus, -a, -um – woolly.

lanceolatus, -a, -um – lance-shaped leaves.

lanosus, -a, -um – woolly.

lanuginosus, -a, -um – woolly, downy.

lapideus, -a, -um – stony.

lasiandrus, -a, -um – pubescent-stamened.

lasianthus, -a, -um – woolly-flowered. lasiocarpus, -a,-um–hairy -or woolly- fruited.

lasiolepis, -is, -is – woolly-scaled.

lasiopetalus, -a, -um – with petals

rough-hairy.

lateralus, -a, -um – lateral, on the side.

lateritius, -a, -um – lateral, on the side.

lateritius, -a, -um – i .l-leaved.

latiglumis, -is, -e – with wide glumes.

latimaculatus, -a, -um – broad-spotted.

latisquamus, -a, -um – broad-scaled.

latissimus, -a, -um – broadest, very broad.

laurifolius, -a, -um – laurel-leaved.

laurinus, -a, -um – laurel-like.

lavandulaceus, -a, -um – lavender-like.

laxiflorus, -a, -um – loose-flowered.

laxus, -a, -um – loose, open.

leianthus, -a, -um – smooth-flowered.

leiocarpus, -a, -um – smooth-fruited.

leiogynus, -a, -um – with a smooth pistil.

leiophyllus, -a, -um – smooth-leaved.

lenticularis, -is, -e - like a lens.

lentiginosus, -a, - -um – freckled.

leopardinus, -a, -um – leopard-spotted.

lepidophyllus, -a, -um – with scaly leaves.

lepidotus, -a, um – with small scurfy scales.

lepidus, -a, -um – graceful, elegant.

leprosus, -a, -um – scrfy.

leptanthus, -a, -um – thin-flowered.

leptocaulis, -is, -e – thin-stemmed.

leptocladus, -a, -um – thin-scaled.

leptophyllus, -a, -um – thin-leaved.

leptopus, -us, -us – thin- or slender-stalked.

leptosepalus, -a, -um – thin-sepaled.

leucanthus, -a, um – white-flowered.

leucocarpus, -a, -um – white-fruited.

leucocaulis, -is, -e-white -stemmed.

leucocephalus, -a, -um – white-headed.

leucodermis, -is, -e – white-skinned.

leuconeurus, -a, -um – white-nerved.

leucophyllus, -a, -um – white-leaved.

leucostachys, -ys, -ys – white-spiked.

leucotrichus, -a, -um – white-haired.

lignosus, -a, -um – like wood.
ligularis, -is, -e – ligulate, strap-shaped.
ligulatus, -a, -um – with a ligule.
limosus, -a, -um – of marshy or muddy places.
linearifolius, -a, -um – with long, slender leaves.
linearis, -is, -e – long and slender.
lineatus, -a, -um – lined, with lines, or stripes.
linguiformis, -is, -e – tongue-shaped.
lingulatus, -a, -um – tongue-shaped.
linifolius, -a, -um – flax-leaved.
lithospermus, -a, -um – with stone-like seeds.
litoralis, -is, -e – growing on the seashore.
lividus, -a, -um – livid, bluish.
lobatus, -a, -um – lobed.
lobularis, -is, -e – lobed.
locularis, -is, -e, with locule.
longebracteatus, -a, -um – long-bracted.
longepedunculatus, -a, -um – long-peduncled.
longestylus, -a, -um – long-styled.
longiflorus, -a, -um – long-flowered.
longifolius, -a, -um – long-leaved.
blongipes, -pes, -pes – long-footed.
longipetalus, -a, -um – long-petaled.
longiscapus, -a, -um – long-scaped.
longispicus, -a, -um – long-spiked.
longistipatus, -a, -um – long-stiped or long-stemmed.
longistylus, -a, -um – long-styled.
lophanthus, -a, -um – crest-flowered.
lorifolius, -a, -um – strap-shaped leaves.
loriformis, -is, -e – strap-shaped.
lucidus, -a, -um – shining, bright.
ludovicianus, -a, -um – of Lousisiana.
lunatus, -a, -um – moon-shaped or crescent.
lupulinus, -a, -um – like hops.
luridus, -a, -um – dim yellow.
luteolus, -a, -um – yellowish.
lutescens (pres. -paart.) – yellowish or becoming yellow.

luteus, -a, -um – yellow.
lyratus, -a, -um – in the form of a lyre.

M

macilentus, -a, -um – lean, meager.
macradenus, -a, -um – long glands.
macranthus, – a, -um – with long or large flowers.
macrocarpus, -a, -um – with long or large fruits.
macrocephalus, -a, -um – large-headed.
macropetalus, -a, -um – large-petaled.
macrophyllus, -a, -um – large-leaved, long-leaved.
macropodus, -a, -um – large-footed.
macrorhizus, -a, -um – with large or long roots.
macrostylus, -a, -um – large-styled.
macrus, -a, -um – long or large.
maculatus, a, -um – spotted, blotched, or stained.
magellanicus, -a, -um – of the strait of Magellan region.
magnificus, -a, -um – magnificent, eminent, distingushed.
magnifolius, -a, -um – with large leaves.
majalis, -is, -e, – flowering in May.
major, -or, -um – larger, greater.
malacoides, -es, -es- soft, mucilaginous.
malacophyllus, -a, -um– soft-leaved.
maliformis, -is, -e- apple-shaped.
malvaceus, -a, -um– malva-like, mallow-like.
malvaeflorus, -a, -um– mallow-flowered.
mamillatus, -a, -um–mamillaris, – is, -e, – having teat-shaped processes.
mammosus, -a, -um – with breasts or nipples.
mammulosus, -a, -um – with small nipples.
mandschuricus, -a, -um– of Manchuria.
manicatus, -a, – um manicate, long-sleeved, covered densely as with thick hairs so that the covering can

be removed as such.

marcidus, -a, -um – faded.

margaritaceus, -a, -um – pearl-like.

margaritus, -a, -um – pearly, of pearls.

marginalis, – is, -e – marginal.

marginellus, -a, -um – somewhat margined.

marianus, -a, -um – of the Mary-land region (U.S.A.).

marilandicus, -a,- um – of the Mary-land region (U.S.A.).

maritimus, -a, -um – growing near the sea.

marmoratus, -a, -um– marbled, mottled.

maroccanus, -a, -um – of Morocco.

marylandicus, -a, -um– of the Mary-land region (U.S.A.).

matronalis, -is -e – pertaining to matrons.

mauritanicus, -a, -um – of Mauretania (N. Africa).

maximus, -a, -um – the largest.

mediopictus, -a, -um – pictured or stripped at the center.

medius, -a, -um – medium, intermediate.

megacarpus, -a, -um – large-fruited.

megalanthus, -a, -um – large-flowered.

megarrhizus, -a, -um – large-rooted.

megastachys, -ys, -ys, – large-spiked.

melancholicus, -a, -um– melancholy, hanging, or drooping.

melanocarpus, -a, -um– black-fruited.

melanococcus, -a, – um -black-berried.

melanotrichus, -a, – um– black-haired.

melantherus, -a, -um – with black anther.

meleagris (noun) – a guinea-fowl.

melleus, -a, -um–pertaining to honey.

mellitus, -a, -um – honey, sweet.

meridionalis, -is, -e – southern.

metallicus, -a, -um – metallic (color or luster).

micans (pres. part.) – glttering,

sparkling mica-like.

micracanthus, -a, -um – small-spined.

micranthus, -a, -um – small-flowered.

microcarpus, -a, -um – small-fruited.

microcephalus, -a, -um – small-headed.

microlepis, -is, -is – small-scaled.

microphyllus, -a, -um – small-leaved.

miliaceus, -a, -um – many-leaved, thousand leaves.

mimus, -a, -um – mimic.

minax, -ax, -ax – threatening, forbidding.

minimus, -a, -um – the smallest.

minor, -or, us – smaller.

minutiflorus, -a, um – minute-flowered.

mirabilis, -is, -e – wonderful, admirable.

mirus, -a, -um– wonderful, admirable.

mirus, -a, -um – wonderful, unusual.

mitis, -is, -e – mild, gentle.

modestus, -a, -um – modest.

moesiacus, -a, -um – of the Balkan region (ancient Moesians).

mollis, is, -e – soft, soft-hairy.

mollissimus, -is, – e –necklace-shaped.

monocarpus, -a, -um – with one carpel.

monocephalus, – a, -um – with one head.

monococcus, -a, -um – one -berried.

monoicus. -a, -um – monoecious.

monopetalus, -a, -um – with one petal.

monophyllus, -a, -um– with one leaf.

monosepalus, -a, -um – with one sepal.

monospermus, -a, -um – one-seeded.

monspellensis, -is, -e – living in caves and mountains.

monstrosus, -a, -um – of Montpelier (France).

monstrosus, -a, -um – monstrous, abnormal.

montanus, -a, -um – of mountains.

montevidensis, – is,   -e – of   Montevideo (Uruguay).

monticolus, -a, -um – inhabiting mountains.

monumentalis, -is, -e – monumental.

mosaicus, -a, -um– partly-colored, as of mosaic.

moschatus, -a, -um – musky, musk-scented.

mucidus, -a, -um – moldy.

mucilaginosus, -a, -um – slimy.

mucronatus, -a, -um – with a short, small, abrupt tip.

multicaulis, -is, -e – with   many stems.

multicostatus, -a, -um – many-ribbed.

multifidus, -a, -um – many-parted.

multiflorus, -a, -um – many-flowered.

multinervis, -is, -e – many-nerved.

multiradiatus, -a, -um – with   numerous rays.

mundulus, -a, -um – armed, fortified.

muralis, -is, -e – growing on walls.

muricatus, -a, -um – muricate, roughened by means of hard points.

murinus, -a, -um – mouse-colored.

musaicaus, -a, -um – musa-like.

muscosus, -a, -um – mossy.

mutabilis, -is, -e – varibale.

muticus, -a, -um – curtailed, blunt.

myriacanthus, -a, -um – myriad-spined.

myriophyllus, -a, -um– myriad-leaved.

myrmecophilus, -a, -um – ant loving.

N

nanellus, -a, -um–very small or dwarf.

nanus, -a, -um – dwarf.

napiformis, -is, -e – turnip-shaped.

narbonensis, -is, -e – of   Narbonne (ancient region of S. France).

natalensis, -is, -e – of   Natal   (S. Africa).

natans (pres. part) – swimming   or floating.

neapolitanus, -a, -um – Neapolitan, of Naples.

nebulosus, -a, -um – nebulous, clouded, indefinite.

neglectus, -a, -um – negleected.

nemoralis, -is, -e–growing in a wood.

nemorosus, -a, -um – growing in a wood or grave.

nemus (noun) -wood.

nephrolepis, -is, -is – kidney scaled.

nerlifolius, -a, -um – nerium-leaved, oleander-leaved.

nervatus, -a, -um – nerved.

nervosus, -a, -um – nerved.

nictitans – (pres.   part) – blinking, winking.

nidulus (noun) – small nest.

nidus (noun) – nest.

nigellus, -a,-um – blackish.

niger, -gra, -grum – black.

nigrescens (pres. part) – blackish, becoming black.

nigrofructus, -a, -um – black-fruited.

nitens (pres. part) – glittering, shining.

nitidus, -a, -um – glittering.

nivalis, -is, -e – snowy, pertaining to snow.

niveus, -a, -um – snow white.

nivosus, -a, -um – snowy,   full   of snow.

nobilior, -or, -us – more noble.

nobilis, -is, -e – noble.

noctiflorus, -a,-um–night- flowering.

nocturnus, -a, -um – of the night, night-blooming.

nodiflorus, -a, -um with flowers.

nodosus, -a, -um – with nodes.

nodulosus, -a,-um–with small nodes.

nolitangere (berb) – do not touvh, touch-me-not.

nonpinnatus, -a, -um – not pinnate.

nonscriptus, -a, -um – not described.

nootkatensis, -is, -e – of   Nootka (Nootka Sound near Vancouver Island).

notatus, -a, -um – marked.

novae-angliae (noun) – of   New England.

noveboracensis, -is, -e – of   New

York.

novi-belgi (noun) – of New Belgium (early name for New York).

nox (noun) – night.

nucifer, -era, -erum – nut bearing.

nudatus, -a, -um – nude, naked, exposed.

nudicaulis, -is, -e – naked-stemmed.

nudiflorus, -a, -um – naked-flowered.

nudus, -a, -um – naked, nude, exposed.

numidicus, -a, -um – of Numidia (ancient country of N. Africa).

nutans (pres. part) – nodding.

nutkatensis, -is, -e – of Nootka (Nootka Sound near Vancouver Island).

nux (noun) -nut.

nyctagineus, -a,-um–night-blooming.

O

obcordatus, -a, -um – inversely cordate.

oblanceolatus, -a, -um – inversely lanceolate.

obliquus, -a, -um – oblique.

oblongatus, -a, -um – oblong.

oblongifolius, -a, -um – oblong-leaved.

obovatus, -a, -um – obovate, inverted ovate.

obtusatus, -a, -um – blunt, rounded.

obtusifolius, -a, -um – obtuse-leaved.

obtusus, -a, -um – obtuse, blunt, rounded.

occidens, – ens, -ens – west.

occidentalis, – is, -e – western.

ocellatus, -a, -um – like an eye.

ocellus – (noun) – small eye.

ochraceus, -a, -um – ochre-colored.

ochroleucus, -a, -um – ochre - colored.

ochroleucus,-a,-um– yellowish-white.

octandryus, -a, -um – with eight stamens.

octoflorus, -a, -um – eight-flowered.

octopetalus, -a, -um – eight-petaled.

odessanus, -a, -um– of Odessa (Black Sea region).

odoratissimus, -a, -um – very fragrant.

odoratus, -a, -um – with an odor.

officinalis, -a, -um – official, medicinal, recognized in Pharmcopoeia.

officinarum (noun) – of the apothecaries.

oleaceus, -a, – um – oily.

oleifer, -era, -erum- – oil-bearing.

oleraceus, -a, -um – herbaceous, oleraceous, garden herb used in cooking.

oliganthus, -a, -um – few-flowered.

oligocarpus, -a, -um – few-fruited.

oligospermus, -a, -um – fewseeded.

olitorius, -a, -um – pertaining to vegetable gardens or gardners.

olivaceus, -a, -um – olive green.

olympicus, -a, -um – of Olympus or Mt. Olympus (Greece).

opacus, -a, -um – opaque, shaded, not transparent.

operculatus, -a, -um – with a lid.

oppositiflorus, -a, -um – with opposite flowers.

oppositifolius, -a, -um – with opposite leaves.

opulifolius, -a, -um– with leaves like those of Opulus.

orbicularis, -is, -e – obicular.

orbiculatus, -a, -um – round or circular, disk-shaped.

orchiodes, -es – es – orchid-like.

orchioides, -es, -es – orchid-like.

orientalis, -is, -e – eastern, oriental.

ornatissimus, -a, -um – very show.

ornatus, -a, -um – adorned, ornate.

ornithocephalus, -a, -um – very show.

ornatus, -a, -um – adorned, ornate.

ornithocephalus, -a, um – like a bird's head.

ornithopodus, ornithoppus, -a; -um – like a bird's foot.

orthobotrys, -ys, -ys – straight-clustered.

orthocarpus, -a, -um – straight-fruit.

orthopterus) -a, -um – strainght-winged.

orthosepalus, -a, -um – straight-
sepaled.

ostrinus, -a, -um – purple.

ovalifolius, -a, -um – oval-leaved.

ovalis, -is, -e – oval.

ovatifolius, -a, -um – ovate-leaved.

ovatus, -am -um – ovate.

ovinus, -a, -um – pertaining to
sheep.

oxycanthus, -a, -um – sharp-spined.

oxygonus, -a – sharp--angled, acte-
angled.

oxyphyllus, -a, - -um – sharp-leaved.

P

pabularius, -a, -um – of fodder or
pasturage.

pachyanthus, -a, -um – thick-
flowered.

pachycarpus, -a, -lum – with thick
pericarp.

pachyphlaeus, -a, -um – thick-
barked.

pachyphyllus, -a, -um – thick-
leaved.

pacificus, -a, -um – of the Pacific, or
regions bordering the Pacific
Ocean.

paleaceus, -a, -um – like chaff.

palestinus, -a, -um – of Palestine.

pallens, -ens, -ens- pale.

pallescens (pres, part.) – becoming
pale.

palliatus, -a, -um – cloacked.

pallidiflorus, -a, -um – with pale
flowers.

pallidus, -a, -um – pallid, pale.

palmatifidus, -a, um – palmately
divided.

palmatus, -a, -um – palmate.

paludosus, -a, -um – marshy.

palus – (noun) – a marsh.

paluster, -tris, -tre – of marshes.

pampinus, -a, -um – tendril.

panduraeformis, -is, -e –
violin-shaped.

paniculatus, -a, -um – in panicles.

paniculiger, -era, -erum – panicle -
bearing.

panis (noun) – bread.

pannosus, -a, -um – ragged, tattered.

papaveraceus, -a, -um – poppy-like.

papilionsceus, -a, -um – like a butterfly.

papillopsus, -a, -um – with small
protuberances.

papposus, -a, -um – with a pappus.

papyraceus, -a, -um – papery.

papyrifer, -era, -erum –
paper-bearing.

paradisiacus, -a, -um – of parks or
gardens.

paradoxus, -a, -um – unusual,
strange.

pardalinus, -a, -um – leopard-like,
spotted.

pardalis (noun) – a panther.

partitus, -a, -um – parted.

parviflorus, -a, -um –
small-flowered.

parvifolius, -fa, -um – very small-
leaved.

parvulus, -a, -um – small.

pascuus, -a, -um – of pastures.

passerinus, -a, -um – pertaining to a
sparrow.

pastoralis, -is, -e – pastoral.

patellaris, -is, -e – circular, disk-
shaped.

patens(pres. part.) – spreading.

patiens, -ens, -ens – patient.

patulus, -a, -um – spreading.

pauciflorus, -a, um – few-flowered.

paucifolius, -a, -um – few-leaved.

pauperculus, -a, -um – poor.

pavonicus, -a, -um – variegated.

pavonius, -a, -um –
peacock-like,fvariegated.

pectinatus, -a, um – comblike, like
the teeth in a comb.

pectinifer,-era,-erum–comb-bearing.

pectoralis, -is, -e, – shaped like a
breastbone.

pedatus, -a, -um – palmately divided
or parted.

pedecarpus, -a, -um – with stalked
fruit.

pedicularius, -a, -um – with a stalk.

peduncularis, -is, -e – with
peduncles.

pedunculatus, -a, -um – with peduncles.

pedunculosus, -a, -um – with many peduncles.

pekinensis, -is, -e,– of peking, China.

pellucidus, -a, -um – with transparent dots.

peltatus, -da, -um – shield -shaped, as a leaf attached by its under side rather than by its margin.

peltifolius, -a, -um – peltate-leaved.

penduliflorus, -a, -um- with pendulous flowers.

pendulinus, -a, -um – somewhat pendulous.

pendulus, -a, -um – pendulous, hanging.

penicillatus, -a, -um – hair-penciled.

pennatifidus, -a, -um – pennatifid.

pennatus, -a, -um – pinnate, feathered.

pensilis, -is, -e – pensile, hanging.

pentalophus, -a, -um – five-winged or five -tufted.

pentandrus, -a, -um - with five stamens.

pentasepalus, -a, -um – with five sepals.

perbellus, -a, -um – very beautiful.

peregrinus, -a, -um – foreign, -exotic.

perennans, -ans, -ans – perennial.

perennis, -is, -e- perennial.

perfoliatus, -a, -um – with stem passing through a leaf.

perforatus, -a, -um – with holes.

perfosus, -a, -um – perfoliate.

pergracilis, -is, -e - very slender.

perpusillus, -a, -um – very small.

persicarius, -a, -um- like a peach.

persicus, -a, -um – like a peach.

persicus, -a, -um – of Persia, like a peach.

persistens. -ens, -ens – persistent.

perspicuus, -a, -um – transparent.

perulatus, -a. -um – wallet-like or pocket-like.

pervenustus, -a, -um–very beautiful.

perviridis, -is, -e – very green, deep green.

petiolaris, -is, -e, pertaining to the petiole.

petiolatus, -a, -um – with petioles.

petiolus -a, um – with petioles.

petraeus, -a, -um – strony.

petrocallis (noun) - rock beauty.

phaeocarpus, -a, -um – dark-fruited.

philadelphicus, -a, -um– of the Philadelphia (Penn, U.S.A.) region.

phleoides, -es, -es, – resembling timothy.

phlogiflourus, -a, -um – flame-flowered, phlox-flowered.

pleniflorus, -a, -um – double-flowered.

pleurostachys, -ys, -ys – side-spiked.

plicatus, -a, -um – folded.

plumarius, -a, -um – with plumes or feathers.

plumosus, -a, -um – feathery.

podocarpus, -a, -um – with stalked fruits.

podophyllus, -a, -um – with stalked leaves.

poeticus, -a, -um – of or pertaining to poets.

politus, -a, -um – many-spined.

polyanthus, -a, -um- many-flowered.

polycarpus, -a, -um – many-fruited.

polycephalus, -a, -um – many-heade.

polylepis, -is, -is – with many scales.

polymorphus, -a, -um – of many forms, variable.

polystachys, -ys, -ys – with many spikes.

polystictus, -a, -um – many-dotted.

pomeridianus, -a, -um – afternoon.

pomifer, -era, -erum – pome-peraing.

pomiformis, -is, -e- shaped like a pome.

ponderosus, -a, -um – heavy, ponderous.

ponticus, -a, -um – of Ponticus (in Asia Minor).

populifolius, a, -um – poplar-leaved.

populneus, – a, -um – pertaining to poplars.

porcinus, -a -um – pertaining to swine.

porphyreticus, -a, -um – purple.

porrifolius, -a, -um- porrum-leaved, leek-leaved.

portoricensis, -is, -e – of Puerto Rico.

protulaceus, -a, -um– portulaca-like, purslane-like.

praealtus, -a, -um – very tall.

praecox, -ox, -ox – precocious, premature, very early.

praemorsus, -a, -um – bitten at the end.

praestans (pres. part.) – distinguished, excelling.

praetextus, -a, -um – bordered.

prasinatus, -a, -um – greenish.

prasinus, -a, -um – grass green.

pratensis, -is, -e – growing in meadows.

pravissimus, a, -um – very crooked.

precatorius, -a, -um – praying, prayerful.

primulinus, -a, -um – primrose -yellow.

princeps, -ceps, -ceps – p r i n c e l y, first.

prismatocarpus, -a, -um – prism-fruited.

proboscideus, -a, -um – proboscis-like.

procerus, -a, -um – tall.

procumbgens (pres. part.) – lying on the ground.

procurrens (pres. part) – extending.

profusus, -a, -um – profuse.

prolifer, -era, -erum – prolific, fruitful.

prolificus, -a, -um – prolific, fruitful.

propendens (pres. part.) – hanging down.

prostratus, -a, -um – lying on the ground.

provincialis, -is, -e – provincial.

pruinosus, -a, -um – h o a r f r o s t ; covered with whitish dust or bloom.

prunifolius, -a, -um – plum-leaved.

pruriens (pres. part.) – itching.

pseudacacius (noun) – false acacia.

pseudonarcissus (noun) – false narcissus.

pseudoplatanus (noun) – false plane tree.

psilostachyus, -a, -um – naked-spiked.

psilostemon, on, -on – smooth-stamened, naked -stamend.

psycodes, -es, -es- fragrant.

pteranthus, -a, -um – with winged flowers.

pterocarpus, -a, -um – with winged fruit.

pubens, -ens, -ens – downy.

pubescens, -ens, -ens – downy, covered with soft hairs.

pubiflorus, -a, -um– downy-flowered.

pubiger, -era, -erum – down-bearing.

pudicus, -a, -um – bashful, retiring, shrinking, modest, chaste.

pugioniformis, -is, -e – dagger-shaped.

pulchellus, -a, -um – pretty, beautiful.

pulcherrimus, -a, -um – very handsome, very beautiful.

pulverulentus, -a, -um – powdered, dust-covered.

pumilus, -a, -um – dwarf, small.

pounctatus, -a, -um – dotted.

pungens, -ens, -ens – prickly, piercing, sharp-pointed.

punicans, -ans, -ans – reddish.

puniceus, -a, -um – purple, crimson.

purgans (pres. part.) – purging.

purpurascens (pres. part.) – p u r - plish, becoming purple.

purpureus, -a, -um – purple.

purpureus, -a, -um – dwarf, small.

pycnacanthus, -a, -um – d e n s e l y spiny.

pycnanthus, -a, -um – densely flowered.

pycnocephalus, -a, -um – t h i c k - headed.

pygmaeyus, -a, -um – dwarf.

pyramidalis, -is, -e – pyramidal.

pyrenaicus, -a, -um– of the Pyrenees.

pyriformis, -is, -e – shaped like a pear.

Q

quadrangularis, -is, -e, quadran-gulatus, -a, - um – four-angled.

quadriflorus, -a, -um – with four flowers.

quadrifolius, -a, -um – with four leaves.

quercifolius, -a, -um – oak-leaved.

quinqueflorus, -a, -um – with five flowers.

quinquefolius, -a, -um – with five leaves.

R

racemosus, -a, -um – in racemes, resembling racemes.

radiatus, -a, -um – radiate, rayed.

radicalus, -a, -um – from the root.

radicans (pres. part) – rooting.

ramosissimus, -a, -um – m a n y - branched.

ramosus, -a, -um – branched.

frapiformis, -is, -e – turnip-shaped.

reclinatus, -a, -um – reclining, bent back.

recurvus, -a, -um – curved back.

redivivus, -a, -um – r e s t o r e d , brought to life.

reflexus, a, -um – bent back.

refractus, -a, -um – curved, broken.

reglis, -is, -e – royal, regal.

regina (noun) – queen.

regius, -a, -um – royal, magnificent, kingly.

religiosu, -a, -um – used for religious purposes, venerated, sacred.

reniformis, -is, -e – kidney-shaped.

repens (pres. part.) – creeping.

reptans (pres. part.) – creepng.

resinosus, -a, -um – resinous.

resupinatus, -a, -um – inverted.

reticulatus, -a, -um – netlike.

retortyus, -a, -um – bent back.

retortus, -a, -um – reflexed.

retrofractis, -is, -e – broken or bent backwards.

retusus, -a, -um – retuse, notched slightly at a rounded apex.

revolutus, -a, -um – revolute, rolled backwards.

rhizophyllus, -a, -um – root-leaved, leaves rooting.

rhombifolius, -a, -um– with rhomble leaves.

rhomboidalis, -is, -e – with rhombic outline.

rhytidophyllus, -a, -um – wrinkl-leaved.

rigens, -ens, -ens – rigid, stiff.

rigidisetae (noun) – stiff bristles.

rigiddus, -a, -um – stiff.

ringens (pres. part). – gaping, open-mouthed.

riparius, -a, -um – growing on side of river.

rivalis, -is, -e – growing on brook-side.

rivularis, -is, -e – growing on brook-side.

roborosus, -a, -um – strong.

robustus, -a, -um – robust, stout.

roridus, -a, -um – dewy, covered with particles which resemble dew.

rosaeflorus, -a, -um – rose -flowered.

roseus, -a, -um – rose-colored.

rostratus, -a, -um – rostrate, beaked.

rosulatus, -a, -um – like a rosette.

rotundifolius, -a, -um – r o u n d - leaved.

rotundus, -a, -um – round.

rubellus, -a, -um – reddish.

rubens, -ens, -ens – red, ruddy.

ruberrimus, -a, -um – very red.

rubescens (pres. part.) – becoming red.

rubicundus, -a, -um – reddish.

rubidus, -a, -um – dark red.

rubiginosus, -a, -um – brownish-red.

ruber, -ra, -rum – red.

rudis, -is, -e – wild, not tilled.

rufidulus, – a, -um – somewhat fufid, reddish.

rufus, -a, -um – fox red.

rugosus, -a, -um – wrinkled.

rugulosus, -a, -um – s o m e w h a t rugose.

rupestris, -is, -e – grwing on rocks.

rupicolus, -a, – um-growing on rocks.

russus, -a, -um – red.

rusticanus, -a, – um – rustic, pertaining to the county.

S

saccatus, -a, -um – saccate, bag-like.

saccharatus, -um – containing sugar.

saccharinus, -a, -um – saccharine, of sugar.

saccharoides, -es, -es, – sweet, like sugar.

saccharum (noun) – sugar.

saccus (noun) sac.

sachalinensis, -is, -e – of Saghalin Island (n. Japan).

sacrorum (noun) – of sacred places.

sagittifolius, -a, -um – arrow-leaved.

sagittalis, -is, -e – arrow-shaped.

sagittatus, -a, -um – arrow-shaped.

salicifolius, -a, -um – willow-leaved.

salicis, -is, -e – of the willow.

salignus, -a, -um – salty.

salsuginosus, -a, -um – fond of salt marshes.

sanctus, -a, -um – holy.

sanguinalis, -is. -e – blood-colored.

sanguineus, -a, -um- blood red.

sanguis (noun) – blood.

sapidus, -a, -um – tasteful, savory, pleasing to taste.

sapientum (noun) – of the wise men or authors.

saponarius, a, -um – soapy.

sarmentosus, -a, -um – sarmentose, bearing runners.

sativus, -a, -um – cultivated.

saxatilis, -is, -e – growing on rocks.

saxifragus, -a, -um – stone-breaking

scaber, -bra, - brum -rough, not smooth.

scaberrimus, -a, -um – very rough.

scandens (pres. part.) – climbing.

scaposus, -a, -um -with scapes.

scariosus, -a, -um – shriveled, thin and not green.

scarletinus, -a, -um – scarlet.

sceleratus, -a, -um – pernicious.

sceptrum (noun) – scepter.

schistaceus, -a, -um – slate gray.

schistosus, -a, -um – schistose, slaty as to tint.

schizopetalus, -a, -um – with cut pe-

tals.

scilloides, -es, -es – like squill.

scissus, -a, -um – split.

sciureus, -a, -um – r e s e m b l i n g squirrel's tail.

sclerophyllus, -a, -um – hard-leaved.

scoparius, -a, -um – broomlike.

scopulinus, -a, -um – rocklike.

scopulorum (noun) – of the rocks.

scorpioides, -es, -es – scorpion-like.

scutatus, -a, -um – buckler-shaped, like a small shield.

scutellatus, -a, -um – shield-shaped.

sebifer, -era, -erum – tallow-bearing.

sebosus, -a, -um – full of tallow or grease.

secalinus, -a, -um – like rye.

secundatus, -a, -um – secund, on the side.

secundiflorus, -a, -um – one-sided, secund-flowered.

secundus, -a, -um – secund, on the side.

segetum (noun) – of corn fields.

semidecandrus, -a, -um – with five stamens, half-of ten stamens.

semiglobosus, -a, -um– half -globose.

semperflorens (pres. part.) – e v e r - blooming.

sempervirens, -ens, -ens – e v e r - green.

sempervivus, -a, -um - ever-living.

senescens (pres. part.) – b e c o m i n g old or gray.,

senilis, -is, -e – senile, old white-haired.

sensibilis, -is, -e – sensible.

sensitivus, -a, -um – sensitive.

sepiarius, -a, -um – of for pertaining to hedges.

septentrionalis, -is, -e – northern.

sericeus, -a, -um – silken.

sericifer, -era, -erum -silk-bearing.

serotinus, -a, -um – late-flowering, fall.

serpens (pres. part) – c r e e p i n g , crawling.

serpentinus, -a, -um – of snakes, serpentine, looping or waving.

serratifolius, -a, -um – with serrate

leaves.

serratus, -a, -um – serrate, saw-toothed.

serrulatus, -a, -um – finely serrate or saw-toothed.

sesquipedalis, -is, -e – one foot and a half long high.

sessiliflorus, -a, -um – with stemless leaves.

sessilis, -is, -e – stemless.

sessilispicatus, -a, -um – with a spike without pedicel.

setaceus, -a, -um – bristle-like.

setifolius, -a, um – bristle-leaved.

setiger, -era, -erum – bristly, bristle-bearing.

setosus, -a. -um – with bristles.

sibericus, -a, -um – of Siberia.

siculus, -a, -um – of Sicily.

signatus, -a, -um – marked, desogmated. attested.

silvaticus, -a, -um – pertaining to woods.

silvestris, -is, -e – pertaining to woods or forest.

similis, -is, -e – similar, like.

simplex, -ex, -ex – simple, unbranched.

simplicifolius, -a, -um – s i m p l e - leaved.

sinensis, -is, -e – Chinese, of China.

sinuatus, -a, -um – sinuate, wavy-margined.

siphiliticus, -a, -um – syphilitic.

smaragdinus, -a, -um – e m e r a l d green.

sobolifer, -era, -erum – b e a r i n g creeping, rooting stems or shoots, sucker-bearing.

somnifer, -era, -erum – sleep-bearing, sleep-producing.

somsnous – (noun) – sleep.

sordidus, -a, -um – dirty.

sparsiflorus, -a, -um – with scattered flowers.

sparsus, -a, -um – scattered.

sparteus, -a, -um – spearlike.

spathulatus, a, -um – with a spathe.

ɪ patiosus, -a, -um – spacious, wide.

speciosus, -a, -um – beautiful, showy, good-looking.

spectabilis, -is, -e – spectacular, remarkable.

speculatus, -a, -um – shining, as if with mirrors.

sphaerocarpus, – a, -um – spherical-fruited.

sphaerocephalus, -a, -um – spherical-headed.

spicatus, a, -um – with spikes.

spiciformis, -is, -e – spike -shaped.

spinescens (pres. part.) – becoming spiny.

spinifer, -era, -erum – b e a r i n g spines.

spinosus, -a, -um – with spines or thorns.

spinulosus, -a, -um – with small spines.

spiralis, -is, -e – spiral.

splendens (pres. part) – glittering, splendid.

spongiosus, -a, -um – spongy.

spontaneus, -a, – um – spontaneous.

spurius, -a, -um – spurious, false, bastard.

squamatus, -a, -um – squamate, with small scalelike leaves or bracts.

squamiger, -a, -um – scale-bearing.

squamosus, -a, -um – squamate, full of scales.

squarrosus, -a, -um- squarrose, with parts spreading or even recurved at ends.

stagninus, -a, -um – growing in a marsh.

stamineus, -a, -um – with prominent stamens.

stans (pres. part.) – erect, standing, upringt.

stellatus, -a, -um – with star, star-like.

stellipilus, -a, -um – with stars, star-like.

stellipilus, -a, -um – with stellate hairs.

stellulatus, -a, -um – somewhat stellate.

stenocephalus, -a, -um – n a r r o w - headed.

stenophyllus -a, -um – n a r r o w - leaved.

sterilis, -is, -e – sterile, infertile.

stipulatus, -a, -um – having stipules.

stipulatus, -a, -um – haing stipules.

stoloniger, -era, -erum – w i t h stolong or runners that take root.

stramineofructus, -a, -um – w i t h straw-colored fruit.

stramineus, -a, -um – straw-colored.

streptocarpus, -a, -um – t w i s t e d - fruited.

streptophyllus, -a, -um – t w i s t e d - leaved.

striatus, -a, -um – with   parallel channels or lines, striate.

strictus, -a, -um– stiff, rigid, bristles.

strobilus, -a, -um – cone.

strumosus, -a, -um – s t r u m o u s , having cushion -like swellings.

stylosus, -a, -um – with   style   or styles prominent.

styracifluus, -a, -um – flowing with storax or gum.

suaveolens, -ens, -ens – s w e e t - smelling.

suavis, -is, -e – sweet, agreeable.

subalpinus, -a, -um – below the al- pine.

subcordatus, -a, -um – almost cor- date.

suberosus, -a, -um – corky.

submersus, -a, -um – submerged.

subovatus, -a, -um – almost ovate.

subterraneus, -a, -um – uner   the ground.

subulatus, -a, -um – awl-shaped.

succidus, -a, -um – sappy.

succulentus, -a, -um – s u c c u l e n t , fleshy.

sudanensis, -is, -e – os the Sudan.

suecius, -a, -um – of Sweden.

suffruticosus, -a, -um – shrubby.

sulcatus, -a, -um – furrowed.

sulfureus, -a, -um – sulfur-colored.

superbus, -a, -um – superb, proud.

supinus, -a, -um – prostrate.

suprafoliaceus, -a, -um- above   the leaf.

suspensus, – a, -um – suspended.

sylvaticus, -a, -um – sylvan,   forest- loving.

sylvestris, -is, -e – growing in woods (forest).

symphoricarpus, -a, um – w i t h fruits together.

syphiliticus, -a, -um – syphilitic.

T

tabacinus, a, -um – like tobacco.

tabulaeformis, -is -e – table-shaped.

tamariscifolius, -a, -um – tama-risk- leaved.

tanacetifolius, -a, -um – tansy-leave.

tardiflorus, -a, -um – late-flowering.

tardus, -a, -um – late.

tartareus, -a, -um – with loose or rough crumbling surface.

tataricus, -a, -um – of Tartary (cent. Asia).

taxifolius, -a, -um – yew-leaved.

tectorus, -a, -um – of   roofs,   or houses.

tectus, -a, -um – concealed, covered.

temulentus, – a, -um – intoxicating.

tenarius, – a, -um – slender.

tenax, -ax, -ax – tenacious, strong.

tenebrosus, -a, -um – of dark or shaded places.

tenellus, -a, -um – flexible, slender.

tener, -a, -um – slender, tender, soft.

tenuiflorus, -a, -um – s l e n d e r flowers.

tenuifolius, -a, -um – slender-leaved.

tenuior, -or, -us – very slender.

tenuis, -is, -e – slender, thin.

terebinthinus, -a, -um – of turpen- tine.

teres, -es -es – terete,   circular   in cross section.

tereticornis, -is, -e ·· with terete or cylindrical horns.

teretifolius, -a, -um – terete leaved.

terminalis, -is, -e – terminal, at the end of a stem or branch.

ternatus, -a, -um– arranged   in threes.

ternifolius, -a, -um – with leaves in threes.

terrestris, -is, -e – of the earth.

tessellatus, -a, -um – tessellate, laid off in squares or in dicelike pattern.

testaceus, -a, -um – light brown, brick-colored.

testiculatus, -a, -um – testiculated, testicled.

testudinarius, -a, -um- like a tortoise-shell.

tetragonolobus, -a, -um – with a four-angled pod.

tetrapterus, -a, -um – four-winged.

tetrastachyus, -a, -um – with four spikes.

textilis, -is, -e – textile, woven.

thelypteroides, -es, -es – t h e l y p terum-like.

thermalis, -is -e – warm, of warm springs.

thuyoides, -es, -es – like thuja (thuya) or arbovitae.

thyoides, -es, -es – See thuyoides.

thyrsiflorus, -a, -um – flowers borne in a thyrsus.

thyrsoideus, -a, -um- thyrse-like.

tigrinus, -a, -um – tiger-striped.

tiliaceus, -a, -um – Tilia-like (like linden or basswood).

tinctorius, -a, -um – a dyer.

tinctus, -a, -um – dyed.

tomentosus, -a, -um – felty, downy.

torminalis, -is, -e, – useful against colic.

torosus, -a, -um – cylindrical with contractions at intervals.

tortilis, -is, -e – twisted.

tortuosus, -a, -um – much twisted.

torulosus, -a, -um – somewhat torose or contracted at intervals.

transalpinus, -a, -um – transalpine.

tremuloides, -es, -es – t r e m b l i n g, quaking.

tremulus, -a, -um – quivering, trembling.

triacanthus, -a, -um – thèree-spined.

triangularis, -is, -e – triangular.

tricephalus, -a, -um- three-headed.

trichocarpus, -a, – um -hairy-fruited.

trichodes, -es, es – pilose.

trichosanthus, -a, -um – with hairy, flowers.

trichospermus, -a, -um – h a i r y -

seeded.

trichotomus, -a, -um – t h r e e - b r a n - ched or three-forked.

tricoccus, -a, -um – three lobed.

tricolor, -or, -or – three-colored.

tricornis, -is, -e – three-horned.

tridens, -ens, -ens– with three teeth.

trifldus, -a,-um – three-cleft or three-parted.

trifoliatus, -a, -um – three-leaved.

trifurcatus, -a, -um – three-forked.

trilobatus, -a, -um – three-lobed.

trimestris, -is,-e – of three sonths, as lasting that time or maturing in it.

trinervis, -is, -e – three-nerved.

trinotatus, -a, -um – t h r e e - m a r k e d or three-spotted.

tripartitus, -a, -um – three-parted.

tripartitus, -a, -um – three-petaled.

triphyllus, -a, -um – three-leaved.

tripterius, -a, -um – three-winged.

trispermus, -a, -um – three-seeded.

tristachyus, -a, -um – three-spiked.

tristis, -is, -e – sad, dull, bitter.

trivialis, -is, -e – common, ordinary, found everywhere.

truncatus, -a, -um – ending abruptly.

tubaeformis, -is, -e–trumpet-shaped.

tubatus, -a, -um – with tubers.

tubiflorus, -a, -um– t r u m p e t - flowered.

tubulosus, -a, -um– tubular, with tubes.

tumidus, -a, um – swollen.

tunicatus, -a, -um – with concentric coats.

turbinatus, -a, -um - top-shaped.

turgidus, -a, -um – turgid, inflated, full.

typhinus, -a, -um – pertaining to fever.

typicus, -a, -um – typical.

U

uliginosus, -a, -um – of wet or marshy places.

umbellatus, -a, -um – with umbels.

umbellifer, -a, -um – bearing umbels.

umbraculifer, -a, -um – bearing an umbrella.

umbrosu, -a, -um – shaded or shade-loving.

uncinatus, -a, -um – hooked at the point, with hooks.

undatus, -a, -um – wavy.

undulatus, -a, -um – with waves.

unguicularis, -is, -e – with claw, tapered to a petiole -like base.

unguipetalus, -a, -um – with clawed petals.

unicornis, -is, -e – one-horned.

uniflorus, -a, -um – one-flowered.

unilabiatus, -a, -um – one-lipped.

unilateralis, -is, -e – one-sided.

unilateralis, -es, -es – uniola-like.

univittatus, -a, -um – with one longitudina stripe.

urceolatus, -a, -um – like a pitcher.

urceolus (noun) – small pitcher.

urens (pres. part.) – burning, stinging.

urninger, -a, -um – pitcher-bearing.

urostachyus, -a, -um – tail-spiked.

ursinus, -a, -um – pertaining to bears, northern (under the Great Bear).

urticaefolius, urticifolius, – a, – um – nettle-leaved.

urticoides, -es, -es – nettle-like.

usitatissimus, -fa, -um – most commonly used.

usneoides, -es, -es – usnea-like.

utilis, -is, -e – useful.

utriculus (noun) – sac.

uvaeformis, -is, -e – like a raceme.

V

vacillans (pres. part.) – swaying.

vagans (pres. part.) – v a g r a n t , wandering.

vaginaeflorus, -a, -um – w i t h flowers in sheath.

vaginans, (pres. part) – sheathing, wrapping around.

vaginatus, -a, -um – sheathed, with a sheath.

valdivianus, -a, -um – of Valdivia (Chile).

valdiviensis, -is, -e – of the region of Valdivia (Chile).

variabilis, -is, -e – variable, of many forms.

varicosus, -a, -um – varicose, with veins or filaments dilated.

bvariegatus, -a, -um – variegated.

variifolius, -a, -um – variable-leaved.

velox, -ox, -ox – rapidly growing.

velutinus, -a, -um – velvety.

venenatus, -a, -um – poisonous.

venenosus, -a, -um – v e n o m o u s , poisonous.

venosus, -a, -um – with veins.

ventralis, -is, -e – facing the center of a flower, opposite of dorsal.

venticosus, -a, -um – v e n t r i c o s e , havifng a swelling or inflation on one side.

venustus, -a, -um – beautiful, handsome, charming.

vermiculatus, -a, -um – wormlike.

vernalis, -is, -e- vernal, flowering in the spring, of spring.

vernicosus, -a, -um – varnished.

vernus, -a, -um – of spring, vernal.

verrucosus, -a, -um – v e r r u c o s e , warted.

verruculosus, -a, -um – very warty or verrucose.

versatilis, -is, -e – movable, swinging freely.

versicolor, -or, -or – variously colored, as of one color blending into another, of changing color.

versiflorus, -a, -um – with different flowers.

versutus, -a, -um – versatile.

verticillatus, -a, -um – w h o r l e d , aranged in a circle about the stem.

vertucosus, -a. -um – with warts, verrucose.

verus, a, -um – the true, genuine, standard.

vesciculosus, -a, -um – with little bladders.

vescus, -a, -um – weak, thin, feeble.

vesicarius, -a, -um – bladder-shaped.

vesicularis, -is, -e – with little bladders or blisters.

vespertinus, -a, -um – blooming in the evening.

vestitus, -a, -um – covered or clothed

with hairs, pubescent.

vexillarius, -a, -um – of the standard petal (pea flower), with a standard.

villosissimus, -a, um – very hairy.

villosus, -a, -um – covered with soft hairs.

viminalis, -is, -e – of osiers, of basket willows.

vinifer, -a, -um – fwine-bearing.

violaceus, -a, -um – violet-colored.

virens (pres. part) – greening.

virescens (pres. part.) – b e c o m i n g green.

virgatus, -a, -um – twiggy.

virginalis, -is, -e – pertaining to young women.

viridiflorus, -a, -um – with green flowers.

viridis, -is, -e, – green.

viridissimus, -a, -um – very green.

viridulus, -a, -um – greenish.

virosus, -a, -um – virulent, poisonous, fetid.

virulentus, -a, -um – virulent, poisonous, fetid.

viscarius, -a, um – viscid, like mistletoe.

viscidulus, -a, -um–somewhat sticky.

viscidus, -a, -um – viscid, sticky.

viscosus, -a, -um – viscid, sticky.

vitaceus, -a, -um – vitis-like, like the grape.

vitellinus, -a, -um – dull yellow approaching red.

vitifolius, -a, -um – v i t i s - l e a v e d, grape-leaved.

vittatus, -a, -um- striped.

vittiger, -a, -um – bearifng stripes.

viviparus, -a, -um – bearing stripes.

viviparus, -a, -um – freely producing asexual propagating parts, as bulbets in the inflorescence.

vixcordatus, -a, -um – nearly heart-shaped.

volatilis, -is, -e – voaltile.

volubilis, -is, -e – twining.

vomicus, -a, -um – emetic.

vomitorius, -a, -um – emetic.

vulcanicus, -a, -um – of Vulcan or of a volcano, volcanic.

vulgaris, -is, -e, – vulgar, common, usual.

vulnerarius, -a, -um – useful for wounds.

vulpes, -pes, -pes – fox.

vulpinus, -a, -um– pertaining to the fox.

X

xanthacanthus, -a, -um – yellow-spined.

xanthinus, -a, -um – yellow.

xanthocarpus, -a, -um – yellow-fruited.

xanthoneurus, -a, -um – yellow-nerved.

xanthorrhizus, -a, -um – yellow-rooted.

Y

yedoensis, -is, -e – of Yedo or Yeddo (Japan).

yunnanensis, -is, -e – of province of Yunnan (china).

Z

zebrinus, -a, -um – zebra-striped.

zeylanicus, -a, -um – Ceylonian, or Ceylon.

zibethinus, -a, -um – like the civet-cat, malodorous.

zizanioides, -es, -es, – zizania-like.

zonalis, -is, -e- zonal.

zonatus, -a, -um – zonal, zoned, banded.

Structures and Specialized Characters

I. THE PLANT

A. Plant Parts

Bud – Immature vegetative or floral shoot or both, often covered by scales.

Flower – Reproductive structure of flowering plants with or without protective envelope, the calyx and/or corolla; short shoot with sporophylls and with or without sterile protective leaves, the calyx and corolla.

Fruit – Matured ovary of flowring plants, with or without accessory parts.

Leaf – A photosynthetic and transpiring organ, usually developed from leaf primordium in the bud; an expanded, usually green, organ borne on the stem of a plant.

Root – An absorbing and anchoring organ, usually initially developed from the radicle and growing downward.

Seed – Matured ovule of seed plants.

Stem – A supporting and conducting organs usually developed initially from the epicotyl and growing upward.

B. Plant Types
(Classification based on habit)

Herb – A usually low, soft, or coarse plant with annual above-ground stems.

Shrub – A much-branched woody perennial plant usually without a single trunk.

Tree – A tall, woody perennial plant usually with a single trunk.

Vine or Liana – An elongate, weak stemmed, often climbing annual or perennial plant, with herbaceous or woody texture.

II. ROOTS

A. Root Parts

Root Cap – Parenchymatous, protective apex of root.

Secondary Root – Lateral root with root cap and hairs, derived from the pericycle.

B. Root Types
(Classification based on position and origin)

Adventitious – Arising from organ other than root; usually lateral.

Secondary – From pericycle within the primary or secondary root; lateral.

C. Root Structural Types

Aerating or Knee – Vertical or horizontal above-ground roots.

Contractile or Pull – Roots capable of shortening, usually drawing the plant or plant part deeper into the soil, usually with a wrinkled surface.

Aerial – Fribrous, adventitious roots, frequently with an adhesive disk; a crampon.

Butters – Roots with board-like or plank-like growth on upper side, presumably a supporting structure.

Fleshy – Succulent roots –

Haustorial – Absorbing roots, within host of some parasitic species.

Moniliform – Elongate roots with regularly arranged swollen areas.

Pneumatophorous – With spongy, aerating roots, usually found in marsh plants.

Fascicled – Fleshy or tuberous roots in a cluster.

Fibrous – With fine, threadlike or slender roots.

Prop or Stilt – Adventitious, supporting roots usually arising at lower nodes.

Tap – Persistent, well-developed primary root.

Tuberous – Fleshy roots resembling stem tubers.

III. STEMS

A. Stem Parts

Bud – An immature shoot.

Internode – A section or region of stem between nodes.

Leaf Scar – A mark indicating former place of attachment of petiole or leaf base.

Lenticel – A pore in the bark.

Prickel – A sharp-pointed outgrowth from the epidermis or cortex of any organ.

Stipular Scar – A mark indicating former place of attachment of stipule.

Terminal Bud Scale Scar Rings – Several marks in a ring indicating former places of attachment of bud scales.

Vascular Bundle or Trace Scar – A mark indicating former place of attachment within the leaf scar of the vascular bundle or trace.

2. Major Stem Parts

Bark – Tissues of plant outside wood or xylem.

Exfoliating Bark – Bark cracking and splitting off in large sheets.

Fissured Bark – Split or cracked bark.

Plated Bark – Split or cracked ark with flat pltes between fissures.

Ringed Bark – Split or cracked bark with flat plates bet   n fissures.

Shreddy Bark – Soft but coarse fibrous bark, usually shallowly fissured.

Smooth Bark – Bark without fissures.

Winged Bark – Bark with one or more thin, flat longitudinal expansions or elongate plates.

Pith – Cantermost tissue of stem, usually soft.

Chambered Pith – Solid core of pith cells absent, only distinct partitions present.

Continuosus Pith – With solid core of parenchyma or pith cells.

Diaphragmed Pith – With solid core of pitch cells and distinct partitions.

Hollow Pith – Disintegrated pith with a large central cavity.

Spongy Pith – Porous, easily compressible pith.

Wood – Xylem consisting of vessels and/or tracheids, fibers, and parenchyma cells.

Annual Ring – Usually one year's growth of wood; spring and summer wood.

Diffuse Porous Wood – Annual rings with vessels or pores more or less evenly distributed.

Non-porous Wood – Annual rings with tracheids only, no vessels produced in spring or summer wood.

Ring Porous Wood – Annual rings with vessels or pores usually in the spring wood, in a well-defined circular band.

B. Stem Types
(Classification based on habit, direction of growth, or position)

Arborescent – Tree-like in a appearance and size.

Ascending – Inclined upward.

Cespitose – Short, much-branched, plant forming a cushion.

Clambering – Sprawling across objects, without climbing structures.

Climbing – Growing upward by means of tendrils, petioles, or adventitious roots.

Columnar – Erect with a stout main stem or trunk.

Decumbent – Reclining or lying on the ground with the tips ascending.

Eramous – With unbranched stems.

Erect – Upright.

Fastigiate–Strictly erect and parallel.

Fruticose – Shrubby.

Geniculate – Abruptly bent at a node, zigzag.

Procumbent, Prostrate, or Reclining – Trailing or lying flat, not rooting at the nodes; *humistrate*.

Ramose – Branched.

Repent – Creeping or lying flat and rooting at the nodes.

Sobliferous – With loosely clumped shoots arising some distance apart from rhizone or under-ground suckers.

Stoloniferous – With runners or propagative shoots rooting at the tip producing new plants; hearing stolons; sarmentose.

Strict – Stiff and rigid.

Supine – Prostrate, with parts oriented upward.

Trailing – Sprawling on ground, usually with adventitious roots.

Twining – Coiling around an object.

Viragate – Wand-like; long, slender, and straight.

C. Stem Structural Types

Bulb – A short, erect, underground stem surrounded by fleshy leaves.

Bulbel – A small bulb produced from the base of a larger bulb.

Bulbil – A small bulb or bulblike body produced on above-ground parts.

Bulbet – A small bulb, irrespective of origin.

Caudex – A short, thick, vertical or branched perennial stem usually subterranean, or at ground level.

Cladode (phylloclade) – A flattened main stem resembling a leaf.

Corm – The enlarged, solid, fleshy base of a stem with scales; an upright underground storage stem.

Cormel – Small corm produced at base of parent corm.

Gulm – Flowering and fruiting stems of grasses and sedges.

Pachycauly – Short, thick, frequently succulent stems, as in some cacti.

Primocane – The first-year non-flowering stem, as in most blackberries; a turion.

Pteriocauly – Winged a stems.

Rhizome – A horizontal underground stem.

Rootstock – A term applied to miscellaneous types of underground stems or parts.

Runner or Stolon – An indeterminate, elongate, above-ground propagative stem, with long internodes, rooting at the tip forming new plants.

Sarcocauly – Fleshy stems.

Scape – A naked flowering stem with or without a few scales leaves, arising from an underground stem.

Sclerocauly – Hard, dryish stems.

Spur – A short shoot on which flowers and fruits or leaves are borne.

Sucker.A short arising below ground or from an old stem, usually fast-growing and adventitious; surculose.

Trendril – Long, slender, coiling branch, adapted for climbing (most tendrils are leaf structures).

Thorn – A sharp-pointed branch.

Tiller – A grass shoot produced from the base of the stem.

Tuber – A thick storage stem, usually not upright.

Turion – An overwintering bud, as in *Lemna*.

Underground Stolon – A determinate, elongate, underground propagative stem with long internodes forming a bulb or tuber at the tip.

IV. BUDS

A. Bud Parts

Bud Primordium – Meristematic tissue that gives rise to a lateral bud.

Flower Primordium – Meristematic tissue that gives rise to a flower.

Leaf Primordium – Meristematic tissue that gives rise to a leaf.

Promeristem – Apical growing or meristematic tissue that gives rise to other bud parts.

Scale – Protective leaf on outside of bud.

B. Bud Types
(Classification based on position and arrangement)

Accessory. Buds lateral to or above axillary buds.

   Collateral – Bud(s) lateral to axillary bud.

   Superposed – Bud(s) above axillary bud.

Axillary or Lateral – In axils of leaves or leaf scars.

Intrapetiolar or Subpetiolar – Axillary bud surrounded by base of petiole.

Pseudo-terminal – Bud appearing apical but is alteral near apex, developing with death or non-development of terminal bud.

Terminal – At apex or end of stem.

C. Bud Structural Types
(Classification based on composition and cover)

Flower – Contains flower primordia; will give rise to one or more flowers.

Leaf – Contains leaf and stem primordia; will give rise to branch with leaves.

Mixed – Contains flower, leaf, and stem primordia; will give rise to branch with leaves and flower(s).

Naked – Shoot and/or flower primordia not surrounded by scales.

Protected – Shoot and/or flower primordia surrounded by scales.

V. LEAVES

A. Leaf Parts

Blade – Flat, expanded portion of leaf.

Leaflet – A distinct and separate segment of a leaf.

Ligule – An outgrowth or projection from the top of the sheath, as in the Poaceae.

Midrib – The central conducting and supporting structure of the blade of a simple leaf.

Sheath – Any more or less tubular portion of the leaf surrounding the stem or culm, as in the Poaceae.

Stipels – Paired scales, spines, or glands at the base of Petiolule.

Midvein – The central conducting and supporting structure of the blade of a leaflet.

Petiole – Leaf stalk.

Petiolule – Leaflet stalk.

Pulvinus – The swollen base of a petiole or petiolule.

Rachilla – Secondary axis of compound leaf.

Rachis – The main axis of a pinnately compound leaf.

Stipules – Paired scales, spines, glands, or blade-like structures at the base of a petiole.

B. Leaf Type
(Classification based primarily on arrangement of leaflets)

Note – This classification is based on discrete segments of leaves or leaflets, but the terms "compound" are equally applicable to segments, divisions, etc. of any structure with a blade; e.g., palmately divided, pinnately cleft etc.

Bifoliolate, Geminate, or Jugate – With two leaflets from a common point.

Bigeminate, Bijuagate – With two orders of leaflets, each bifoliolate;

doubly paired.

Hipalmately Compound – With two orders of leaflets, each palmately compound.

Bipinnately Compound – With two orders of leaflets, each pinnately compound.

Biternate – With two orders of leaflets, each ternately compound.

Compound – With leaf divided into two or more leaflets.

Decompound – A general term for leaflets in two or more orders-bi-, tri-ate-pinnately, palmately, or ternately compound.

Imparipinnately Compound – *Oddipinnately* compound, with a terminal leaflets.

Paripinnately Compound – *Evenpin-nately* compound, without a terminal leaflet.

Pinnately Compound – With leaflets arranged oppositely or laternately along a common axis, the rachis.

Simple – With leaf not divided into leaflets.

Tergeminate – With three orders of leaflets, each bifoliolate, or with geminate leaflets ternately compound.

Ternately Compound – With leaflets in three's.

Trifoliolate – With leaflets three, pinnately compound with terminal petiolule longer than lateral; or palmately compound with petiolules equal in length.

Triplamately Compound – With three orders of leaflets, each palmately compound.

Tripinnately Compound–With three orders of leaflets, each pinnately compound.

Palmately Compound– With leaflets from one point at end of petiole.

Palmate-pinnate – With first order leaflets palmately arranged, second order pinnately arranged.

Unifoliolate – With a single leaflet with a petiolule distinct from the petiole of the whole leaf, as in *Cercis*.

C. Leaf Structural Types

Pract – Modified leaf found in the inflorescence.

Brecteole or Prophyllum – Small leaf, usually on a pedicel.

Chaff or Pale – Scale or bract at base of tubular flower in composites.

Epetiolulate – Without petiolule, leaflet sessile.

Epicalyx – Group of leaves resembling sepals below the true calyx.

Exstipellate – Without stipels.

Fly Trap – Hinged, insectivorous leaf, as in *Dionaea*.

Glume – Bract, usually occurring in pairs, at the base of the grass spikelet.

Incomplete – Leaf without one or more parts; blade, petiole, stipules.

Lemma – Outer scale subtending grass floret.

Palea – Inner scale subtending grass florest.

Complete – Leaf with blade, petiole, and stipules.

Cotyledon – Embryonic leaf.

Elaminate – Without blade.

Epetiolate – Without petiole, leaf sessile.

Phyllary – One of the involucral leaves subtending a capitulum, as in composites.

Phyllodium – Flattened blade-like petiole or midrib.

Pitcher – Ventricose to tubular insectivorous leaf, as in *Sarracenia*.

Scale – Small, non-green leaf on bud and modified stem.

Spathe An enlarged bract enclosing an inflorescence.

Spine – Sharp-pointed petiole, midrib, vein, or stipule.

Sporophyll – A spore learing leaf.

Storage Leaf – Succulent, fleshy leaf.

Tendril – Usually a coiled rachis or twining leaflet modification.

Tentacular – Glandular-haired or tentacle-bearing insectivorous leaf, as in *Drosera*.

D. Petiole and Petiolule Structural Types

Channelled or Canaliculate – With a longitudinal groove.

Inflated – Swollen or thickened, as in *Eichhornia*.

Pericladial – With a sheathing base, as in the Apiaceae.

Petiolate – With a petiole.

Petiolulate – With a petiolule.

Phyllodial – Flattened and blade-like.

Pulvinal – With a swollen base, as in the Fabaceae.

Sessile or Absent – Without petiole or petiolule.

Winged – With flattened blade-like margins.

E. Stipule and Stipel Types
(Classification based primarily on attachment or function)

Adnate – With stipule attached to petiole.

Basal – With stipules attached near base of petiole.

Interpetiolar – With connate stipules from two opposite leaves.

Lateral – With stipules adnate to petiole and free part of stipules located along the petiole.

Median – With stipules adnate to petiole with free part of stipules near middle of petiole.

Photosynthetic – Blade-like and green.

Sheathing or Protective – Enclosing a bud or flower.

Stipellate – With stipels.

Stipulate – With stipules.

Vestigial – Minute; a remnant.

VI. INFLORESCENCES

A. Inflorescence Parts

Bract – Modified, usually reduced, leaf in the inflorescence.

Bracteole or Bractlet – A secondary or smaller bract.

Cupule – Fused involucral bracts subtending flower, as in *Quercus*.

Involucel – Small involucre; secondary involucre.

Epicalyx or Calycle – A whorl of bracts below but resembling a true calyx.

Flower – Modified reproductive shoot of angiosperm.

Phyllary – Individual bract within involucre.

Involucre – A group or cluster- of bracts subtending an inflorescence.

Pedicel – Individual flower stalk.

Peduncle – Main stalk for entire inflorescence.

Perigynium – Sac-like bract subtending the pistillate flower, as in *Carex*.

Rachilla – Central axia fo a grass or sedge spikelet.

Rachis – Major axis within an inflorescence.

Ray – Secondary axis in a compound inflorescence.

Scape – Naked peduncle.

Spathe – A sheathing leaf subtending or enclosing an inflorescence.

B. Inflorescence Types
(Classification based primarily on arrangement and development)

Note – Inflorescene types are essentially secondary arrangements. The primary arrangement (alternate, Opposite, or whorled) of the individual flowers should be indicated; and the tertiary arrangement of sessile-flowered inflorencences should be noted; e.g., spikelets racemose of heads umbellate. *Determinate inflorescences* have the central flower mturing first with the arrest of the elongation of the central axix; *indeterminate inflorescences* have the lateral or lower flowers maturing first without the arrest of the elongation of the central axix.

1. Inflorescences with sessile Flowers

Ament or Catkin – A unisexual spike or elongate axis with simple dichasia that fall as a unit after flowering or fruiting.

Capitulum or Head – A determinate or indeterminate crowded group of sessile or subsessile flowers on a compound receptacle or torus.

Glomerule – An indeterminate dense cluster of sessile or subsessile flowers.

Hypanthodium – An inflorescence with flowers on wall of a concave capitulum as in *Ficus*.

Spadix – Unbranched, indeterminate, elongate inflorescence with sessile flowers.

Spikelet or Locusta – A small spike; the basic inflorescence unit in grasses and sedges.

2. Unbranched Inflorescences with Pedicellate Flowers

Cincinnus – A tight, modified helicoid cyme in which pedicels are short on the developed side.

Corymb – A flat-topped or convex indeterminate cluster of flowers.

Cymule – A simple, small dichasium.

Helicoid Cyme or Bostryx – A determinate inflorescence in which the branches devlop on the side only, appearing simple.

Raceme – Unbranched, indeterminate inflorescence with pedicelled flowers.

Scorpiodid Cyme or Rhipidium – A zigzag determinate infloréscence with branches developed on oposite sides of the rachis alternately.

Simple Cyme or Dichasium – A determinate, dichotomous inflorescence with the pedicels of equla length.

Umbel – A determinate or inderminate flattopped or convex inflorescence with the pedicels arising at a common point.

3. Branched Inflorescences with Pedicellate Flowers

Compound Corymb – A branched corymb.

Compound cyme – A branched cyme.

Compound Umbel – An umbel with primary rays or peduncles arising at a common point with a secondary umbel arising form the tip of the primary rays; a branched umbel.

Panicle– Branched inflorescance with pedicelled flowers.

Thyrse – A many-flowered inflorencence with an indeterminate central axis and with many opposite lateral dichasia.

Verticilaster – Whorled dichasia at the nodes of an elongate rachis.

4. General Inflorescence Terms and Types

Cyathium – A pseudanthium subtended by an involucre, frequently with petaloid glands, as in *Euphorbia*.

Monochasium – A cymose Inflorescence with one main axis.

Pleiochasium – Compound dichasium in which each cymule has three.

Scapose – With a solitary flower on a leafless peduncle or scape, usually arisimg from a basal rosette.

Secound – One-sided arrangement.

Pleiochasium – compound dichasium in which each cymule has thres lateral branches.

Pseudanthium – Several flowers simulating a simple flower but composed of more than a single axis with subsidiary flowers.

Solitary – One-flowered, not an inflorescence.

Umbellet – The secondary umbel in a compound umbel.

VII. FLOWERS

A. Flower Parts

Accessory Organs – The Calyx and corolla.

Androecium – One or more whorls or groups of stamens; al stamens in flower.

Androgynophore – The stipe or

column on which stamens and carpels are borne.

Calyx – The lowermost whorl of modifixed leaves, sepals.

Carpel – The female sporophyll within flower; flora organ that bears ovules in angiosperms; unit of com pound pistil.

Clinanthium – The compound receptacle of the composite head.

Corolla – The whorl of petals located above the sepals.

Disc – A discoid structure developed from receptacle at base of ovary or form stamens around the ovary.

Essential Organs – The andreecium and gynoecium.

Gynoecium or Pistil – the whorl or group of carpels in the center or at the top of the flower; all carpels in a flower.

Cynophore – The stipe of a pistil or carpel.

Hypanthium – The fused of coalesced bassal portion of floral parts (sepals, petals, stamens) around the ovary.

Pedicel – the flower stalk.

Perianth – An aggregation of tepals or combined calyx and corolla.

Polyphore – A receptacle or torus bearing many distinct carpels, as in *Rosa*.

Receptacle or Torus – The region at end of pedicel or on axis to which flower is attached.

Sepal – A calyx member or segment; a unit or the calyx.

Stamen – The male sporophyll within the flower; the floral organ that bears pollen in angiosperme.

Tepal – A perianth member or segment; term used for perianth parts undifferentiated into distinctive sepals and petals.

Whorl – A cyclic or acyclic group of sepals, or petals, or stamens, or carpels.

B. Flower Types
(Classification based on evolutionary flower-pollinator relationships-from)

Actinomorphic – Flowers with radial symmetry and parts arranged at one level; with definite numbers of parts and size; e.g. *Anemone, Caltha.*

Amorphic or Paleomorphic – Flowers without symmetry; usually with an indefinite number of stamens and carpels, and usually subtended by bracts or discolored upper leaves; e.g., *Salix discolor, Echinope ritro* (mostly fossil forms).

Haplomorphic – Flowers with parts spirally arranged at a simple level in a semispheric or hemispheric form; petals or tepals colored; parts numerous; e.g., *Nymphaea, Magnolia.*

Plemorphic – Actinomorphic with numbers or parts reduced; e.g., *Tripoganda.*

Storemorphic – Flowers 3-dimensional with basically radial symmetry; parts many or reduced, and usually regular; e.g, *Narcissus, Aquilegia.*

Zygomorphic – Flowers with bilateral symmetry; parts usually reduced in number and irregular; e.g., *Cypripedium, Salvia.*

VIII. PERIANTH AND HYPANTHIUM

A Perianth Parts

Anterior Lobes – The lobes away from axis, toward the subtending bract, abaxial lobes.

Anterior Ridges, Lines, Grooves – The lines, grooves, ridges in or on the dorsal side, abaxial, within the perianth.

Hood – A cover-shaped Perianth part, usually with a turned down margin.

Horn – A curved, pointed and hollow protuberance from the perianth.

Hypogynium – Perianth-like structure of bony scales subtending the ovary, as in Scleria and other members of the Cyperaceae.

Base – Bottom or lower portion.

Beard– A tuft, line or zone of trichomes.

Bristle – A stiff, strong trichome, as in the perianth of some members of the Cyperaceae.

Callosity – A thickened, raised area. which is usually hards; a *callus.*

Carina – Keel.

Claw – The long, narrow petiole-like base of a sepal or petal.

Corona – A crown; any outgrowth between the stamens and corolla which may be petaline or staminal in orighin.

Dorsal Side – Back or abaxial side, or the lower side of a perianth part.

Faucal Area – The throat area.

Fringe – The modified margin of petal, sepal, tepal or lip.

Palate – The raised area in the throat of a sympetalous corolla.

Pappus – Bristly or scaly calyx in the Asteraceae.

Petal – A corolla member or segment; a unit of the corolla.

Keel – The two united petals of a papilionaceous flower; any structure ridged like the bottom of boat.

Ligule – The strap-shaped portion of a ray or ligulate corolla.

Lip or Labellum – Either of two variously shaped parts into which a corolla or Calyx is divided, usually into an upper and lower lip, as in the Lamiaceae and Orchidaceae.

Limb – Expanded portion of corola or Calyx above the tube,throat or claw.

Lobe – Any, usually rounded, segment or part of the perianth part in the Poaceae; hyaline scales at base of ovary in the Poaceae.

Spur – A tubular or pointed projection from the perianth.

Standard, Banner, or Vexillum – The uper, usually wide petal in a papilionaceopus corolla.

Tepal – A member or segment of perianth in which the parts are not differentiated into distinct sepals and petals.

Posterior Lobe – The lobe next to axis, away from the subtending bract; adaxial lobe.

Posterior Ridges, Lines, Grooves – The lines, grooves, ridges in or on the ventral side, adaxial, within the perianth.

Pouch or Sac – A bag-shaped structure.

Scale – Small, scarious to coriaceous flattened bodies within the perianth, as in the Cyperaceae and Asteraceae.

Throat – An open, expanded tube in the perianth.

Tube – The cylindrical part of the perianth. :

Ventral Side – Top side or upper side of a perianth part.

Wing – Lateral petals, as in the Fabacea; a flattened extension, appendage or projection from a perianth part.

B. Hypanthium Parts

Base – Bottom or lower portion of the hypanthium.

Limb – Free, flared portion of the hapanthium.

Neck – Narrowed portion of hypanthium, between the base and a flared limb.

Tube or Casing – Cylindrical part or the hypanthium.

C. Perianth Types

Achlamydeous – Without perianth.

Apetalous – No petals or corolla.

Apopetalous or choripetalous – With separate petals.

Aposepalous or Chorisepalous – With separate sepals.

Asepalous – No sepals or calyx.

Chlamyudeous – With perianth.

Dichlamydeous– With perianth composed of distinct calyx and corolla.

Homochlamydeous – With perianth composed of similar parts, each part a tepal.

Sympetalous – With fused petals.

Synsepalous – With fused sepals.

D. Perianth and Hypanthium Structural Types

Actinomorphic – With radially arranged perianth parts; ray-like figure.

Bilabiate – Two-lipped, with two unequal divisions.

Caloarate – Spurred.

Calceolate – Slipper-swhaped, as in the corolla of *Cypripedium.*

Campanulate–Bell-shaped; with flaring tube about as broad as long and a flearing limb.

Carinate – Keeled.

Corniculate -

   Papilionaceous– With large posterior petal (banner or standard) two lateral petals (Wings) and usually two connate lower petals (keel); as in the Fabaceae.

   Personate – Two-lipped with the upper arched and the lower protruding into corolla throat.

   Rotate – Wheel-shaped, with short tube and wide limb at right angles to tube.

Coronate–Tubular or flaring perianth or staminal outgrowth; petaloid appendage.

Cruciate – Four separate petals in cross form.

Cucullate – Hooded.

Galeate – Helmet-shaped, as one sepal in *Aconitum.*

Gibbous – Inflated on one side near the base.

Globose – Round.

Infundibular – Funnel-shaped.

Ligulate or Ray – Strap-shaped.

Saccate – Pouch-like.

Salverform, Trumpet-shape – With slender tube and limb nearly at right angles to tube.

Subglobose – Almost round or spherical.

Tubular – Cylinirical.

Unguiculate – Clawed.

Urceolate – Urn-ahaped.

Ventricose – Inflated on one side near the middle.

IX. ANDROECIUM

A. Androecial Parts

Stamen – Male sporophyll within the flower; floreal organ that bears pollen in angiosperme.

Staminodium – Sterile stamen, may be modified as a nectary or petaloid structure.

Staminal Disc – A fleshy, elevated cushion formed from coalesced staminodia or nectaries.

B. Androecial Types
(Classification based primarily on fusion of parts)

Apostemonous – With separate stamens.

Diadelphous – With two groups of stamens connate by their filaments.

Cynandrial or Cynostemial – With fused stamens and carpels (stigma and style) as in the Orchid sceae.

Monadelphous – With one group of stamens connate by their filaments.

Petalostemonous – With filaments fused to corolla, anthers free.

Polydelphous – With several Groups of stamens connate by their filamerts.

Syngenesious – With fused anthers.

C. Stamen parts

Anther – Pollen-bearing portion of stamen.

Filament – Stamen stalk.

D. Stamen Structural Types
(Classification based primarily on structure of filament and anther)

Note – In this classification intermediate types of stamens do occur; shapes, apices, and bases of anthers should be described separately and independently of stamen type.

Appendicular– Typical stgamen with a variously-shaped or modified, protruding connective, as in *Viola*.

Laminar – Leaf-like stamen without a distinct anther and filament but with embedded or superficial microsporangia, as in *Degeneria.*

Petalantherous – With a terminal anther and distinctly petaloid filament, as in *Saxifraga*.

Petaloid – Petal-like stamen without distinct anther and filament but with marginal microsponangia, as in Magnolia *nitida.*

Filantherous or Typcial – Stamen with distinct anther and filament with or without thecal appendages, as in *Rhexia* or *Vaccinium*.

E. Anther parts

Connective – Filament extenaion between thecae.

Locule – compartment of an anther.

Pollen Grain – Young male gametophyte.

Pollen Sac – Male sporangium.

Theca – One half of anther containing two pollen sacs or male sporangia.

F. Anther Types
(Classification based on dehiscence)

Longitudinal – Dehisciang along along axis of theca.

Extrorse – Dehiscing longitudinally outward.

Introrse – Dehiscing longitudinally inward.

Latrorse – Dehiscing longitudinally and laterally.

Poricidal – Dehiscing through a pore at apex of theca.

Transverse – Dehiscing at right angles to long axis of theca.

Valvular – Dehiscing through a pore covered by a flap of tissue.

G. Anther Attachment

Dyads – Grains occurring in clusters of two.

Filiform – Thread-like.

Monad – Grains occurring singly.

Pollinia – Grains occurring in uniformcoherent masses.

Polyad – Grains occurring in groups of more than four.

Tetrads – Grains occurring in groups of four.

X. GYNOECIUM

A. Gynoecial Parts

Carpel – Female sporophyll within flower, floral organ that bears ovules in angiosperms.

Carpopodium – Short, thick, pistil, late stalk.

Locule – Overy cavity.

Ovary – Ovule-bearing part of pistil.

Placenta – Ovule-earing region of oveary wall.

Stigma – Pollen-receptive portion of pistil.

Style – Attenuatged, non-ovule-bearing portion of pistil between stigam and ovary.

B. Gynoecial Types
(Classification based on fusion)

Apocarpous – With carpels separate.

Semicarpous – With ovaries of adjacent carpels partly fused, stigmas and styles separate.

Syncarpous – With stigmas, styles, and ovaries completely fused.

Synovarious – With ovaries of adjacent carpels completely fused. Styles and stigmas separate.

Synstylovarious – With ovaries and styles of adjacent carpels completely fused, stigmas separate.

Unicarpellous or Stylodious – With solitary, free carpel in gynoecium.

C. Carpel Parts

Funiculus – Stalk by which ovule is attached to placenta.

Locule – Ovary cavity.

Ovary – Ovule-bearing part of carpel in simple ovary.

Ovule – Embryonic seed consisting of

integument (s) and nucellus.

Placenta – Ovule-bearing region of ovary wall.

Stigma – Pollen receptive portion of carpel.

Stipe– Podogyne, Carpopodium, Basal stalk.

Style – Attenaated portion of carpel between stigma and ovary.

D. Carpel Types
(Classification based on presence or absence of style and stips)

Astylocarpellous – Without a style and a stipe.

Astylocarpepodic – Without a style. With a stipe.

Stylocarpepodic – With a style and stipe.

Stylocarpellous – With a style and without a stipe, the normal carpel.

E. Stigma Types
(Classification based on shape)

Capitate – Head-like.

Clavate – Club-shaped.

Crested or Cristate – With fa terminal ridge or tuft.

Decurrent – Elongate, extending downward.

Diffuse – Spread ovar a wide surface.

Discoid – Disc-like.

Fimbriate – Fringed.

Lineate – In lines, stigmatic surface linear.

Lobed – Divided into lobes.

Plumose – Feather-like.

Terete – Cylinirical and elongate.

F. Style Types
(Classification based primarily on shape)

Astylous – Style absent.

Conduplicate – Folded with a longitudinal groove.

Cristate – Crested.

Eccentric – Off-center style.

Fimbriate – Fringed.

Flabellate – Fan-shaped.

Geniculate – Bent abruptly.

Gynobasic – Attached at base of ovary in central depression.

Heterostylous – With styles of diffrent sizes or lengths or shapes within a species.

Homostylous – With styles of same sizes or lenghts and shapes.

Involute – With margine infolded longitudinally, with groove present.

Petaloid – Petal-like.

Stylopodic – With a stylopodium or discoid base, as in the Apiaceae.

Terete – Cylindrical and elongate.

Tubercuhte – With hard, swollen, persistent base or tubercle.

Umbraculate – Umbrella-shaped, as in *Sarracenia*.

G. Ovule Parts

Chalaza – End of ovule opposite micropyle.

Embryo Sac – Female gaemtophyte.

Integuments – Outer covering of ovule; embryonic seed coat.

Micropyle – Hole through integument(s).

Nucellus– Femlae sporangium within ovule; megasporangium in seed plants.

Raphe – Longitudinal ridge on outer integument.

H. Ovule Types
(Classi fioation based on orientation of ovule body in relation to the funiculus and micropyle)

Amphitropous – With body bent or curved on both sides so that the microphyle is near the medially attached funiculus.

Antropous – With body completely inverted so that funiculus is attached basally near adjoining micropyle area.

Campylotropous – With body bent or curved on one side so that micropyle is near medially attached funiculus.

Hemianatropous or Hemitropous – With body half-inverted so that funiculus is attached near middle with micropyle terminal and at right angles.

Orthotropous or Atropous – With straight body so that funicular attachment is at one end and micropyle at other.

XI. FRUITS

A.Fruit Parts

Carpophore – Flora axis extension between adjacent carpels, as in the Apiaceae.

Ectocarp or Exocarp – Outermost layer of pericarp.

Endocarp – Innermost differentiated layer of pericarp.

Funiculus – Seed stalk.

Mericarp – A portion of fruit that seemingly matured as a separate fruit.

Mesocarp – Middle layer of pericarp.

Pericarp – Fruit wall.

Placenta – Region of attachment of seeds on inner fruit wall.

Replum – Persistent septum after dehiscence of fruits, as in the Brassiceae.

Retinaculum, Jaculator or Echma – A persistent indurated, hook-like funiculus in the fruits os Acanthaceae.

Rostellum or Beak – Persistent stylar base on fruit.

Seed – A matured ovule.

Septum or Dissepiment – Partition.

B. Fruit Structural Types
(Classification based primarily on origin, texture, and dehiscence; types grouped as simple, aggregate, multiple, accessory)

1. Simple Fruits
(Fruit derived from the ovary of a solitary pistil in a single flower)

a. Dry Indehiscent Fruit Types
(Fruits that do not split open at maturity)

Achene – A one-seeded, dry, indehiscent fruit with seed attached to fruit wall at one point only, derived from a one-loculed superior ovary.

Balausta – Many-seeded, many-loculed indehiscent fruit with a tough, leathery pericarp, as in *Punica*.

Balausta – Many-seeded, many-loculed indehiscent fruit with a tough, leathery pericarp, as in *Punica*.

Calybium – A hard one-loculed dry fruit derived from an inferior ovary, as in *Quercus*.

Nutlet – A small nut.

Samare – A winged, dry fruit.

Capsule, Indehiscent – Dry fruit derived from a two-or more loculed ovary, as in *Peplis*.

Caryopsis or Grain – A one-seeded dry, indehiscent fruit with the seed coat adnate to the fruit wall, derived from a one-loculed superior ovary.

Cypsela – An achene derived from a one-loculed, inferior ovary.

Nut – A one-seeded, dry, indehiscent fruit with a hard pericarp, usually derived from a one-loculed ovary.

Utricle – A small, bladdery or inflated, one-seeded, dry fruit.

b. Dry Dehiscent Fruit Types
(Fruits that split open at maturity)

Capsule – Dry, dehiscent fruit derived from a compound ovary of 2 or more carpels.

Diplotegium – A pyxis derived from an inferior ovary.

Follicle – A dry, dehiscent fruit derived from one carpel that splits along one suture.

Legume – A usually dry, dehiscent fruit derived from one carpel that splits along two sutures.

Loment – A legume that separates transversely between seed sections.

Silicle – A dry, dehiscent fruit derived from two or more carpels that dehisce along two sutures and which has a persistent pertition after dehiscence and is as broad as, or broader, than long.

Silique – A silicle type fruit that is

longer than broad.

C. Capsule Types
(Classification based on type of dehiscence)

Acrocidal Capsule – One that dehisces through terminal slites, or fissures, as in *Staphylea*.

Anomalicidal or Rputring Capsule – One that dehisces irregularily, as in *Ammannia*.

Bascidal Capsule – One that dehisces through basal slits or fissures, as in some species of *Aristolochia*.

Circumscissle Capsule or Pyxis – One that dehisces circumferentially, as in *Plantago*.

Denticidal Capsule – One that dehisces apically, leaving a ring of teeth, as in *Cerastium*.

Indehiscent Capsule – One that does not dehisce at maturity, as in Peplis.

Loculicidal Capsule – One that dehisces longitudinally into the cavity of the locule, as in *Epilobium*.

Operculate Capsule – One that dehisces through pores, each of which is covered by a flap, cap, or lid, as in *Papaver*.

Poricidalo Capsule – One that dehisces through pores, as in *Triodanis*.

Septicidal Capsule – One that dehisces longitudinally through the septa. as in *Penstemon*.

Valvular or Septifragal Capsule – One with Valves breaking away from the septa, as in *Ipomoea*.

d. Schizocarpic Fruit Types
(Fruits derived from a simple, two-or more-locular compound ovary in which the locules separate at fruit maturity simulating fruits derived from the ovaries of simple pistils)

Schizocarpic Achenes – Separating achenes which are one-seeded dry, indehiscent fruits with seed attached to fruit wall at one point only, derived from a superior ovary, as in sidalcea.

Schizocarpic Carcerules – Separating carcerules which are dry, few-seeded indehiscent locules, as in *Althaea*.

Schizocarpic Serries – Separating berries which have a fleshy pericarp, as in *Phytolacca*.

Schizocarpic Mericarps, Cremocarp, or Carpopodium – Separating mericarps which are dry, seed-like fruits derived from an inferior ovary, as in the Apiaceae.

Schizocarpic Follicles – Separating follicles which are dry, dehiscent fruits derived from one carpel, splitting along one suture, as in Apocynaceae.

Schizocarpic Samaras – Separating samaras which are winged, dry fruits as in *Acer*.

e. Fleshy Fruit Types

Amphisarca – A berry-like succulent fruit with a crustaceous or woody rindd as in *Lagenaria*.

Berry – Fleshy fruit, with succulent pericarp, as in *Vitis*.

Drupe – A fleshy fruit, with succulent pericarp, as in *Prunus*.

Drupelet – A small drupe, as in *Rubus*.

Hesperidium – A thick-skinned septate berry with the bulk of the fruit derived from glandular hairs, as in *Citrus*.

Peop – A berry with a leathery non-septate rind derived from an inferior ovary, as in *Cucurbita*.

Pyrene – Fleshy fruit with each seed surrounded by a bony endocaryp, as in *Ilex*.

2. Aggregate Fruit Types (Coocarpium)
(A group of separate fruits developed from carpels of one flower)

Achenecetum – An aggregation of achenes, as in *Ranunculus*.

Baccacetum or Etaerio – An aggregation of berries, as in *Actaea*.

Drupecetum – An aggregation of

drupelets, as in *Rubus.*

Follicetum – An aggregation of follicles, as in *Caltha.*

Samaracetum – An aggregation of samaras, as in *Liriodendron.*

3. Multiple Fruit Types
(Fruits on a common axis that are usually coalesced and derived from the ovaries of several flowers)

Bibacca – A fused double berry, as in *Lonicera.*

Sorosis – Fruits on a common axis that are usually coalesced and derived from the ovaries of several flowers, as in *Morus.*

Syconium – A syncarp with the achenes borne on the inside of a hollowed-out receptacle or peduncle, as in Ficus.

4. Accessory Fruit Types
(Fruits derived from simple or compound ovaries and some one-ovarian tissues, as the hypanthium; Classification arranged alphabetically; types of accessory structures given in parentheses below.)

Bur (Involucre) – Cypsela enclosed in dry involucre, as in *Xanthium.*

Goenocarpium (Various Structures) – Multiple fruit derived from ovaries, floral parts, and receptacles or many coalesced flowers, as in *Ananas.*

Diclesium (Calyx) – Achene or nut surrounded by a persistent calyx, as in *Mirabalis.*

Pseudocarp (Receptiacle) – An aggregation of achenes embedded in a fleshy receptacle, as in *Fragaria*

Pseudodrupe (Involucre) – Two-four loculed nut surrounded by a fleshy involucre, as in *Juglans.*

Glans (Involucre) – Nut subtended by a cupulate, dry involucre, as in *Quercus.*

Hip or Cynarrhodion (Receptacle and Hypanthium) – An aggregation of achenes surrounded by an urceolate receptacle and hypanthium, as in *Rosa.*

Rome (Receptacle and Hypanthium) – A berry-like fruit, adnate to a fleshy receptacle, with cartilaginous endocarp, as in *Malus.*

XII. SEEDS AND SEEDLINGS

A. Seed Parts

Aril – Outgrowth of uniculus, raphe, or integuments; or fleshy integuments or seed coat, a *sarcotesta.*

Chalaza – Funicular end of seed body.

Embryo – Young sporophyte consisting of epicotyl, hypocotyl, radicle, and one or more cotyledons.

Endosperm – Food reserve tissue in seed derived from fertilized polar muclei; or food reserve derived from megametophte in gymnosperms.

Hilum – Funiculoar scar on seed coat.

Micropyle – Hole through seed coat.

Perisperm – Food reserve in seed derived from diploid nucellus or integuments.

Raphe – Ridge on seed coat formed from adnate funiculus.

Seed Coat – Outer protective covering of seed.

B. Seed Types
(Classification based on type of nourishing tissue)

Cotylespermous – With food reserve in cotyledon, derived from zygote.

Endospermous or Albuminous – With food reserve in endosperm or albumen, derived from fertilized polar nuclei.

Hypocetylespermous or Macropodial – With food reserve stored in hypocotyl, derived from zygote.

Perispermous – With food reserve in perisperm, derived from diploid nucellus or integuments.

C. Embryo Parts

Coleoptile – Protective sheath around epicotyl in grasses.

Cotyledon – Embryonic leaf or leaves

in seed.

Epicotyl – Apical end of embryo axis, that gives rise to shhoit system.

Hypocotyl – Embryonic stem in seed, located below cotyledons.

Plumule – Embryonic leaves in seed derived from epicotyl.

Radicle – Basal end of embryo axis that gives rise to root system.

D. Embryo Types
(Classification based primarily on embryo shape, size, and position)

Bent – Foliate embryo with expanded and usually thick cotyledons in an axile position bent upon the hypocotyl in a jacknife position.

Dwarf – Axial embryo variable in size relative to seed, small to nearly total size of seed; seeds 0.2-2 nm. long.

Folded – Foliate embryo with cotyledons usually thin and extensively expanded and folded in various ways.

Investing – Axial embryo usually erect with thick cotyledons overlapping and encasing the the somewhat dwarfed hypocotyl; endosperm wanting or limited.

Lateral–Basal or baso-lateral embryo, discoid or lenticular, usually surrounded by copious endosperm.

Linear – Axial embryo several times longer than broad, straight, curved or coiled; cotyledons not expanded; endosperm present or absent.

Broad – Basal, globular or lenthicular embryo in copious endosperm.

Capitate – More or less basal head like or turbinate embryo in copious endosperm.

Micro – Axial embryo in minute seed, less than 0.2 nm. long; minute and undifferentiated to almost total size of seed.

Peripheral – Peripheral embryo large and elongate, arcuate, annular, spirolobal, or straight; cotyledons narrow or expanded; perisperm central or lateral.

Rudimentary – Basal, small non-peripheral embryo in small to large seed; relatively undifferentiated; endosperm copious.

Spatulate – Foliate, erect embryo with variable cotyledons, thin to thick and slightly expanded to broad.

Spatulate – Foliate, erect embryo with variable cotyledons, thin to thick and slightly expanded to broad.

E. Aril Structural Types and Selected Seed Surface Features

1. Aril Structural Types

Arillate – General term for an outgrowth from the funiculus, seed coat or chalaza or a fleshy seed coat.

Carunculate – With an excrescent outgrowth from integuments near the hilum, as in *Euphorbia*.

Fibrous – With 'stringy or cord-like seed coat, as mace in *Myristica*.

Funicular – With a persistent elongate funculus attached to seed coat, as in *Magnolia*.

Sarcous – With the seed coat fleshy.

Strophiolate – With elongate aril or strophiole in the hilum region.

2. Special Seed Surface Features

Alate – Winged.

Circumalate – Winged circumferentially.

Comose -With a tuft of trichomes.

Coronate – With a crown.

Crested – With elevated ridge or redges, raphal.

Umbonate – With a distinct projection usually from the side.

Verrucose – Warty.

F. Seedling Parts
(Specialized parts only)

Cataphyll – Rudimentary scale leaf produced by seedling, usually in cryptocotylar species.

Collet – External demarcation between hypocotyl and root.

Eophyll – Term applied to first few

leaves with green, expanded lamina developed by seedlings; transitional type leaves developed before formation of adult leaves.

Metaphyll – Adult leaf.

G. Seedling Types
(Classification based on position of cotyledons in germination)

Cryptocotylar or Hypogeous – With the cotyledons remaining inside the seed, seed usually remaining below ground.

Phanerocotylar or Epigeous – With the cotyledons emergent from seed, usually apearing above ground.

GENERAL CHARACTERS AND CHARACTER STATES

I. POSITION AND ARRANGEMENT

A. Location or Environmental Position

(Classification based on position of organs or parts in their surrouding environment)

1. General

Aerial or Epigeous – Above the ground or water; in the air.

Emmergent – With part (s) of plant aerial and part (s) submersed; rising out of the water above the surface.

Epipetric – Upon rock.

Epiphytic – Upon another plant.

Floating – Upon the surface of the water.

Submersed – Beneath the surface of the water.

Subterranean or Hypogeous – Below the surface of the ground.

Surfcial or Epigeous – Upon or spread over the surface of the ground.

2. Special
(Selected location terms)

Aerocaulous – With aerial stems.

Aerophyllous – With aerial leaves.

Amphicarpous – With fruits in two environments; e.g., aerial and subterranean.

Emersifolious–With emergent leaves.

Epirhizous – With roots upon another plant.

Flotophyllous – With floating leaves.

Geoflorous – With subterranean flowers.

Petrorhizous – With roots on rock.

Submersicaulous – With submersed stems.

Surcarpous – With fruits on surface of ground.

B. Position
(Classification based on location parts or organs with respect to other dissimilar parts, or organs)

1. General

Apical or Terminal – At the top, tip, or end of a structure.

Basal or Radical – At the bottom or base of a structure.

Discontinuous – Basal and lateral, basal and terminal, or lateral and terminal; not continuous.

Continuous – Basal, lateral, and terminal.

Lateral or Axillary – On the side of a structure or at the nodes of the axis.

2. Special
(Classification based on positional terms usually applicable to individual parts)

a. Androecial Position

Acrocaulous – With terminal branches.

Basicaulous – With basal branches.

Caulous – With branches more or less evenly spaced along trunk.

Subacrocaulous – With branches at or near tip of main stem.

Subbasicaulous – With branches at or near base of main stem.

Zonocaulous – With branches interminttently spaced along main stem.

C. Cotyledons Position

Accumbent or pleurorhiza – Reclinate with cotyledon edges against hypocotyl.

Incumbent or Notorhizal – Reclinate with sides of cotyledons against hypocotyl.

d. Flower, Fruit, Inflorescence, Infructescence Position

Acrocaulous – At the tip of the stem.

Amphiflorous, Amphicarpous – Flowers or fruits above and below ground, as in *Amphicarpum*.

Axillary – In axil of leaf.

Basicaulous – Near base of stem.

Cauline or Caulous – On old woody stem.

Epiphyllous – From a phyllocled or peculiar bract, as in *Trilia*.

Geoflorous, Geocarpous – Flowers or fruits below ground, as in *Amphicarpum*.

Infrafoliar – On the stem between the leaves, as in the Arecaceae.

Leaf-opposed – On stem opposite the base of the leaf, as in *Alchemilla*.

Suprafoliar – On the stem above the leaves, as in the Arecaceae.

Terminal – At or near tip of branch.

e. leaf Position

Acroramous – Leaves terminal, near apex of branch.

Aphyllopodic – Without bleadebearing leaves at base of plant.

Basiramous – Leaves on lower part of branch.

Cauline or Ramous – Leaves more or less evenly distributed on stem or branch.

Phyllopodic – With blade-bearing leaves at base of plant.

Radical -Leaves basal, near ground, usually from caudex or rootstock.

f. Ovary Position

Inferior – Other floral organs attached above ovary with hypanthium adnate to ovary.

Half-inferior – Other floral organs attached around ovary with hypanthium adnate to lower half of ovary.

Superior – Other floral organs attached below ovary.

g. Ovule Position
(Based on position of ovule in locule and orientation of the micropyle and raphe.)

Epitropous, dorsal – Ovule pendulous or hanging, micropyle above, raphe dorsal (away from ventral bundle).

Heterotropous – Ovule erect, micropyle below, raphe dorsal (away from ventral bundle).

Hypotropous, ventral – Ovule erect, microphle below, raphe ventral (toward ventral bundle).

Epitropous, ventral – Ovule pendulous or hanging, micropyle above, raphe ventral (toward ventral bundle.)

Pleurotropous, drosal – Ovule horizontal, micropyle toward ventral bundle, raphe above.

Pleurotropous, ventral – Ovule horizontal, micropyle toward ventral bundle, raphe below.

h. Perinath and Androecium Position
(Classification based on insertion of Floral Parts-Corolla, Calyx, and Androecium-the androperianth)

Epigyny – The condition in which the sepals, petals, stamens are attached to the floral trube above the ovary with the ovary adnate to the tube or hypanthium.

Epihyperigyny – The condition in which the sepals, petals, stamens are attached to the floral tuge or hypanthium surrounding the ovary; a combination perigyny and partly inferior ovary.

Epihypogyny – The condition in which the sepals, petals, stamens are attached about half-way from the base of the ovary to the partly adnate hypanthium tube; half-inferior insertion of parts.Attached to the floral or hypanthium cup above the ovary with the lower part of the hypanthium completely adnate to the ovary.

Hypanepigyny – The condition in which the sepals, petals, stamens are attached to the elongate floral tube or hypanthium above the inferior ovary, as in *Oenothera*.

Hypogyny – The condition in which the sepals, petals, stamens are attached below the ovary.

Perigyny – The condition in which the sepals, petals, stamens are attached to the floral tube or hypanthium surrounding the ovary with the tube or hypanthium free form the ovary.

1. Placenta Position (Placentation)

Axile – With the placentae along the central axis in a compound ovary with septa.

Basal – With the placenta at the base of the ovary.

Free-central – With the placenta along the central axis in a compound ovary without septa.

Laminate – With the placenta over the inner surface of the ovary wall.

Marginal or Ventral – With the placenta along the margin of the simple ovary.

Parietal – With the placentae on the wall or intruding partitions of a unilocular compound ovary.

Pendulous, Apical, or Suspended – With the placenta at the top of the ovary.

j. Radicle Position

Antitropous – With radicle pointing away from hilum.

K. Stamen Position.

Allagostemonous – Having stamens attached to petal and torus alternately.

Antipetalous – Opposite the petals.

Antisepalous – Oposite the sepals.

Epipetalous – With stamens attached to or inserted upon petals or corolla.

Episepalous – With stamens attached or inserted upon sepals or Calyx.

Syntropous – With radicle pointing toward hilum.

Cryptantherous – With stamens included.

Diplostemonous – With aments in two whorls, outer opposite the sepals, inner opposite petals.

Obdiplostemonous – With stamens in two whorls, outer opposite petals, inner opposite the sepals.

Phaneranterous – With stamens exserted.

1. Style Position

Gynobasic – At the base of an invaginated ovary.

Lateral – At the side of an ovary.

Subapical – At one side near apex or ovary.

Terminal or Apical – At the apex of the ovary.

C. Arrangement
(Classification based on location of parts in relation to each other)

Alternate – One leaf or other structure per node.

Clustered, conglomerate, Agglomerate, Crowded, Aggregate – Parts dense, usually irregularly over-lapping each other.

Symmetry of arrangement broken, with uneven lengths of internodes.

Loose, Distant, or Scattered – Parts widely separated from one another, usually irregularly.

Continuous – Symmetry of arrangement even, not broken.

Decussate – Opposite leaves at right angle to preceding pair.

Equitant – Leaves 2-ranked with overlapping bases, usually sharply folded along midrib.

Fasciculate – Leaves or other strtuctures in a cluster form a common point.

Geminate or Binate – paired; in pairs.

Imbricate – Leaves or other structures overlapping.

Opposite – Two leaves or other structures per node, on opposite sides of stem or central axis.

Polystichous – Leaves or other structures in many rows.

Rosulate – Leaves in a rosette.

Secund or Unilateral – Flowers or other structures on one side of axis.

Tetrastichous – Leaves or other structures in four rows.

Whorled, Radiate, or Verticillate – Three or more leaves or other structures per node.

2. Special
(Classification based on arrangement with special terms applicable to individual plant parts)

a. Stamen Arrangement

Didymous – With stamens in two equal pairs.

Didynamous – With stamens in two unequal pairs.

Tetradynamous – With stamens in two groups, usually four long and two short.

Tridynamous – With stamens in two equal groups of three.

b. Thecal Arrangement
(The thecae in this classification can be conjuctive or disjunctive)

Divergent – Thecae or anther cells divaricate or separated from one another at an acute anjle to the connective or fillament.

Parallel – Thecae or anther cells along side of the connective cells longitudinal to each other.

Oblique – Thecae or anther cells lower on one side of connective than the other.

Transverse or Explanate – Thecae anther cells with maximum divergence of about $90°$ from the connective of filament.

D. Orientation
(Classification based on arrangement of parts in relation to vertical angle of divergence from a central axis or point)

1. General

Acroscopic – Facing aspically.

Agglomerate, Conglomerate, Crowded, or aggregate – Dense structures with varied angles of divergence.

Antrorse – Bent or directed upward.

Assurgent – Directed upward or forward.

Basiscopic – Facing basally.

Connivent – Convergent apically without fusion.

Contorted – Twisted around az central axis; twisted.

Declinate – Directed or curved downward.

Deflexed – Bent abruptly downward.

Dextrorse – Rising helically from right to left, a characteristic of twining stems.

Inflexed – Bent abruptly inward or upward.

patent – Spreading.

Pendulous – Hanging loosely freely.

Reclinate – Bent down upon the axis, no angle of divergence.

Reflexed – Bent or turned downward.

Retrorse – Bent or directed downward.

Salient, Porrect, or Projected – Pointed outward, usually said of teeth.

Sinistrorse – Rising helically from left to right, a characteristic of twining stems.

Twining – Twisted around a central axis.

2. Special
(Classification based on stated degrees of divergence)

Appressed or adpressed – Pressed closely to axis upward with angle of divergence $15°$ or less.

Ascending – Directed upward with an angle of divergence of $16\text{-}45°$.

Depressed – Pressed closely to axis downward with angle of divertgence of 166-165⁰.

Divergent, Patent, or Divaricate – More or less horizontally spreading with angle of divergence of 15⁰ or less up or down from the horizontal.

Horizontally – Spreading outward at 90⁰ from vertical axis or plane.

Inclined – Ascending at 46-75⁰ angle of divergence.

Reclined – Descending at 106-135⁰ angle of divergence.

Resupinate – Inverted or twisted 180⁰, as in pedicels in the Orchidaceae.

E. Transverse Posture
(Classification based on position of ends of single sturcture in relation to its center or transverse axis)

Applanate or Plane – Flat, without vertical curves or bends.

Arcuate – Curved like a crescent, can be downward or supward.

Cernuous – Drooping.

Flexuous – With a series of long or open vertical curves at right angles to the central axis.

Geniculate – Abruptly bent vertically, usually near the base.**Incurved** – Curved inward or upward.

Lorate – With elongate vertical waves in the margins or sides at right angles to the longitudinal axis.

Recurved – Curved outward or downward.

Squarrose – Usually sharply curved downward or outward in the apical region, as the bracts of some species of *Aster*.

Undulate – With a series of vertical curves at right angles to the central axis.

F. Longitudinal Posture
(Classification based on position of the sides of a single structure in relation to its central axis)

Conduplicate – Longitudinally folded upward or downward along the central axis so that ventral and/or dorsal sides face each other.

Geniculate – Abruptly bent horizontally, usually in series.

Induplicate – Having margins bent inward and touching margin of each adnacent structure.

Involute – Margins or outer portion of sides rolled inward over upper or ventral surface.

Plicate – With a series or longitudinal folds; plaited.

Revolute – Margins or outer portion of sides rollled outward or downward over lower or dorsal surface.

Rolled – Sides enrolled, usually loosely, over upper or lower surfaces.

Sinuate – Long horizontal curves in the body of the structure parallel to the centralo axis.

Straight – Without a curve, bend, or angle.

Tortuous – Irregularly twisted.

Valvate – Sides enrolled, adaxially or abasially so that margins touch.

G. General Structural Position
(Pertains to regional locations on a structure)

1. General

Abaxial – Away from the axis. the lower surface of the leaf; dorsal.

Adaxial – Next to the axis; facing the stem; ventral.

Apical – At or near the tip.

Basal – At or near the bottom.

Central – In the middle or middle plane of a structure.

Cicrcumferential – At or near the Circumference; surrounding a rounded structure, of an outer face of organ; lower side of leaf; abaxial.

Marginal – pertaining to the border or edge.

Medial – Upon or along the longitudinal axis.

Peripheral – On the outer surface or edge.

Proximal – Near the point of origin or attachment.

Subbasal – Near the base.

Distal -Away from the point of origin or attachment.

Dorsal – Pertaining to the surface most distant form the axis; back.

Ventral – Pertaining to the surface nearest the axis; innewrface of an organ; the upper surface of the leaf; adaxial.

2. Special
(Selected terms for location on a structure)

Acrocaulous – At tip of stem.

Basipetiolar – At the base of the petiole.

Centroramous – At the center of the branch.

Dorsilaminar – On dorsal side of blade.

Laterospermous – On the side of the seed.

pericarpous – Around the fruit.

Suprarhizous – On top of the root.

Ventristipular – On ventral side of stipule.

H. Embryonic Position
(Position of immature organs or parts)

1. Aestivation or Prefloration
(Classification based on position of embryonic perianth parts. Calyx and corolla any have different aestivation types).

Alternate – Having structure in two rows or series os that the inner structure has its margins overlapped by a margin from each adjacent outer sturcture.

Cochleate – Having one hollow or helmet-shaped structure which encloses or covers the others.

Contorted – Having several structures in a whorl or close spiral with one margin covering the margin of an adjacent structute.

Imbricate – Having margins overlapping.

Induplicate – Having margins bent inward and touching margin of adjacent structure.

Quincuncial – Having five structuresl, two of which are exterior, two interior, and fifth with one margin covering interior structure and other margin coverd by that of one of the exterior structures.

Convolute – Having one leaf of perianth part rolled in another, usually twisted apically.

Valvate – Having margins of adjacent structures touching at edges only.

Vexillate – Having one structure larger than others which is folded over smaller enclosed structures.

2. Cotyledon Ptyxis
(Classification based on position of cotyledons in seed)

Conduplicate – Cotyledons folded lengthwise along midrib with one cotyledon covering other and inner cotyledon covering hypocotyl.

Contortuplicate – With weirdly folded corrugat cotyledons.

Diplecolobal – With incumbent cotyledons folded two or more times.

Spirolobal – With incumbent cotyledons folded times.

3. Ptyxis
(Classification based on rolling or folding of individual embryonic leaves and arangement of embryonic leaves within a structure)

Circinate – With lamina rolled from apex to base with apex in center of coil.

Conduplicate – With lamina folded once adaxially along midrib or midvein.

Implicate – With lamina folded or curved transversely near the apex.

Convolute – With one lamina enrolled in another lamina enrolled in another lamina.

Corrugate – With lamina irregularly folded in all directions, wrinkled.

Curvative or arcuate – With lamina folded transversely into an are.

Reclinate – With lamina folded or curved backwards from near its base so th at embryonbic blade is parallel to its petiole, hypocotyle, or stem.

Involute – With lamina margins enrolled adaxially.

Planate or Plain – With lamina flat, without folds or rolls.

Plicate – With many longitudinal folds in lamina.

Revolute – With lamina margins enrolled abaxially.

Supervolute – With lamina with one edge tightly enrolled covering the first, loosely convolute.

II. NUMBER AND SIZE

(Pertains to selected terms dealing with numbers.)

Ancipital – Two-edged.

Bicafrpellate – Two-carpelled.

Bidentate – Two-toothed.

Biflorous – Two-flowered.

Bifoliate – Two-leaved.

Bifoliate – Two-lipped.

Bilocular -Two-locular.

Binate – Twinned.

Biseriate – Twinned.

Bisexual -Both sexes in same flowere (monoclinous, perfect).

Diadelphous -With two stamens per flower.

Diandrous – With tow stamens per flower.

Dichasium – Cymose inflorescence in which each axis produces a pair of lateral axes.

Monadelphous – With one group of stamens connate by their filaments.

Monocarpellate – One-carpelled.

Monocarpellate – One-carpelled.

Monocephelaous – One-headed, as in composites.

Monochasium – Cymose inflorescence in which each central axis bears one lateral axis.

Monochlamydeous – With perianth composed of similar parts; each being a tepal; with one whorl of accessory parts.

Monocotyledonous   -With   one cotyledon.

Monoecious – With staminate and carpellate flowers on same plant.

Monophyllous – One-leaved.

Multicellular – Many-celled.

Dichlamydeous – With two perianth parts, a distinct calyx and corolla.

Dicotyledonous – With two cotyledons.

Diclinous – Having the stamens and carpels in separate flowers, imperfect, either monoecious or dioecious.

Dioecious – With staminate and carpellate flowers on separte plants.

Dipetrous – Two- winged.

Dyad – Pollen grains   occurring in clusters of two.

Monad – Pollen grains occurring singly.

Multicipital – With many axes or stems from one rootstock or caudex.

Multicostal – Many-ribbed.

Multilocular – Many-locular.

Multiseriate – Many-rowed; in many series.

Multistriate – Many-lined.

Pentagonal – five-angled.

Pentandrous -With five stamens.

Polyad – Pollen grains in clusters or more than fot ·.

Polyandrous -Many-stamened.

Polycarpellate – Many-carpellate.

Polycephalous – Many-headed, as in composites.

Polydelphous – With several groups of stamesn connate by their filaments.

Tetrad – Pollen grains in clusters of four.

Trifoliate – Three-leaved.

Tetragonal – four-angled.

Trifoliolate -With three leaflets.

Tetrahedral – Having the form of a tetrahedron.

Tetralocular – Four-locular.

Tetrandrous -With four stamens.

Tricarpellate – Three-carpellate.

Triflorous – Three-flowered.

Trigonous – Three-angled.
Triquetrous – Three-angled with the sides usually concave.
Unilocualr – One-locular.
Uniseriate – One-rowed; in one series.
Unisexual – With only one sex in each flower.
Ampliate – Enlarged; dilated.
Angustate – Narrow.
Anisocarpous – With unequal carpesl.
Anisocotylous – With unequal cotyledons.
Anisolateral – With unequal sides.
Anisopetalous – With unequal petals.
Anisophllous – With unequal leaves.
Anisostylous – With unequal styles.
Depauperate – Small and usually poorly developed.
Dilated – Widened; expanded.
Dwarf – Very small.
Gigantic – Very large.
Heterandrous – With stamens of different sizes and/or shapes.
Heterobalsty – With juvenile foliage distinctly different from adult foliage in size or shape.
Heterocarpous -With carpels of different sizes and/or shapes.
Heterocladous – With stems of different sizes and/or shapes.
Heteropetalous – With petals of different sizes and/or shapes.
Heterophyllous – With leaves of different sizes and/or shapes.
Heterosepalous – With sepals of different sizes and/or shapes.
Heterostichus – With unequal rows.
Homandrous – With stamens of same size and shape.
Homocarpous – With carpels of same size and shape.
HYpophyllous – With small leaves, as bracta scales, cataphylls.
Inequilateral – With unequal sides.
Isocotylous – With cotyledons of same size and shape.
Isodynamous – With equally developed structures.

Isopetalous – With petals of same size and shapes.
Isophyllus – With leaves of same size and shape.
Isosepalous – With sepals of same size and shape.
Isostichous – With equal rows.
Latiflorous – With broad-flowers.
Leptophyllous – With leaves to 25 wq. mm. in size .
Major – Greater in size.
Minor – Smaller in size.
Minute – Very small.
Nanophyllous – With leaves to 225 sq. mm. in size.
Platycanthous – With flat and usually large spines.
Reduced – Decreased in size.
Robust – Large.

C. Cycly
(Pertains to number of whorls of floral parts, leaves, or stems)

Acarpous – No carpels or carpellate whorl; no pistil.
Achlamydeous – Without perianth.
Apetalous – No petals or corolla.
Aphyllous – Without leaves, no whorls of leaves.
Arihizous – Without roots, no whorls of roots.
Asepalous – No sepals or calyx.
Astemonous or Arandrous – No stamens or androecium.
Chlamydeous – With perianth.
Complete – With four types of floral parts.
Dichlamydeous– With perianth composed of distinct calyx and corolla.
Dicyclic – Two-whorled.
Homochlamydeous – With perianth composed of similar parts, each part a tepal.
Incomplete – One or more types of floral parts absent.
Monocyclic – One-whorled.
Oligotaxy – Reduction in number of whorls.
Pentacyclic – Five-whorled.
Pleiotaxy – Increase in number of whorls.

Polycyclic – Many-whorled.

Tetracyclic – Four-whorled.

Tricyclic – Three whorled.

D. Merosity

(Pertains to number of parts within whorls of floral parts, leaves, or stems)

Dimerous – Whorl with two members.

Heteromerous or Anissomerous – With different number of members in different whorls.

Isomerous – With same number of members in different whorls.

Monomerous – Whorl with one member.

Oligomerous – With reduction in number of members within whorl.

Pentamerous – Whorl with five members.

Polymerous – Whorl with many members.

Pseudomomnomerous – Whorl seemingly with one member which is a fusion product to two or more parts.

Tetramerous – Whorl with four members.

Trimerous – Whorl with three members.

E. Fusion

(Pertains to fusion of members within and between whorls of floral parts)

1. General

Adherent – With unlime parts of organs joined, but only superficially and without actual histological continuity.

Adnate – With unlike parts or organs integrally fused to one another with histological continuity.

Coalesced – With like or unlike parts or organs incompletely separated; parti ally fused in a more or less irregular fashion.

Coherent – With like parts or organs joined, but only superficially and without actual histological continuity.

Contiguous – Touching but not adnate, connate, adherent, or coherant.

Distinct – With like parts or organs unjoined and separate from one another.

Fasciated – Unnaturally and often monstrously connate or adnate, the coalesced parts often unnaturally proliferated in size and/or number; e.g., inflorescence of *Celosia*.

Free – Unlike parts or organs unjoined and separate from one another.

2. Special

(Selected terms pertaining to fusion)

Anthocarpous – Having a body of combined floral and fruit parts, as in multiple fruits.

Apocarpous – With separate carpels.

Apopetalous or Choripetalous – With separate petals.

Aposepalous or Chorisepalous – With separate sepals.

Apostemonous – With separate stamens.

Column, Gynostemium or Gynandrium – With fused stamens and carpels (stigma and style) as in *Orchis*.

Conjugate – Fused pairs, as the fruits or *Lonicera*.

Diadelphous – With two groups of stamens connate by their filaments.

Hypanthium – Fused floral parts forming an envelope around the ovary.

Monadelphous – With one group of stamens connate by their filaments.

Petalostemonous – With filaments fused to corolla, anthers free.

Polydelphous – With several groups of stamens connate by their filaments.

Sympetalous – With fused petals.

Syncarpous – With fused carpels.

Syncotyly – Cotyledons coalesced, forming a funnel or trumpet.

Synsepalous – With fused sepals.

Syngenesious – With fused anthers.

3. Hypanthium Adnation
(Based on fusion with ovary)

Absent – No hypanthium present.

Adnate – Hypanthium competely fused to ovary.

Free – Hypanthium surrounding but competely free from the ovary.

Free-adnate – Hypanthium fused with ovary and having a free limb around or above ovary.

Partly adnate – Hypanthium adnate to part of the ovary and with no free limb or tube.

F. Division

Bifid – Cut or divided into two lobes or parts.

Bifurcate – Divided into two forkes or branches.

Cleft – Cut 1/4-1/2 of distance of middle or structure.

Dichotomous – Divided into two equal parts.

Dimidiate – Divided into unequal halves.

Dissected – Irregularly cut into numberous seqments.

Divided – Cut 3/4 to almost entire distance to middle of structure.

Furcate – Forked.

Incised – Margins sharply and deeply cut, usually jaggedly.

Lacerate – Irregularly cut, appearing torn.

Laciniate – Cut into closely parallel ribbonlike or straplike projections.

Lobed – Round-toothed, cut 1/8-1/4 of distance to middle of structure.

Palmatifid – Cut palmately.

Palmatisect – Sectioned or divided palmately into distinct segments.

Parted – Cut 1/2-3/4 of distance to middle of structure.

Pectinate – Having closely parallel toothlike projections; comblike.

Pedate, Bipalmate – palmately cleft or divided with lateral lobes cleft

or divided.

Pinnatifid – Cut pinnately.

Pinnatisect – Sectioned or divided pinnately into distinct segments.

Quadrifid – Cut or divided into four lobes or parts.

Septate – Divided by internal partitions into locules or cells.

Serrated – Cut into sawlike teeth.

Trifid – Cut or divided into three lobes or parts.

Trifurcate – Divided into three forks or branches; three-forked.

III. SHAPES

A. Shapes-Plane and solid

1. Symmetric Figures.

A. **Elliptic** – With widest axis at midpoint of structure and with margins symmetrically curved.

Plane L/W

a. Narrowly elliptic more than 6:1-3:1
b. Elliptic 2:1-3:2
c. Widely elliptic 6:5
d. Circular 1:1
e. Oblate 5:6
f. Transversely elliptic 2:3-1:2
g. Narrowly transversely elliptic 1:3-1:6 or more

Solid L/D

a. Narrowly ellipsoid more than 6:1-3:1
b. Ellipsoid 2:1-3:2
c. Broadly ellipsoid 6:5
d. Spheroid 1:1
e. obloid 5:6
f. Transversely ellipsoid 2 : 3-1 :2
g. Lenthicular 1:3-1:6 or more

B. **Oblong.** With wides axis at midpoint of structure and with margins essentially parallel.

Plane L/W

a. Linear more than 12:1
b. Narrowly oblong 6:1-3:1 or lorate
c. Oblong 2:1-3:2

d. Widely oblong 6:5
e. Square 1:1
f. Transversely widely oblong 5:6
g. Transversely narrowly oblong 1:3-1 :6
h. Transversely narrowly oblong 1:3-1 :6
i. Transversely linear 1:12 or more

Solid L/D

a. cylindric or terete more than 12:1
b. Narrowly oblong 6:1-3 :1
c. Oblong 2:1-3:2
d. Broadly oblong 6:5
e. Cubical 1:1
f. Transversely broadly oblong 5:6
g. Transversely oblong 2:3-1 :2
h. Transversely narrowly oblong 1:3-1 :6
i. Transversely Cyllindrical or terate 1:12 or more

C. Ovate. With widest axis below middle and with margins symmetrically curved; egg-shaped.

a. Lanceolate more than 6:1-3: 1
b. Ovate 2:1-3:2
c. Widely ovate 6:5
d. Very widely ovate 1:1
e. Widely depressed ovate 5:6
f. Depressed ovate 2:3-1:2

a. Lanceoloid more than 6:1-3:1
b. Ovoid 2:1-3:2
c. Broadly ovoid 6:5
d. Very broadly ovoid 1:1
e. Broadly depressed ovoid 5:6
f. Depressed ovoid 2:3-1 :2

D. Obovate Inversely ovate.

a. Oblanceolate more than 6:1-3:1
b. Obovate 2:1-3 :2
c. Widely obovate 6:5
d. Very widely obovate 1:1
e. Widely depressed obovate 5:6
f. Depressed obovate 2:3-1:2

a. Oblanceolid more than 6:1-3:1
b. Obovoid 2:1-3:2
c. Broadly obovoid 6:5

d. Very broadly obovoid 1:1
e. Broadly depressed obovoid 5:6
f. Depressed obovoid 2:3-1:2

E. Rhombic. With widest axis at midpoint of structure, and with margins; elliptic but margins straight and middle angled.

Plane L/W

a. Narrowly rhombic more than 6:1-3:1
b. Rhombic 2:1-3:2
c. Widely rhombic 6:5
d. Quadrate rhombic 1:1
e. Transversely widely rhombic 5:6
f. Trnasversely rhombic 2:3-1:2
g. Narrowly transversely rhombic 1:2-1:6 or more

a. Narrowly rhomboid more than 3:1
b. Rhomboid 2:1-3:2
c. Broadly rhomboid 6:5
d. Quadrate rhomboid 1:1
e. Transversely broadly rhomboid 5:6
f. Transversely rhomboid 2:3-1:2
g. Narrowly ·transversely rhomboid 1:3-1:6 or more

F. Trullate. With widest axis below middle and with straight margins; ovate but margins straight and angled below middle; trowel-shaped.

plane L/W

a. Narrowly trullate more than 6:1-3:1
b. Trullate 2:1-3:2
c. Widely trullate 6:5
d. Very widely trullate 1:1
e. Widely depressed trullate 5:6
f. Transversely depressed trullate 2:3-1:2

Solid L/D

a. Narrowly trulloid more than 6:1-3:1
b. Trulloid 2:1-3:2
c. Broadly trulloid 6:5
d. Very broadly trulloid 1:1
e. Broadly depressed trulloid 5:6
f. Transversely depressed trulloid 2:3-1:2

G. Obtrullate. Inversely trullate.

Plane L/W

a. Narrowly obtrullate more than 6:1-3:1
b. Obtrullate 3:2-2:1
c. Widely obtrullate 6:5
d. Very widely obtrullate 1:1
e. Widely depressed obtrullate 5:6
f. Transversely depressed obtrullate 2:3-1:2

Solid L/D

a. Narrowly obtrulloid more than 6:1-3:1
b. Obtrulloid 3:2-2:1
c. Broadly obtrullid 6:5
d. Very broadly obtrulloid 1:1
e. Broadly depressed obtrulloid 5:6
f. Transversely depressed obtrulloid 2:3-1:2

H. Triangular. With three sides and three angles.

Plane L/W

a. Linear-triangular more than 12:1
b. Narrowly triangular 6:1-3:1
c. Triangular 2:1-3:2
d. Widely deltate 6:5
e. Deltate 1:1
f. Shallowly deltate 5:6
g. Shallowly triangular 2:3
h. Very shallowly trangular 1:3-1:6 or more.

Solid L/D

a. Subulate more than 12:1
b. Narrowly Pyramidal 6:1-3:1
c. Traingular 2;1-3:2
d. Broadly deltoid 6:5
e. Deltoid 1:1
f. Shallowly deltoid 5:6
g. Shallowly Pyramidal 2:3-1:2
h. Very shallowly pyramidal 1:3 1:6 or more.

I. Obtriangular. Inversely triangular.

Plane L/W

a. Linear-obtriangular or narrowly cuneate more than 12:1
b. Cuneate 6:1-3:1
c. Obtriangular 2:1-3:2 or widely cuneate
d. Widely obdelate 6:5
e. Obdeltate 1:1
f. Shallowly obdeltate 5:6
g. Shallowly obtraingular 2:3-1:2
h. Very shallowly otriangular 1:3-1:6 or more

Solid L/D

a. Linear-obpyramidal or narrowly cuneiform more than 12;1
b. Cuneiform 6:1-3:1
c. Obpyramidal or broadly cuneiform 2:1-3:2
d. Broadly obadeltoid 6:5
e. Obdeltoid 1;1
f. Shallowly obdeltoid 5:6
g. Shallowly obpyramidal 2:3-1:2
h. Very shallowly obpyramidal 1:3-1:6 or more

2. Special Plane Figures-Outline

Acicular – Needlelike, round or grooved in cross section.
Auriculiform – Usually obovate with two small rounded, basal lobes.
cordiform – Heart-shaped.
Dimidiate – Inequilateral with one-half wholly or nearly wanting.
Falcate – Scimitar-shaped.
Dimidiate – Inequilateral with one-half wholly or nearly wanting.
Falcate – Scimitar-shaped.
Filiform– Threadlike, usually flexuous.
Hastiform – Triangular with two flaring basal lobes.
Lunate – Crescent-Shaped, with acute ends.
Lyrate – Lyre-shaped; pinnatifid with large terminal lobe and smaller lower lobes.
Obcordiform – Inversely cordiform.
Panduriform – Fiddle-shaped; obovate with sinus or indentation on

each side near base and with two small basal lobes.

Peltiform – Rounded with petiole attached to center of blade or apparently to laminar tissue.

Rectangular – Box-shaped, longer than wide.

Reniform– Kidney-shaped, with shallow sinus and widely rounded margins.

Runcinate – Oblanceolate with lacerate to parted margins.

Sagittiform – Triangular-ovate with two straight or slightly incurved basal lobes.

Spathulate or Spatulate – Oblong or obovate apically with a long attenuate base.

Acerose – Needle-shaped; sharp.

Annular – Ring-like.

Arcuate – Bent like the arc of a circle.

Botuliform – Sausage-shaped.

Capillate – Hair-shaped.

Capitate – Head-like.

Clavate – Club-shaped.

Cochleate – Snail-shaped.

Compressed or Complanate – Flattened.

Conical – having figure of true cone.

Coroniform -Crown-shaped.

Cotyliform -Cup-shaped.

Crateriform – Shallow cup-shaped as the involucre of some species of *Quercus*.

Cruciform or Cruciate -Cross-shaped.

Cylindric -Long-tubular.

Cymbiform – Boat-shaped.

Discid – Orbicular with convex face.

Dolabriform – Axe-shaped.

Excentric – One-sided; off-center.

Falcate or Seculate – Sickle-shaped.

Fistulose – Solow, as a culm without pith.

Flabelliform – Fan-shaped.

Fusiform – Spindle-shaped; broadest in middle and tapering to each end.

Half-terete – Flat on one side, terete on other; semicricular in in cross section.

Hippocrepiform – Horseshoe-shaped.

Lenticular – Biconvex, usually elongate and flattish.

Lingulate – Tongue-shaped, plano-convex in cross section.

Meniscodal – Thin and concave-convex.

Napiform – Turnip-shaped.

Navicular – Boat-shaped.

Nodiform or Nodulose – Knotty or knobby, as the roots of most of the Fabaceae.

Obconic – Inversely conical.

patelliform -Knee-shaped; disk-shaped.

pisiform – Pea-shaped.

Pyriform – Pear-shaped.

Rectangular – Boxlike, longer than wide.

Spiral – Twisted like a corkscrew.

Stellate – Star-shaped.

Strombiform –Elongate snail-shaped.

Terete – Cylindrical.

Torose – cylindrical with contractions at intervals.

Turbinate – Top-shaped; obconic.

Turgid – Tumid or swollen.

Umbilicate – Depressed in the center.

Umbonate – Round with a projection in center.

Umbraculiform – Umbrella-shaped.

Vermiform – Worm-shaped.

B. Apices and Bases
(Leaves, petals, sepals, scales, bracts or other flattened structures)

1. Apices and Bases with Sinuses:

Auriculate – Lobe rounded; sinus depth variable; outer margin concave, inner convex or straight.

Cordate – (Apex obcordate) Lobe rounded; sinus depth 1/8-1/4 distance to midpoint of blade; margins convex and/or straight.

Cleft – Lobe rounded; sinus depth 1/4-1/2 distance to midpoint of blade; margins convex and/or straight.

Emarginate – Lobe rounded; sinus depth 1/16-1/8 distance to midpoint of blade; margins straight or convex.

Hastate – Lobe pointed and oriented outward or divergent in relation to petiole or midrib; sinus depth variable; margins variable.

Lobate – Lobe rounded. sinus depth variable; outer and inner margins concave.

Obtuse – Margins straight to convex, forming a terminal angle more than 90^0.

Reniform – Lobe rounded; sinus depth variable; outer margin convex or straight, inner concave.

Retuse – Lobe rounded; sinus depth to 1/6 distance to midpoint of blade; margins convex.

Rounded – Margins and apex forming a smooth are. .

Sagittate – Lobe pointed and oriented downward or inward in relation to petiole or midribe; sinus depth variable; margins variable.

Truncate – Cut straight across; ending abruptly almost at right angles to midrib or midvein.

2. Apices and Bases without Sinuses:

Acute (Base cuneate) – Margins straight to convex forming a terminal angle $45^0 - 90^0$

Acuminate (Base narrowly cuneate) – Margins straight to convex forming a terminal angle of less than 45^0

Caudate (Base attenuate) – Acuminate with concave margins.

Cuspidate – Acute but coriaceous and stiff.

Spinose of Pungent – Acuminate but coriaceous and stiff.

3. Apices with Midrib, Midvein or Vein Extension :

Apiculate – More than 3:1 l/W, usually slightly curled and flexuous.

Aristate -More than 3:1 l/w, usually prolonged, straight and stiff.

Cirrhous – More than 10:1 l/w, coiled and flexuous.

Mucronate – Less than 3:1 l/w, straight and stiff.

Mucrounulate – 1:1 l/w, or broader than long; straight.

Muticous – Without a vein extension, awn or hair.

Piliferous – More than 20:1 l/w, hairlike, flexuous.

4. Specialized Bases and Leaf Attachments:

Amplexicaul – Completely clasping the stem.

Clasping – Partly surrounding the stem.

Connate or Connate-perfoliate – Having bases of opposite leaves fused around the stem.

Decurrant – Extending along stem downward from leaf base.

Ligulate – Having a tongue-like outgrowth at base of blade or top of sheath.

Oblique – Having an asymmetrical base.

Ocreolate – Having a stipular tube surrounding stem above insertion of petiole or blade.

Ocreolate – Diminutive of ocreate; usually applied to bract bases.

Peltate – Usually havinbg petiole attached near the centesr on the underside of blade.

Perfoliate – Having base completely surrounding stem.

Petiolate – with a petiole.

Sessile – Without a petiole.

Sheathing – Having tubular structure enclosing stem below apparent insertion of blade or petiole.

Surcurrent – Extending along stem-supward from leaf base.

C. Margins
(Leaves, petals, sepals, bracts, scales or other flattened structure)

Note – For precison in margin description the type (as described below), the symmetry of the individual tooth, the margins of the individual tooth, the apex of the individual tooth, the type of sinus (rounded or angled), the number of teeth per unit of margin measurement, the spacing (reqular or irregular) of the teeth, the nature of teeth (simple or compound in two or more size groups) should be indicated or described.

Margin Types:

Aculeate – Prickly.

Bicrenate or Doubly-crenate – With smaller rounded teeth on larger rounded teeth.

Biserrate or Doubly-serrate – With sharply cut teeth on the margins of larger sharply cut teeth.

Ciliate – With trichomes protruding from margins.

Cleft – Indentations or incisions cut 1/4-1/2 distance to midrib or midvein.

Crenate – Shallowly ascending round-toothed, or teeth obutuse; teeth cut less than 1/8 Way to midrib or midvein.

Crenulate – Diminutive of crenate, teeth cut to 1/16 distance to midrib or mid-vein.

Crispate – Curled; margins divided and twisted in more than one plane.

Dentate – Margins with rounded or sharp, coarse teeth that poin outwards at right angles to midrib or midvein, cut 1/16 to 1/8 distance to midrib or midvein.

Denticulate – Diminutive of dentate, cut to 1/16 distance to midrib or midvein.

Divided – Indentations or incisions cut 3/4-almost completely to midrib or midvein.

Entire – Withotu indentations or incisions or margins; smooth.

Erose – Irregularly, shallowly toothed and/or lobed margins; appearing gnawed.

Filamentose or Filiferous – With coarse marginal fibers or threads.

Fimbriate – Margins fringed.

Fimbriolate – Minutely fimbriate.

Incised – Marging sharply and deeply cut, usually jaggedly.

Involute – Margins rolled inward.

Lacerate – Margins irregularly cut, appearing torn.

Laciniate – Margins cut into ribbon-like segments.

Lobed – Large, round-toothed, cut 1/8-1/4 distance to midvein.

Palmatifid – Cut palmately.

Parted – Indentations or incisions cut 1/2-3/4 distance to midrib.

Pinnatifid – Cut pinnately.

Repand – Sinuate with indentions less than 1/16 distance to midrib or midvein.

Retrorsely Crenate – rounded teeth directed toward base.

Retrorosely Serrate – Sharp or pointed teeth directed toward base.

Revolute – Margins rolled under.

Serrate – Saw-toothed; teeth sharp and ascending, but cut 1/16-1/8 distance to midrib or midvein.

Serrulate – Diminutive of serrate, but cut to 1/16 distance to midrib or midvein.

Sinuate – Margins shallowly and smoothly indented, wavy in a horizontal plane, without distinctive teeth or lobes, indented 1/16-1/8 distance to midrib or mid-vein.

Undulate – Margins shallowly and smoothly indented, wavy in a vertical plane.

IV.
SURFACE-VENATION-TEXTURS

A. Configuration of Surface
(Classification based on patterns of configuration)

Aciculate – Finely marked as with pin pricks, fine lines usually raniomly arranged.

Alate – Winged.

Alveolate – Honey-combed.

Areolate – Divided into many angular or squarish spaces.

Bullate – Puckered or blistered.

Canaliculate – Longitudinally grooved, usually in relation to petioles or midribs.

Cancellate or Clathrate – Latticed.

Corrugate – Redged.

Costate – Coarsely ribbed.

Fenestrate – With windowlike holes throught the leaves or other structures.

Flexuous – Coarsely undualte with folds at right to long axis.

Fovelate – Pitted.

Plicate or Plaited – Fluted, longitudinally folded.

Punctate – Covered with minute impressions or depressions.

Pustulate – With scattered blister-like swellings.

Reticulate -Netted.

Ribbed – With longitudinal nerves.

Ringed – With old bud scale scar rings.

Rugous – Covered with coarse reticulate lines.

Ruminate – Coarsely wrinkled, appearing as chewed.

Scarred – With old leaf base, stipular and/or branch scar regions.

Smooth or Pnane – Without configuration.

Striate – With longitudinal lines.

Striate – With longitudinal grooves.

Tortuous – Having the surface variously twisted.

1. General

Acrodromous – With two or more primary or strongly developed secondary veins diverging at or above the base of the blade and running in convergent arches toward the apex over some or all of the blade lengh, the arches not basally curved.

Actinodromous – With three or more primary veina diverging radially form a single point at or above the base of the blade and running toward the margin, reaching it or not.

Brochidodromous – With a single primary vein, the secondary veins not termina ting at the margin but joined together in a series of prominent upward arches or marginal loops on each side of the primary vein.

Campylodromous – With several primary veins or their brancehs diverging at or close to a single point and running in strongly developed, basally recurved arches which converge toward the apex, reaching it or not.

Cladodromous – With a single primary vein, the secondary veins not erminating at the margin and freely remified toward it.

Craspedodromous, Mixed – With a single primary vein, some of the secondary veins terminating at the margin and an aproximately equal number otherwise.

Craspedodromous, Simple – With a single primary vein, all of the secondary veins and their branches terminating at the margin.

Eucamptodromous – With a single primary vein, the secondary veins curved upward and gradalluy diminishing distally within the margin and interconnected by a series of cross-veins without forming con-spicous marginal loops.

Hyphodromous – With a single primary vein and all other venation absent, rudimentary, or concealed within a coriaceous or fleshy blade.

Palinacrinodromous – Actinodromous, the primary veins with one or more subsidiary radiations above the primary one.

Parallelodromous – With two or more primary veins originating beside one another at the blade base and running more or less parallel to the apex where they converge.

Reticulodromous – With a single primary vein, the secondary veine not terminating at the margin and losing their identities near the margin by repeated branching, yielding a dense reticulum.

Semicraspedodromous – With a

single primary vein, the secondary veins branching just within the margin, one braqnch form each terminating at the other forming a marginal loop and joining the superadjacent secondary vein.

2. Special

(Traditional Classification)

Dichotomous – With veins branching or forking in pairs equally.

Netted or Reticulate – With veins forming a network.

Prinnately Netted – With secondary veins arising from midrib or midvein.

Palmately or Digita⸱   ⸱ Netted – With three or more p⸱ mary veins arising from a common point.

Parallel – With veins extending from base to apex, essentially parallel.

Penni-parallel – With veins extending from midrib to margins, essentially parallel.

ANATOMY

A

Abaxial – That surface of any structure which is remote or turned away from the axis.

Absciss – Layer of meristematic cell just outside cork layer, to whom fall of leaves, floral parts, fruits and certain branches is due.

Abscission – The separation of parts.

Abscission layer – Layer at base of leaf stalk in woody dicotyledons and gymnosperms, in which the parenchyma cells become separated from one another through dissolution of middle lamella before leaf-fall.

Accessory bud – An additional axillary bud; a bud formed on a leaf.

Accessory cells – Auxiliary cells.

Acicular – Like a needle in shape; sharp pointed.

Acrogenous – Increasing in growth at summit or apex.

Acropetal – Development of organs in succession towards apex, the oldest at base, youngest at tip *e.g.*, leaves on a shoot.

Acuminate – Drawn out into long point; tapering; pointed.

Adaxial – Turned towards the axis.

Adenine – A compound occurring in many cells; $C_5H_5N_5$.

Adventitious – Tissues and organs arising in abnormal positions; secondary.

Aerenchyma – Cortex of sumberged roots of certain swamp plants; aerating cortical tissue in floating portions of some aquatic plants.

Aerial – Roots growing above ground, from stems.

Aerophyte – A plant growing attached to an aerial portion of another plant; epiphyte.

Agranular – Without granules.

Air-cells – Air spaces in plant tissue.

Alburnum – Sap-wood or splint-wood, soft white substance between inner bark and true wood; outer young wood of dicotyledon.

Aleurone – Protein grains found in general protoplasm, and used as reserved food material; aleurone.

Aleuroplast – Colourless plastid, storing protein.

Alkaloid – Basic nitrogenous organic substance with poisonous or medicinal properties produced in certain plant species as caffeine, morphine, nicotine etc.

Ameiosis – Occurrence of only one division in meiosis instead of two.

Ameiotic – Parthenogenesis in which meiosis is suppressed.

Amino acids – Compounds containing amino (NH_2) and carboxyl (COOH) groups, constituents of proteins, synthesized in autotrophic organisms.

Amitosis – Direct cell-division and cleavage of nucleus without thread-like formation of nuclear material.

Amphicribral – Amphiphloic.

Amphicribral bundle – Concentric bundles are amphicribral, when phloem surrounds xylem, *e.g.*, in some ferns.

Amphiphloic – With phloem both peripheral and central to xylem.

Amphivasal – With primary xylem surrounding or on two sides of amphicribral, amphiphloic, periphloic.

Amphivasal bundle – Concentric

bundles are amphivasal, when xylem surrounds phloem, *e.g.*, in *Dracaena*.

Amyloplast – A leucoplast or colourless starch forming granule in plants; amyloplastid.

Amylum – Vegetable starch.

Anaphase – A stage in mitosis during divergence of daughter chromosomes; the stages of mitosis upto division of chromatin into chromosomes; kataphase.

Anatomy – The science which treats of the structure of plants and of animals, as determined by dissection.

Angienchyma – Vascular tissue.

Angiosperm – Having seeds in a closed case, the ovary.

Angular – Collenchyma with cellwalls thickened in the angles of the cells.

Annual ring – One of the rings, seen in transverse sections of dicotyledons, indicating the secondary growth during a year.

Annular – Ring-like, certain vessels in xylem, owing to rint-like thickenings in their interior.

Anomaly – Any departure from type characteristics.

Anthocyanin – One of the blue, reddish or violet pigments of flowers, leaves, fruits and stems.

Anticlinal – Line of division of cells at right angles to surface of apex of a growing point.

Apical – At tip; cell at tip of growing point; meristem.

Apical meristem – Growing point (zone of cell division) at tip of root and stem in vascular plants, having its origin in a single cell (initial), *e.g.*, Pteridophyta, or in a group of cells (initials), *e.g.*, Spermatophyta.

Apposition – The formation of successive layers in growth of a cell wall.

Aquatic – Living in water. An aquatic plant.

Arachnoid – Consisting of entangled hairs.

Artefact – An appearance, or apparent structure, due to preparation and not natural.

Articulate – Jointed; articulated.

Astrosclereid – A multiradiate sclereid or stone cell.

Autoplast – Chloroplast.

Axial – Axis or stem.

Axillary – Growing in axil, as buds.

Axis – The main stem or central cylinder.

B

Bark – The tissues external to the vasculr cambium, collectively; pholem, cortex and periderm; outer dead tissues and cork.

Bast – The inner fibrous bark of certain trees.

Bicollateral – Having the two sides similar, *e.g.*, a vascular bundle with phloem on both sides of sylem, as in Cucrbitaceae and Solanaceae.

Bifacial – Leaves with distinct upper and lower surfaces; dorsiventral.

Bilateral – Having two sides symmetrical about an axis.

Blind pit – A cell wall pit which is not backed by a complementary pit.

Biometry – Application of mathematics to the study of living things.

Bordered pit – A form of pit developed on walls of tracheids and wood-vessels, with overarching border of secondry cell-wall.

Botany – The branch of biology dealing with plants; phytology.

Brachysclereid – A stone-cell.

Branch gaps – Gaps in the vascular cylinder of a main stem, subtending branch traces.

Branch traces – The vascular bundles connecting those of a main stem of those of a branch.

Brownian movements – (R. Brown. Scottish botanist). The passive vibratory movements of fine particles when suspended in a fluid.

Bud – A rudimentary shoot, or

flower.

Bulbous – Like a bulb.

Bulliform – Thin-walled cell which cause rolling, folding or opening of leaves by turgor changes.

Bundle-sheath – A layer of large parenchymatous cells surrounding vascular tissue of leaf-vein.

C

Callose – An occasional carbohydrate or periodic component of plant cell walls; as on sieve-plates.

Calles – Tissue that forms over cut or damaged plant surface; deposit of callose on sieve-plates.

Cambial – Pertaining cambium cells.

Cambium – The tissue from which secondary growth arises in stems and roots.

Canada balsam – Gum commonly used, dissolved in xylene, for making permanent microscopical preparations. The object is placed in a thin layer of balsam solution between cover-slip and slide. The balsam dries hard, and because its refractive index is like that of proteins and other constituents of biological objects, it makes them very transparent.

Capitate – Swollen at tip.

Carbohydrates – Compounds of carbon, hydrogen and oxygen, aldehydes or ketones constituting sugars, or condensation products thereof.

Carotene – A yellow pigment synthesized by plants.

Carotenoids – Pigments occurring in plants, including carotenes xanthophylls and other fat-soluble pigments.

Cassparian band – (R. Caspary. German botanist). A cork or wood-like strip encircling radial walls of endodermis cells; Casparian strip.

Cauline – Vascular bundles not passing into leaves.

Cell – A small cavity or hollow; a unit mass of protoplasm, usually containing a nucleus or nuclear material; originally, the cell-wall.

Cell Division – Division of cell] both cytoplasm and nucleus, into two.

Cell lineage – Developmental history in terms of descent by cell division of later cells from earlier cells. The cell lineage of an organ traces the succession of cells, from the zygote onwards, which culminates in the group-of cells constituting that organ.

Cell membrane – A bimolecular layer of lipoids and proteins enveloping the protoplasm of a cell; plasma membrane; ectoplast and tonoplast of a plant cell.

Cell plate – Equatorial thickening of spindle fibres from which partition wall arises during division of plant cells.

Cell sap – Fluid in vacuoles os plant cell.

Cell theory – Theory, initiated by Schleiden and Schwann, 1838-39; taht all animals and plants are made up of cells and their products, and that growth and reproduction are fundamentally due to division of cells.

Cellular – Consisting of cell.

Celluose – A carbohydrate forming main part of plant cell-walls. $(C_6H_{10}O_5)x$.

Cell-wall – Investing portion of cell.

Cement – A uniting substance, as between cells or animals.

Central body – Centrosome.

Central cylinder – Stele.

Centrifugal – Turning of turned away from centre of plants or axis.

Centriole – The central particle of the centrosome; the centrosome itself.

Centripetal – Turning or turned towards centre or axis.

Centromere – The part of the chromosome located at the point lying on the equator of the spindle at metaphase, and dividing at anaphase controlling chromosome activity.

Centrosome – A cell-organ, the centre of dynamic activity in mitosis, consisting of centriole and attraction-sphere.

Chitin – Nitrogen-containing polysaccharide with long fibrous molecules, forming material of considerable mechanical strength and resistance to chemicals. Present in cuticle.

Chlorenchyma – Tissues collectively, or stem tissue, or mesophyll, containing chlorophyll.

Chlorophyll – The green colouring matter found in plants and in some animals.

Chloroplast – A minute granule or plastid containing chlorophylls a and b, found in plant cells exposed to light.

Chloroplast pigment – Chlorophylls, carotene, and xanthophyll.

Chonodriosomes – Mitochondria, chondriomites, chondrioconts, Chondriospheres, chondrioplast.

Chromatid – A component of tetrad in meiosis; a half chromosome between early prophase and metaphase in mitosis, or between diplotene and second metaphase in meiosis.

Chromatin – A substance in the nucleus which contains nucleic acid, proteids, and stains with basic dyes.

Chromatophore – A coloured plastid of plants and animals; a colourless body in cytoplasm and developing into a leucoplast, chloroplast or chromoplast.

Chromomere – One of the chromatin granules of which a chromosome is formed and whcih corresponds to a gene.

Chromoplast – A coloured plastid or pigment body; coloured plastid other than a chloroplast; chromoplastid.

Chromosome – One of deeply staining bodies, the number of which is constant for the cells of a species, into which the chromatin resolves itself during karyokinesis, and meiosis.

Cluster-crystals – Globular aggregates of calcium oxalate crystals in plant cells; sphaeraphides.

Collateral – Side by side; bundles with xylem and phloem in the same radius.

Collenchyma – Parenchymatous peripheral supporting tissue with cells more or less elongayted and thickened, either at the angles or on walls adjoining interecellular spaces, or tangentially.

Colloid – A gelatinous substance which does not readily diffuse through an animal or vegetable membrane.

Columnar – Cell longer than broad.

Companion cell – A narrow cell, retaining its nucleus, derived, from a cell giving rise also to a sieve-tube element, in phloem of angiosperms.

Complementary – Non-suberized cells loosely arranged in cork tissue and forming air passages.

Concentric – Having a common centre.

Concentric bundle – The vascular bundle, with one tissue surrounding the other i.e., amphicribral and amphivasal.

Conjunctive – A tissue, mesocycle and pericycle in a stele.

Cork – A tissue derived usually from outer layer of cortex in woody plants.

Cork-cambium – Phellogen.

Corpus – Body; core of apical meristem within the tunica.

Cortex – The extrastelar fundamental tissue of the sporophyte.

Cortical – Pertaining the cortex.

Cristae – Folds of the inner membrane of a mitochondrion.

Crystalloids – A protein crystal found of certain plant cells.

Crystal sand – A deposit of minute crystals of calcium oxalate, as in Solanaceae.

Cuticle – An outer skin or pellicle.

Cuticularisation – Cutinisation in external layers of epidermal cells.

Cutin – A mixture of substances associated with cellulose, found in external layers of thickened epidermal cells of plants.

Cutinisation – The deposition of cutin in cell-wall, thereby forming cuticle.

Cyanin – The blue pigment.

Cystolith – A mass of calcium carbonate occasionally of silica, formed on ingrowths of epidermal cell walls in some plants.

Cytokinesis – In a dividing cell, division of cytoplams as distinct from division of nucleus.

Cytology – Study of cells.

Cytoplasm – Substance of cell-body exclusive of nucleus.

Cytosine – A cleavage product of nucleic acids, $C_4H_6N_3O$.

D

Deoxyribose nucleic acid, DNA – Stable nucleic acid component of kinetoplasts, chromosomes, bacterial cells and phages, which consists structurally of two spirals linked transversely and constitutes a pattern or template for replication.

Dermatogen – The young or embryonic epidermis in plants.

Dextrose – Grape sugar or glucose, the end product of starch digestion, $C_6H_{12}O^6$.

Diakinesis – The later prophase stage of meiosis, between diplotene and prometaphase; movement of chromosomes between metaphase and telophase.

Diarch – With two xylem and two phloem bundles; root in which protoxylem bundles meet and form a plate of tissue cross cylinder with phloem bundle on each side.

Diastase – An enzyme which acts principally in converting starch into sugar.

Dicotyledon – A plant with two seed-leaves.

Dictyosome – Unit of Golgi apparatus several of which occur as discrete bodies in cells of plants.

Dictyostele – Amphiphloic siphonostele that is broken up by crowd ded leaf gaps into a net work of distinct vascular strands (meristeles), each surrounded by a endodermis. Present in stems of certain ferns.

Diploid – Having the 2n number of chromosomes.

Diplotene – Stage in meiosis at which bivalent chromosomes split longitudinally.

Discoid – Flat and circular; disc-shaped.

DNA – Deoxyribonucleic acid, a compound consisting of a large number of nucleotides attached together in single file to form a long strand.

Dorsiventral – With upper and lower surfaces distinct; bifacial.

Duct – Any tube which conveys fluid or other substance.

Duplication – A translocated chromosome fragment attached to one of normal set.

Dynamic – Producing or manifesting activity.

E

Ecology – Study of the relations of plants, particularly of plant communities, to their surroundings.

Ectoplast – The protoplasmic film or plasma-membrane just within the true wall of a cell.

Elaioplast – A plastid in a plant cell, which forms or helps to form oil globules.

Embryo – A young organism in early stage of development.

Emergence – An outgrowth from sub-epidermal tissue.

Endarch – With central protoxylem, or with several surrounding a central pith.

Endodermis – Innermost layer of cortex in plants; layer surrounding pericycle.

Endoplasmic reticulum (ER) – Complex mesh work of tubular channels, often extended into slit like cavities (*cisternae*) together with more or less flattened vesicles, all bounded by unit membranes, occurring in the cytoplasm of many eucaryotic cells; usually only visible by electron microscopy.

Enzyme – A catalyst produced by living organism and acting on one or more specific substrates.

Epiblema – The outermost layer of root tissue; piliferous layer.

Epicotyl – That portion oc seedling stem above cotyledons.

Epidermis – The outermost protective layer of stems, roots and leaves.

Epiphyte – Plant which lives on surface of other plants.

Epithelium – Layer of cells living schizogenously formed secretory canals and cavities, *e.g.*, in resin canals of pine.

Ergastic – Life-less cell inclusions, as fat, starch.

Ergastoplasm – Endoplasmic reticulum.

Essential oils – Volatile oils, composed of various constituents and contained in plant organs, with characteristic odour.

Eumeristem – Meristem composed of isodiametric thin-walled cells.

Eumitosis – Typical mitosis.

Exarch – With protoxylem strands outside metaxylem, or in touch with pericycle.

Excentric – One sided.

Exodermis – A specialized layer below the piliferous layer.

Extracellular – Occurring outside the cell.

Extracortical – Not within the cortex.

Extranuclear – Situated outside the stele.

Extraxylary – Occurring outside the stele.

Extraxylary – On the outside of the xylem.

F

Fibre – Elongated plant cell for mechanical strenth.

Fibre – Fibres of a nature intermediate between that of libriform fibres and of tracheids.

Fibrous – Composed of fibres.

Fibrovascular – Bundle of vascular tissue surrounded by non-vascular fibrous tissues.

Filament – A thread-like structure.

Flower – Specialized reproductive shoot, consisting of an axis (*receptacle*), on which are inserted four different sorts of organs.

Fructose – Fruit sugar.

G

Gland – Superficial dischargin secretion externally, e.g. glandular hair, nectary, hydathode; or embedded in tissue, occurring as isolated cells containing the secretion, or as layer of cells' surrounding intercellular space (secretary cavity) into which secretion is discharged, *e.g.*, resin canal of pine.

Gland cell – An isolated secreting cell.

Glandular – With glands.

Glandular tissue – Tissue or massed cells, parenchymatous and filled with granular protoplasm, adapted for secretion of aromatic substances in plants.

Globoid – A spherical body in aleurone grains, a double phosphate of calcium and magnesium.

Globose – Spherical.

Globule – Any minute spherical structure.

Glucose – The grape sugar, dextrose.

Glycogen – A carbohydrate storage product of plants.

Golgi apparatus or complex – (C. Golgi, Italian, histologist). Cell-constituents, localised or diffused, often consisting of separate elements, the Golgi bodies, batonettes, dictyosomes or pseudo-chromosomes, containing lipo-protein, and concerned with cel-lular synthesis and secretion.

Grana – Minute particles consisting of a pile of thin double platelets, probably containing chlorophyll, in chloroplasts.

Granum – Singular of grana.

Ground tissue – Conjunctive parenchyam.

Growing point – A part of plant body at which cell-division is localised, generally terminal and composed of meristematic cells.

Guanine – A purine base found in some plants; $C_6H_5ON_5$.

Guard cells – Two cells surrounding stomata of aerial epidermis of plant tissue.

Gum – An exudation of certain plants and trees; vegetable mucilage.

H

Hair – Any epidermal filamentous outgrowth consisting of one or more cells, varied in shape.

Halophyte – A shoreplant; plant capable of thriving on salt impregnated soils.

Haploid – Having a single set of un-paired chromosomes in each nucleus.

Haustorium – Specialized organ of parasitic plant, *e.g.*, *Cuscuta*, which penetrates into and withdraws food material from tissues of host plants.

Heart-wood – The darker, harder, central wood of trees; duramen.

Helix – A spiral.

Hemicellulose – One of several polysaccharides, chemically unrelated to cellulose, occurring as cell-wall constituents in cotyledons, endosperms, and woody tissues, and serving as reserve food.

Herb – Plant with no persistent parts above ground, as distinct from shrubs and trees.

Herbaceous – Being a herb.

Hereditary – Transmissible from parent to offspring, as characteristics.

Hexarch – Having six radiating vascular strands; *e.g.*, roots.

Hilum – Nucleus of starch grain.

Hinge-cells – Large epidermal cell wich, by changes in turgor, control rolling and unrolling of a leaf.

Histogens – Tissue producing zones or layers, plerome, periblem, dermatogen, and calyptrogen.

Histology – The science which treats of the detailed structure of anima of plant tissues; microscopic morphology.

Histone – A basic protein constituent of cell nuclei.

Homologous – Resembling in structure and origin; chromosomes with the same sequence of genes.

Hormones – Substances normally produced in cells and necessary for the proper functioning of other distant cells to which they are conveyed end of the body as a whole.

Hydathode – An epidermal structure specialized for secretion, or for exudation of water, water stoma.

Hydrophyte – An aquatic plant.

Hypocotyl – Part of seedling stem below cotyledons.

Hypodermis – The cellular layer lying beneath the epidermis.

I

Idioblast – Plant cell containing oil, gum, calcium carbonate or other product and which differs from the surrounding parenchyma.

Intercalary – (Of a meristem) situated between regions of permanent tissue, *e.g.*, at base of nodes and leaves in many monocotyledons.

Intercellular – Among or between cells.

Interfascicular – Situated between the vascular bundles.

Internode – The part between first and second mitotic divisions; inter-kinesis.

Interxylary – Between xylem

strands; interxylary phloem.

Intracellular – Within a cell.

Intrafasicular – Within a vascular bundle.

Intrastelar – Within the stele of a stem or root; ground tissue, bundles.

Intraxylary – Within wood or xylem.

Inulin – A carbohydrate occurring in rhizomes and roots of many plants; dahlia starch; $(C_6H_{10}O_5)_x$.

Invertase – A plant enzyme which converts cane sugar into dextrose and laevulose.

Irritability – *e.g.*, protoplasmic movements. A universal property of living things.

Isobilateral – A form of bilateral symmetry where a structure is divisible in two planes at right angles.

Isobilateral – Having equal diameters; cells or other structures.

Isolateral – Having equal sides; leaves with palisade tissue on both sides.

K

Karyokinesis – Indirect cell-division; mitosis.

Karyolymph – Nuclear sap.

Kataphase – The stages of mitosis from formation of chromosomes to dividion of cell; anaphase.

L

Lacunate – Collenchyma with cell-walls thickened where bordering intercellular spaces.

Lamella – Any thin plate.

Latex – A milky, or clear, sometimes coloured, juice or emulsion of diverse composition found in some plants, as in spurges, rubber trees.

Laticifer – Any tatex containing cell, series of cells or duct.

Laticiferous – Conveying latex; cells, tissue, vessels.

Leaf-gap – Gap in vascular cylinder of stem, a parenchymatous region associated with leaf-traces.

Leaf trace – Scar marking position where a leaf was formerly attached to the stem.

Leaf scar – Vascular bundles extending from stem bunbles of leaf-base.

Leuticel – Ventilating pore in angiosperm stems or roots.

Leptotene – Stage in early prophase of first division of meiosis.

Leucoplastids – Colourless plastids from which amylo-, chloro- and chromoplastids arise.

Leucoplasts – Colourless granules of plant cytoplasm.

Lignin – Climbing plants found in tropical forests, with long woody rope like stems of anomalous anatomical structure.

Libriform – Resembling bast.

Lignification – Wood formation; thickening of plant cell-walls by deposition of lignin.

Lignin – A complex substance which, associated with cellulose, causes the thickening of plant cell-walls, and so forms wood.

Lipoid – Resembling a fatty substance.

Lumen – Central cavity of a plant cell.

Lysosomes – Particle in cytoplasm, smaller than mitochondria, consisting of a membrane enclosing several enzymes.

M

Macerate – To wear or to isolate parts of a tissue of organ.

Macrosclereids – Relatively large columnar sclereids, as in coat of certain seeds.

Matrix – Ground substance of connective tissue.

Medullary – In region of medulla.

Medullary rays – A number of strands of connective tissue extending between pith and pericycle.

Meiosis – Process of reduction division of germ-cell chromosomes from diploid to haploid number at maturation.

Meristele – See *dictyostele*.

Meristem – Tissue formed of cell all capable of diversification, as found at growing points; meristematic tissue.

Meristematic – Consisting of meristem; tissue cells of growing point.

Mesarch – Xylem having metaxylem developing in all directions from the protoxylem, characteristic of ferns.

Mesophyll – The internal parenchyma of a leaf.

Mesophyte – A plant thriving in temperate climate with normal amount of moisture.

Messenger RNA – RNA molecule that conveys from the DNA the information that is to be translated into the structure of a particular polypeptide molecule.

Metabolic – Chemical change, constructive and destructive, occurring in living organism.

Metaphase – The stage in mitosis or meiosis in which chromosomes are split up in equatorial plate.

Metaphloem – The phloem of secondary xylem.

Metaxylem – Secondary xylem with many thick walled cells.

Micron – One thousandth part of a millimetre symbol: .

Microscope – The ordinary microscope of the laboratory is a compound microscope with two sets of lenses (objective and eyepiece) which magnify the object in two steps.

Microtome – Machine for cutting extremely thin sections of tissue (usually 3 to 20).

Microtomy – The cutting of thin sections of objects, as of tissues, or cells, in preparing specimens for microscopic or ultramicroscopic examination.

Middle lamella – The layer derived from the cell plate, and covered on both sides by cellulose in formation of the wall of a plant cell.

Mid-rib – The large central vein of a leaf, constitution of the petiole.

Mitochondria –Granular, rodshaped, or filamentous self-replicating organellae in cytoplasm, consisting of an outer and inner membrane containing phosphates and numerous enzymes, varying in different tissues and functioning in cell respiration and nutrition; chondriosomes.

Mitochondrion – Singular of mitochondria.

Mitosis – Indirect or karyokinetic nuclear division, with choromosome formation, spindle formation, with or without centrosome activity.

Mitotic – Produced by mitosis.

Monocotyledonous – Having one cotyledon.

Morphogenesis – The development of shape; origin and development of organs or parts of organisms.

Morphology – The science of form and structure of plants and animals, as distinct from consideration of functions.

Mucilage – A substance of varying composition, hard when dry, swelling and slimy when moist, produced in cell-walls of certain plants.

Mucilaginous – Composed of mucilage.

Mycorrhiza – Association of fungal mycellium with roots of a higher plant.

N

Nectar – Sweet substance secreted by special glands, nectaries, in flowers and certain leaves.

Nectary – A group of modified sub-epidermal cells of no definite position in a flower, less commonly in leaves, secreting nectar; a nectar gland.

Nodal – Pertaining a mode or nodes.

Node – The knob or joint of a stem at which leaves arise.

Nuclear membrane – Delicate membrane bounding a nucleus, formed for surrounding cytoplasm.

Nuclear sap – Karyolymph.

Nucleic – Plural of nucleus.

Nucleic – Acids containing phosphorus, found in nuclei of cells.

Nucleolus – A dense rounded mass in a cell-nucleus, consisting of protein and ribonucleic acid granules, and functioning in RNA and protein synthesis controlled by a special region or nucleolar organiser in the chromosome; plasmosome or a karyosome.

Nucleoplasm – Reticular nuclear substance; karyoplasm.

Nucleoprotein – A compound of protein and nucleic acid, a constituent of cell nuclei.

Nucleotide – Compound formed from sugar (with 5 carbon atoms) phosphoric acid, and a nitrogen-containing base (purine or pyrimidine).

Nucleus – Complex spheroidal mass essential to life of most cells.

O

Obtuse – With blunt or rounded end.

Oil gland – A gland whcih secretes oil.

Oil immersion objective – Objective of light microscope, space between which and the cover slip over the object examined is filled with drop of oil of same refractive index as glass; system used for highest magnification with the light microscope.

Ontogeny – The whole course of development during an individual's life-history.

Organ – Multicellular part of a plant which forms a structural and functional unit, *e.g.*, leaf.

Organelle – A persistent structure with specialized function forming part of a cell, *e.g.* a mitochondrion.

Osteosclereid – A sclereid with both ends knobbed.

Oxalates – Salts of oxalic acid, occurring as metabolic by-products in various plant tissues.

P

Pachytene – Stage in prophase of first division of meiosis, follow zygotene.

Palisade – Tissue, the layer or layers of photosynthetic cells beneath the epidermis of many foliage leaves.

Papain – A proteolytic enzyme in fruit juice of the tree *Carica papaya*.

Parasite – An organism living with or within another to its own advantage in food or shelter.

Parenchyma – Plant tissue, generally soft and of thin-walled relatively undifferentiated cells, which may vary in structure and function as pith, or mesophyll, etc.

Parenchymatous – Pertaining or found in parenchyma; a kind of cell.

Passage cells – Thin-walled endodermal or exodermal cells of root, which permit passage of solutions.

Pectic – Substances in cell walls and cell sap of plants, including pectic acid and its salts, pectin and pectose; enzymes; pectosinase, pectase, and pectinase, which hydrolyse pectic substances.

Pectose – A carbohydrate constituent of plant cell-walls, converted into pectin and cellulose by action of pectosinase.

Pedicel – Stalk of individual flowers of an inflorescence.

Pentarch – With five alternating xylem and phloem groups.

Periblem – Layers of ground of fundamental tissue between dermatogen and plerome of growing points.

Periclinal – System of cells parallel to surface of apex of a growing point.

Pericycle – The external layer of stele, the layer between endodermis and conducting tissues.

Periderm – The outer layer of bark; phellogen, phellem and phelloderm collectively; epiphloem.

Permanent tissue – Tissue consisting of cells which have completed their period of growth and subsequently change little until they lose their protoplasm and die.

Permeability – Of membrane, extent to which molecules of a given kind can pass through it.

Phellem – Cork; cork and non-suberous layers forming external zone of periderm; phellem.

Phelloderm – The secondary parenchymatous suberous cortex of trees, formed on inner side of cork cambium.

Phellogen – The cork-cambium of tree stems arising as a secondary meristerm and giving rise to cork cambium.

Phioem – Bast tissue; the soft bast of vascular bundles, consisting of sieve-tube tissue.

Phloem parenchyma – Thin-walled parenchyma associated with sieve tubes of phloem.

Phloem sheath – Pericycle, together with inner layer of a bundle sheath later consists of two layers.

Phloic – In green plants, synthesis of organic compounds from water and carbon dioxide using energy absorbed by chlorophyll from sunlight.

Phyllocalde – A green flattened or rounded stem functioning a leaf as in *Cactus*; flattened axillary bud, as in *Ruscus*; phyllocladium, cladode, cladophyll.

Phyllode – Winged petiole with flattened surfaces placed laterally to stem, functioning as leaf.

Phylogeny – Evolutionary history.

Physiology – Study of the processsess which go on in living organism.

Pigment – Colouring matter in plants.

Pigment cell – A chromatophore.

Piliferous – Bearing or producing hair; outermost layer of root or epiblema which gives rise to root hairs.

Pit – A depression formed in course of cell-wall thickening of plant tissue.

Pit chamber – The cavity of a bordered pit bwlow the overarching border.

Pit-fields – Areas of depressions in primary cell-walls.

Pith – Central part of an organ of plant; central core of usually parenchymatous tissue in those stems in which the vascular tissue is in the form of a cylinder.

Pit membrane – Middle lamella of plant cell-wall forming floor of pits of adjacent cells.

Plasmalemma – Plasma membrane (external plasma membrane in plants).

Plasma membrane–The membrane forming the surface of cytoplasm and consisting of a bimolecular phospholipid layer between an inner and outer layer of protein molecules.

Plasmodesma – Cytoplasmic threads penetrating cell wall and forming intercellular bridge; *plural* plasmodesmata.

Plastid – A cell-boyd other than nucleus or centrosome.

Plastidome – In a cell, the plastids as a whole; cytoplasmic inclusions which give rise to plastids.

Plastochondria – Mitochondria.

Plerome – The core or central part of an apical meristem.

Polyarch – Having many protoxylem bundles.

Pore – A minute opening or passage, as of sieve plates, stomata etc.

Primary – First; principal; original.

Primary meristem – Ground meristem, procambium and protcderm.

Procambium – The tissue from which vascular bundles are developed.

Promeristem – Meristerm of growing point, and primary meristem.

Prometaphase – Stage bwtween prophase and metaphase in mitosis and meiosis.

Prophase – The preparatory change,s the first stage in mitosis, or in meiosis.

Proplastid – An immature plastid, as in meristematic cells.

Protein – A nitrogenous compound of cell prctoplasm; a complex substance characteristic of living matter and consisting of aggregates of amino-acids, and generally containing sulphur.

Protoderm – The outer cell layer of apical meristerm; primordial epidermis of plants; superficial dermatogen.

Protophloem – The first phloem elements of a vascular bundle.

Protoplasm – Living cell substances; cytoplasm and karyoplasm.

Protoxylem – Pertaining or consisting of protoplasm.

Protoplast – Protoplasm of a plant cell.

Protoxylem – Primary xylem lying next pith of stems.

Pteridophyta – Division of plant kingdom comprising ferns, horsetails, club-mosses etc.

Purine – Bases — adenine and guanine.

Pyrimidine – Bases — thymine and cytosine.

Q

Quantasomes – Regulary arranged sub-units observed by electron microscopy in thylakoid lamellae. Estimated each contains about 300 chlorophyll moleculws whcih are believed to function as a photosynthetic units in absorption of light quanta.

Quiescent centre – Inactive or passive region of cells.

R

Raphides – Needle shaped crystals of calcium oxalate occurring in bundles in certain plant cells.

Radial – Pertaining to radius.

Ray – A parenchymatous band penetrating from cortex towards centre of stem.

Replication – Duplication of a molecule or aggregate by copying from a pre-existing molecule or structure of the same kind of mitochondria, chloroplast etc.

Reticulate – Like network; thickening of cell-wall.

Rhizome – Underground stem, bearing buds in axils of reduced scale-like leaves.

Rhytidome – The outer bark.

Ribonucleic acid – RNA, a nucleic acid containing adenine, guanine, cytosine and uracil, in nucleolus, mitochondria, and ribosomes, and taking part in cytoplasmic protein synthesis.

Ribosomes – Spherical granules or microsomal particles containing ribo-uncleic acid, on nucelar membrane and membranes of endoplasmic reticulum, and taking part in protein synthesis.

RNA – Ribouncleic acid, a molecule consisting of large number of nucleotides attached together of form a long strand one nucleotide thick. Each nucleotide contains the sugar ribose, and one of the four different bases found in DNA except that uracil replaces thymine.

Root – Descending protion of plant, fising it in soil, and absorbing moisture and nutrients.

Root-cap – A protective cap of tissue at apex of root.

Root-hairs – Unicellular epidermal outgrowths from roots, of protective and absorbent function.

Rosette – A cluster of crystals, as in certain plant cells.

S

Saprophyte – The plant that obtains organic matter in solution from dead and decaying tissues of plants or animals, *e.g. Monotropa.*

Saprophytic – A plant wich lives on dead and decaying organic matter.

Sap-wood – The more superficial, softer wood of trees; alburnum.

Scalariform – Ladder-shaped; vessels or tissues having bars like a ladder.

Scalariform thickening – Internal thickening of wall of a xylem vessel or tracheid, in the form of more of less transverse bars, suggestive of ladder rungs.

Scape – Leafless flowering stem arising from ground level, *e.g.*, *Canna*.

Schizogenous – Cavity originaing by separation of cells, *e.g. Citrus*.

Sclereid – Any cell with a thick lignified wall; a sclerenchymatous cell; a stone cell.

Sclerencyma – Plant tissue of thickened end of the hard cells or vessels.

Scleroid – Hard.

Sclerotic – Containing lignin.

Secondary – Arising not from growing point, but from other tissue.

Secondary cortex – Phelloderm.

Secondary wood – Wood formed from cambium.

Sepel – One of the parts forming calyx of dicotyledonous flowers.

Septum – Partition or wall.

Shoot – Stem of a vascular plant derived from the plumule.

Sieve area – Perforated area of cell wall of sieve elements, with groups of pores surrounded by callose.

Sieve cell – A phloem cell having perforated areas of cell-wall; a cell of sieve tubes.

Sieve elements – The conducting part of phloem; sieve cells and sieve tube cells.

Sieve plate – Part of the wall of a sieve cell, containing simple or compound sieve areas; the perforated and thcikened end of a sieve tube cell.

Sieve tubes – Phloem vessels, long slender structures consisting of elongayted cells placed end to end, forming lines of conduction.

Somatic cells (Soma) – The cells of an organism, other than the germ cells.

Spermatophyta – Seed plants. Division of plant kingdom providing dominant flora of present day, including most trees, shrubs, herbs, grasses etc.

Spindle – A structure formed of a chromatin fibres during mitosis.

Spiral – Thickening of cell wall.

Spireme – Thread-like appearance of nuclear chromatin during prophase of mitosis.

Spongy – Parenchyma of mesophyll.

Sporophyte – Phase of life cycle of plants which had diploid nuclei, and during whcih spores are produced.

Stamen – Organ of flower which produces pollen grains.

Starch – The common carbohydrate formed by plants and stored in seeds. ($C_6H_{10}O_5)_x$.

Starch sheath – Endodermis with starch grains.

Stele – A bulky strand or cylinder of vascular tissue contained in stem and root of plants, developed from plerome.

Stellate – Star shaped hair.

Stem – Main axis of a plant.

Steroids – Complex hydrocarbons, chemically similar, occurring in plants and animals.

Sting – Stinging hari and cell.

Stoma – A small orifice: minute openings with guard cells, in epidermis of plants, especially on under surface of leaves, or, the stomatic pores only.

Stomata – Plural of stoma.

Stone cells – Sclerotic cells are rounded sclerenchymatous elements, as found in pear; brachysclereids.

Suberisation – Modification of cell-walls due to suberin formation.

Suberin – The waxy substance developed in a thickened cell-well, characteristic of cork tissues.

Subsidiary cells – Additional modified epidermal cells lying outside guard cells.

Substomatal – Hypostomatic.

Sucrose – Cane sugar $C_{12}H_{22}O_{11}$.

Sulcate – Furrowed, grooved.

Superficial – On a near the surface.

T

Tabular – Flattened, as certain cells.

Tanning – Group of astringent substances of wide occurrence in platns, dissolved in cell sap; particularly common in the bark of trees, unripe fruits, leaves and galls. Complex organic compounds containing phenols, hydroxy-acids, or glucosides.

Tap root – Root system with a prominent main root, directing vertically downward and bearing smaller lateral roots.

Taxonomy – Study of the classification of organisms according to their resemblances and differences.

Telophase – Terminal stage of mitosis or meiosis during which nuclei revert to resting stage.

Terminal – Situated, at the end, as terminal bud at end of twig.

Tetrarch – With four protoxylem bundles.

Thylakoid – In photosynthetic organisms, vesicle, wall of which bears photosynthetic pigments. Thylakoids vary in form arrangement in different groups of organisms.

Tissue – The fundamental structure of which animal and plant organs are composed; an organization of like cells.

Tissue culture – A technique for maintaining fragments of animal or plant tissue or separated cells alive after their removal from the organism.

Tonoplast – A vascular membrane; a plastid with distinct vacuole walls; a special form of vacuole-producing plastid.

Torus – Thickened centre of a bordered pit membrane.

Trachea – Spiral or annular vascular tissue of plants; wood vessel.

Tracheid – One of the cells with spiral thickening or bordered pits, conducting water and solutes,a nd forming wood tissue.

Tracheophyta – Division including all vascular plants (Pteridophyta and Spermatophyta).

Transection – Cross section; transverse section.

Transfer RNA – A relativley small molecule of RNA, whose function is to place the amino acids that will be linked into a polypeptide molecule in the specific sequence specfied by a molecule of *Messenger RNA*.

Transfusion Tissue – Tissue of empty cells with pitted and occasionally, internally thickened walls, and protein containing, parenchyma cells, accompanying vascular tissue in leaves of most gymnosperms, lying on either side of the vascular bundles of the single vein.

Triarch – Having three xylem bundles uniting to form the woody tissue of root.

Trichoblast – A cell, of plant epidermis which develops into a root hair.

Trichome – An outgrowth of plant epidermis, either hairs or scales; a hair tuft.

Tunica – Apical meristematic cell giving rise to protoderm.

Tunica – An interpretation of the shoot apex which recognizes two tissue zones in the promerister, *tunica* consisting of one or more peripheral layers, in which planes of cell division are predominantly anticlinal, enclosing, *corpus*, or central tissue of irregularly arranged cells in which the planes of cell division vary.

Tylosis – Development of irregular cells in a cell cavity; a cellular intrusion into vessel through pit of parenchyma cells.

U

Ultrastructure – Structure, at the molecular or electron microscopical level.

Unciform – Shaped like a hook or barbed.

Uncinate – Unciform; hook like.

Unit membrane – The common form, as seen in the electron microscope, of the membranes organelles of the cells : i.e., of the plasma membrane and of the membranes organells such as mitochondria, nucleus and endoplasmic reticulum.

V

Vacuole – One of spaces in cell protoplasm containing air, sap, or partially digested food.

Vascular – Containing, or concerning vessels.

Vascular bundle – A group of special cells consisting of two parts, xylem or wood and phloem or bast portion; many have in addition a thin strip of cambium separating the two parts.

Vascular cylinder – Stele.

Vascular plant – Plant possessing vascular system, member of Tracheophyta. (Pteridophyta and Spermatophyta).

Vascular system – Plant tissue consisting mainly of xylem and phloem which forms a continous systme throughout all parts of higher plants. It functions in conduction of water, mineral salts, and synthesized food materials and for mechanical support.

Vascular tissue – Specially modified plant-cells, usually consisting of either tracheae or sieve cells for circulation of sap.

Veins – Strands vascular tissue of leaf.

Velamen – A specialized moisture absorbing tissue.

Ventral – Situated on lower surface.

Vesicle – Small globular or bladder-like air space in tissues.

Vessel – Any tube or canal with properly defind walls.

W

Water stomata – Pores on surfaces of leaves for excretion of water; hydathodes.

Wax – A substance soluble in fat solvents.

Wood – The hard substance of a tree stem, xylem of vascular bundles.

Wood vessel – An element of tracheal tissue, a long tabular structure formed by cell-fusion.

X

Xanthophylls – Yellow or brown carotenoid pigments found in plastids.

Xerophyte – A plant growing in desert or alkaline or physiologically dry soil.

Xylary – Pertaining xylem.

Xylem – Woody tissue; lignified portion of vascular bundle.

Xylem-parenchyma – Short lignified cells surrounding vascular cells or produced with other xylem cells toward, the end of the growing season.

Xylem-ray – Ray or plate of xylem between two medullary rays.

Z

Zygotene – Stage in prophase of first division of meiosis following leptotene, in which pairing (synapsis) of homogous chromosomes occurs with formation of bivalents.

Experimental Taxonomy

Agamospermy – Includes those forms of apomixis in which seeds are produced by one of a wide variety of asexual processes. Agamospermy does not recessarily produce clones because meiotic segregation and recombination can occur in certain kinds of agamospermy.

Agamic Complex – Complex of hybrid and polyploid forms reproducing by agamospermy so that each original form comes to be represented by an apomictic population. These are extremely difficult taxonomically because each "population" is actually the extension of a single individual of hybrid origin

Allele (Allelomorph) – One of the alternative forms of a particular gene which can occur at a locus. When more than two alleles are known for a given locus and gene, they are called *multiple alleles*.

Allopolyploid (alloploid) – A polyploid in which the original chromosome sets (genomes) have been derived from different sources as the result of different genomes. In a strict allotetraploid only bivalents are formed at meiosis and no multivalent associations.

Amphidiploid (literally both diploid)–The allotetraploid formed by doubling the chromosomes of a sterile F1 hybrid which contains two haploid genomes which are incapable of effective pairing at meiosis. Fertility is restored after chromosome doubling because each genome has become duplex, bivalents are formed at meiosis, and balanced chromosome assortment ensues.

Aneuploid – Chromosome numbers which are not exact multiples of an original haploid number, with one or more chromosomes missing ($2n - 1$) or in addition ($2n + 1$) to the euploid number.

Aneuploid series. A series of aneuploid numbers in a group of individuals, or species, each chromosomes number of the series differing from the next by the addition (or subtraction) of one or two chromosomes (e.g., $2x$, $2x + 1$, $2x + 2$, $2x + 4$, $2x + 5$, etc.).

Apomixis – The substitution of some form of asexual reproduction for the sexual process *amphimxis* in an organism derived from sexually reproducing ancestors. Apomixis should not be confused with *amixis*, the asexual forms of reproduction which are common in procaryotic organisms.

Autodeme – A deme composed of predominantly self-fertilizing individuals.

Autopolyploid (autoploid) – A polyploid in which all of the chromosome sets (genomes) have arisen from the same source, each chromosome is capable of pairing (autosyndesis) with a corresponding chromosome in each of the other sets. Easily applied to experimentally induced polyploids; e.g., a normal diploid is doubled to produce an autotetraploid, this can then be crossed with the original diploid to produce an autotriploid and so on. The same events can and do occur in nature as the result of spontaneous chromosome doubling. In natural populations, however, both within and between related species, every possible degree of similarity from essential identity to marked genic and even structural divergence may have occurred between the parents of the original diploid individual in which the chromosome doubling occurred.

B

Basic number – x. The number of chromosomes in the original genome (haploid set) from which a polyploid or a group of polyploid forms or species is known or postulated to have arisen.

Biological Species – Grant (1971) ; "The sum total of the races that interbreed frequently or occasionally with one another, and that intergrade more or less continuously in their phenotypic characters". Mayr (1963); "Groups of interbreeding natural populations that are reproductively isolated from other such groups". Dobzhansky (1970): "A Mendelian population is a community of individuals of a sexually reproducing species within which matings take place. A biological species is that most inclusive Mendelian population; it is integrated by the bonds of sexual reproduction and parentage".Taxonomic species are sometimes equivalent to biological species but for a variety of reasons, some good, some bad, are often less inclusive.

C

Cenospecies = Coenogamodeme – All the ecospecies so related that they may exchange genes among themselves to a limited extent through hybridization. This is equivalent in most cases to the *biological species*.

Clinal Variation – A sequence of morphologically interrelated forms which have evolved in response to an environmental gradient and form a more or less continuous series, a *cline*. The populations in the series are *clinodemes*. Clinal variation is in contrast to introgression but easily confounded with it.

Clone – All the members of a population of cells or cells or organisms derived asexually from a single individual by vegetative (mitotic) reproduction.

Comparium=Syngamodeme–All the cenospecies between which hybridization is possible either directly or through intermediaries. Gene exchange between cenospecies is impossible.

D

Deme Terminology – The suffix "-deme" as originally proposed and correctly employed is defined as "a term, always used in this terminology with a pre-fix, denoting any group of individuals of a specified taxon. The nature of the association of the group of individuals or the kinds of populational unit which they may represent must always be specified by the addition of an appropriate prefix.Second order, compound prefixes may also be used to specify relationships more precisely. The word *deme* used alone simply means "a group of individuals of a specified taxon" and shows no other relationships unless this is specified in the context (Davis and Heywood, 1963; Heslop-Harrison, 1967).

Topodeme – A deme occurring in a specified geographical area.

Ecodeme – A deme occurring in a specified kind of habitat.

Phenodeme – A deme which differs from others phenotypically.

Genodeme – A deme which differs from other genetically.

Cytodeme – A deme which differs from others cytologically.

Plastodeme – A deme which does not differ from others genotypically, but does differ phenotypically.

Gamodeme – A deme that is a panmictic breeding unit.

Autodeme – A deme composed of predominantly self-fertilizing individuals.

Endodeme – A gamodeme composed of predominantly endogamous (closely inbreeding) dioeclous plants or animals.

Clinodeme – One of a series of demes which collectively show a variational trend and thus collectively from a *cline.*

Dibasic polyploidy – Allopolyploids with chromosome numbers which result from combinations of two genomes with different basic numbers.

Diploid (2n) – The chromosome complement of a zygote, two complete sets of chromosomes one from each parental gamete – the zygotic number. The 2n diploid chromosome number is the normal somatic chromosome number of any organism which has arisen from a zygote, the organism itself is often referred to as diploid.

Disomic – The normal diploid (2n) condition in which a given chromosome is represented twice in the complement.

Dominance and Recessiveness – A typical Mendition dominant gene is defined as one that exproses in the F_1 hybrid between two parents and the character that is suppressed in the F_1 is referred to as recessive. Both genes express themselves in the F_2 generation in a 3 : 1 ratio dominant to recessive.

Dysploid – Diverse aneuploid chromosome numbers which do not fall into a seires or regular pattern.

E

Ecospecies = Hologamodeme – All the ecotypes so related that they are able to exchange genes freely without loss of fertility or vigor in the offspring. In many groups this unit is approximately equivalent to the *taxonomic species.*

Ecotype = Ecological Race = Ecogenodeme – A population differing genetically from others which occupies a specific ecological habitat.

Effective Population Size – The number of individu: of a population which are actuary reproducing in any generation, the *breeding population.*

Epistasis – This is the general term for the interaction of non-allelic genes and is commonly employed in the analysis of quantitative inheritance. The *epistatic* effect of an individual gene refers to the manifestation of its effect over that of non-allelic genes. The converse, *hypostatic*, refers to the effect upon the expression of the non-allelic genes in such a relationship.

Euploid – A chromosome number which is an exact multiple of a given haploid number.

F

Fitness – Success in producing viable and fertile offspring.

Fitness versus Flexibility – Fitness is the degree of adaptation of an individual or a population to the environment in which it lives. The ultimate measure of fitness in terms of population genetics is the relative ability to reproduce over generations. Flexibility is the capability of response to change, ie., the adjustment of adaptation to environmental change. "

Fitness places a premium on homogeneity with minimal expressed variability; flexibility depends upon a wide range of free or potential genetic variation. The best compromise between these two conflicting demands upon a population appears to be a wide range of variation coupled with a large modal class of individuals which are relatively uniform and near the optimum in adaptive characters.

Founder Principle (Ernst Mayr) –

If a small population invades a new area and becomes the basis of a new evolutionary line, divergence from the original ancestral population will be accelerated because the founder population includes (as the result of sampling) only a limited portion of the ancestral gene pool.

G

Gene and Locus – The *gene* is the ultimate unit of Mendelian inheritance, a precise sequence of nucleoitide triplets (codons), which occupies a specific chromo-somal locus, segregates and recombines as a unit, and which can mutate to various allelic forms which have different effects in inheritance.The locus (*plural, loci*) is the specific point in the linear order of genes in the chromosome which is occupied by a particular gene.

Genetic Drift – This is the loss or the fixation of a gene without or in spite of selection because of the "accidents of sampling". Genetic drift has been too often invoked to account for circumstances for which the investigator had no ready explanation. The random fixation (gene frequency = 1.0) or loss ($f = 0.0$) is most likely to occur when one of two conditions exists; (1) the population size is very small, 10-30 individuals, or (2) the effect of the given gene is very small, in which case the effect of fixation either way would be difficult to detect.

Genecology – Study of the genetic basis of ecological differentiation. "A turning-point in the study of evolutionary taxonomy".

Gene Flow – Changes in gene frequency which result from the migration of genes or gametes into a breeding population from a different population.

Gene Frequency – The proportion of a given allele at a specific locus which occurs with reference to all other alleles at this locus in a given breeding population. Usually expressed in values from 0 to 1.0; A gene frequency of 0.5 indicates that half the loci in the population carry the specified allele.

Gene Pool – The total genetic potentialities in the sum total of the genes in a breeding population from which the genes of the next generation are derived.

Genome – A complete haploid set of chromosomes; the haploid chromosome complement of a diploid race or species.

Genotype and Phenotype – The genetic constitution of an individual organism is the *genotype*. This means the total genetic endowment of the organism present in the original zygote. The *phenotype* is the expressed characteristics of an individual organism. This means all the characters of the organism which result from the interaction of its genetic endowment (genotype) with the environment during development and throughout the life of the individual. (Note : Genotype and phenotype are often used in a much restricted sense for purposes of analysis. Genotype designates the genic or chromo-somal constitution of an individual in a particular context; phenotype the character expression in such an individual).

H

Haploid (monoploid, n) – One complete set of chromosomes, such as the complement in the nucleus of one of the gametes. The *n* number of chromosomes in the haploid = monoploid = gametic number of chromosomes.

Hardy-Weinberg Law – The principle that both gene and genotype frequencies are constant from generation to generation provided that there is no migration, mutation, or selection with respect to the gene in question, i.e., the population is in Hardy-Weinberg equilibrium. The principle is based upon a large random mating population.

Heterosis – Superiority of heterozygous genotypes in one or more characters over the corresponding homozygotes, the phenotypic result of gene interaction in the heterozygotes.

Heterostyly – Combined morphological, physiological, and genetic mechanism in certain angiosperms which promotes cross-breeding. Flowers of two, some-times three kinds which differ in the relative positions of the styles (and anthers) are produced by separate individuals. Successful pollination and fertilization usually occurs only between flowers which differ, e.g., long styled by short styled or the converse, thus cross-pollination is ensured.

Heterozygote – A zygote or individual carrying two different alleles of a given gene (e.g., Aa).

Homoploid series – A group of related species (e.g., the species of *Pinus*, in all of them $2n = 24$) all of which have the same chromosome number.

Homozygote – A zygote or individual formed form the fusion of gametes each of which carries the same allele of a given gene (e.g. AA).

Hybrid – In taxonomy, the result of a cross between two taxa, often *interspecific*. In genetics, the result of a cross between any two individuals or races which differ genetically.

Hybrid Swarm – A population of interspecific or interracial hybrids and their segregates and intercrossed derivatives. Formation of such a population can be the first step in introgressive hybridization, but is not the actual process of introgression which aways involves successive backcrossing.

Hybrid Zone – Geographic area or region in which races of species meet and hybridize.

I

Inbreeding – A breeding system in which sexual reproduction involves the mating and union of gametes of individuals which are more closely related than would occur with random mating. *Selfing, sib-mating,* and *back-crossing* (e.g., parent-offspring matings) are examples of close inbreeding. *Self-pollination* is rather frequent in plants, especially among annuals. It may be obligate (e.g., cleistogamy) or faculative and is often aided by floral adaptations which enhance the opportunities for selfing.

Introgressive Hybridization – A procession of successive hybridizations in nature which causes migration of genetic material from one species into another. An initial interspecific hybridization is followed by a series of successive backcrosses to one of the parents (it can occur often called *introgression.*

Isolating Mechanism – Any of a number of intrinsic (genetic or cytogenetic) extrinsic (geographic, ecological environments) mechanisms or circumstances which prevent interbreeding and thus isolate two parts of a once continuous population.

L

Linkage – The non-random, "linked" inheritance of groups of genes and the characters they control which happen to be spatially associated on the same chromosome (T.H. Morgan, 1910). The degree of linkage is relative to the physical distances between the genes on the chromosome. These distances are inferred from the frequencies of crossing-over between the genes as determined from the associations of phenotypic characters in breeding experiments. The cross-over (map) distances between genes can be affected by factors other than physical distance, e.g., structural modification of the chromosomes, or genetic effects on cross-over frequency.

M

Mendelian Population – An interbreeding group of individuals sharing a common gene pool.

Modifier Genes and Genetic Background – Genes which are known or assumed to modify the primary effects of other genes are often called *modifier genes*. The effect of any gene or gene complex under study always occurs in the context of all the other genes which are present in the genotype. These other genes constitute the *genetic background* or *residual genotype*.

Monosomic – An individual in which the diploid chromosome complement lacks one member of one of the chromosome pairs, $2n - 1$. Thus one member of the complement is monoploid or simplex. There are as many different monosomic types possible as the number of chromosome pairs.

Mutation – Occurrence of a heritable change in the genotype of an organism which was not inherited from its ancestors; the ultimate source of all genetic variability. (cf. recombination). "All genetic changes, except those due to recombination, are mutations" (Dobzhansky, 1970). A mutation may be as minute as the substitution of a single nucleotide pair in the DNA molecule, as great as a major change in chromosome structure or number.

N

Natural Selection – The preservation of favorable individual differences and variations, and the destruction of those which are injurious is called Natural Selection or the Survival of the Fittest' – Darwin.

Natural selection is the "steering mechanism" of evolution. It may vary in type, kind, and intensity and may have varied results.

Nullisomic – Aneuploid in which both members of a chromosome pair are missing, $2n - 2$. This is not found at the diploid level, but can occur in polyploids, in which the genetic material of the missing chromosomes may be duplicated elsewhere in the chromosome complement so that no essential functions are lost.

O

Oligogenes and Polygenes – Terms for the genes assumed to correspond respectively to the qualitative and quantitative character effects. There are many genes which are observed to have only relatively small effects, but other genes (oligogenes) have not only major effects but also minor effects on other characters and act as "polygenes". The polygenic inheritance of recent literature is essentially equivalent to the *multiple factor* inheritance of earlier authors.

Outbreeding (Cross-breeding) – A breeding system in which sexual reproduction involves the mating and union of gametes of different individuals (requires cross-pollinationin seed plants). The common mode of sexual reproduction in higher animals. In plants outbreeding may be obligate or faculatative. The necessity or tendency for outbreeding may be reinforced by *dioecism, self-incompatibility, self-sterility, heterostyly* and other mechanisms.It is usually aided by wind, insect, or other pollinating agents. *Out-breeding* may imply mating between individuals which are less closely related than would occur with random mating.

Overdominance– Overdominance occurs when the heterozygote (Aa) exceeds both homozygotes (AA or aa) in the genotypic value of the character under study. The expression of the heterozygote falls outside the range of the two homozygotes.

P

Panmictic Unit = Gamodeme – A local population whose individual members have the same chance of mating and producing progeny -- the smallest Mendelian population.

Panmixis – The interbreeding of all the actively reproducing members of a population in such a way that mating is at random, i.e., each individual has the same probability of mating with any other individual.

Penetrance and Expressivity – *Penetrance* is the frequency in percent of dominants or of recessive hymozygotes which are actually expressed phenotypically. *Expressivity* refers to the strength or weakness in phenotypic of geno-type and environment.

Polyploid (ploid) – Any individual (or a cell) which contains three or more complete sets of chromosomes. Various combinations of words or numbers with "-ploid" indicate the number of haploid sets of chromosomes.
triploid = $3n$ hexaploid = $6n$
tetraploid = $4n$ Octoploid = $8n$
pentaploid = $5n$ 10-ploid = $10n$, etc.

Polyploidy (ploddy) – Variations in chromosome number involving more than the diploid number of complete chromosome sets.

Polyploid series ($2x$, $3x$, $4x$, $6x$, ...). A group of related races or species with various chromosome numbers all of which are multiples of a single basic number (e.g., the euploid series indicated).

Postzygotic Isolation – This results form mechanisms which reduce the viability or fertility of hybrid zygotes or individuals.

Prezygotic Isolation – This results from any geographical, ecological, or temporal circumstance or behavioral, structural, or genetic incompatibility which prevents the formation of hybrid zygotes.

Q

Qualitative and Quantitative Characters – These terms designate the extremes of a continuous array of a character expressions and gene effects. Qualitative characters have been described as differences in kind: major effects upon color, shape, hyper-development or abortion of certain organs, etc., but even such characters have quantitative attributes. Quantitative characters are those which can only be described in matrical terms: numbers, lengths, areas, volumes, etc., and in which the net effect of a single gene among the many affecting the same character is small in relation to the magnitude of non-heritable, environmentally induced modification. Quantitative characters exhibit *continuous variation*; qualitative characters produce *discontinuous variation*.

R

Recombination – The constant reassortment of chromosomes, crossover segments; and therefore genes which produces new genotypes and thus new phenotypes in each generation of sexually reproducing organisms. Recombination is the immediate source of the variation observed in populations, the raw material upon which the processes of evolution and speciation act.

Recurrent Mutation – Precisely the same mutation may occur again and again at the same genetic locus, either in nature or under the influence of a given physical or chemical agent (*a mutagen*). The frequency of such mutations under a given set of conditions is the *mutation rate*.

Reproductive Biology – A collective

term applied to all of the many mechanisms, agents, and adaptations involved in the reproductive process of a species. In higher plants, for example, reproductive biology includes floral structure, mode and/or agent of pollination, self or cross in compatibility, various types of sterility, time and season of flowering, and similar characteristics. Although fundamental to an understanding of the dynamics of any nutural population, reproductive biology is inadequately known in most plant species.

S

Segmental allopolyploid – A polyploid in which part of the chromosome complement is essentially autopolyploid in origin, part is allopolyploid. One of the frequent and more easily recognized intermediate situations between classical autopolyploidy and allopolyploidy.

Segregation and Independent Assortment – *Segregation* is the separation of allelic pairs, the genes received from the parents, and their distribution to different cells, the gametes. (*Independent assortment* describes the independence of non-allelic genes (or non-homologous chromosomes) in the segregation process; non-alleles on different chromosomes segregate randomly with respect to each other, cf. *linkage*.

Selection – "Anything tending to produce systematic, heritable change between one generation and the next" -- G.G. Simpson.

Self-Incompatibility – Genetically controlled physiological interactions which prevent or inhibit (*partial self-incompatibility*) the completion of self-pollination and/or self-fertili-zation in certain perfect flowered or monoecious plants (e.g., species of *Nictoiana, Hemerocallis, Trifollium*) and thus

promote cross-breeding. The same genotypic combinations which cause *self-incompatibility* also cause *cross-incompatibility* reactions when they occur in separate individuals which may be otherwise unrelated.

Supergene – A gene complex which is more or less permanently linked and transmitted as a unit is sometimes called a *supergene*. It is often difficult to distinguish such gene complexes from individual genes by experimental means.

T

Tetrasomic – An aneuploid in which a single chromosome pair is duplicated and thus is tetraploid, $2n + 2$. This is more likely to occur in polyploids.

Transgressive Variation – When the range of variability found in a segregating generation such as the F2 exceeds the range found in either parent or the F1, the variation is *transgressive*.

Trisomic – Aneuploid in which one of the individual chromosomes is present in triplicate $(2 + 1)$. As with monosomics there are as many possible trisomic types as there are different chromosomes in the haploid genome. A secondary trisomic occurs when the extra chromosome is an isochromosome.

Types of Selection Compared

Centripetal, stabilizing or normalizing selection – The selection that occurs in overy normal population eliminating extreme variations, maintaining the adaptation of the group somewhere near its optimum for the environment in which it lives.

Balancing selection – a type of stabilizing selection in which two or more discontinuous forms are maintained in the same interbreeding population (balanced polymorphism) because each form is of advantage to the population part of the time, or in part of the environmental circumstances under which the population lives, such as seasonal fluctuations in temperature.

Centrifugal, disruptive or diversifying selection – selection away from the original made of a population in two or more direction either as a response to a more or less sudden change in the original environment (short duration) or as a form of balancing selection which produces a comples polymorphism.

Linear or directional selection – contained selection in a given direction producing an evolutionary trend, a sequence of a adaptive changes in response to an environmental gradient.

U

Unit Characters and Pleiotropy – Specific genes often produce spectacular effects upon specific characters. Further study has shown that most genes affect many different characteristics even when one of the effects is more obvious than the others. Also some genes (pleiotropic) have major effects upon a number of seemingly unrelated characters. Since the action of most genes is related to enzyme systems and metabolic pathways, these results are to be expected.

V

Viability and Lethality – All genes affect in some direct or indirect way the survival, health, and longevity of the organism. Relative *viability* can be measured by comparison with a standard or *"normal"* population. Even *lethality*, in terms of gene action, is not absolute but relative, because death caused by the action of a specific gene or gene complex may be *haplontic* in spore or gamete, or *diplontic* in the zygote, or delayed to a late embryonic stage or even to early maturity.

EVOLUTION

A

Abiotic Selection– Selection primarily by the physical environment, usually unidirectional. (See parallel evolution).

Adaptation – The result of natural selection; the evolution of a character, or group of characters that favor survivial under certain selective forces.

Adaptive Radiation – Evolutionary diversification within a group of related species or other taxa brought about by adaptations to different selective forces through time or migration.

Adaptive zone – The "place" or "niche" of an organisam in the evolutionary reticulum of its, commumity; the set of conditions to which the ff spring of an organism will be preadapted.

Advanced – An organism, character to condition presumed to be more specialized or evolved than another to which it is compared.

Adventitious Embryony – The production of an embryonic sporophyte by mitotic divisions from the tissues of the parental sporophyte without an intervening game tophytic generation.

Agamospermy – The formation of seeds without fertilization; pollen of male gametes may be necessary to stimulate division of the zygote.

Allogamy – Cross-fertilization.

Allometry – Different growth rates in different parts of the same organism.

Allopatric – Populations or species with separate and nonoverlapping geographic ranges.

Allopatric Speciation – The differentaion of geographically isolated populations into separate species.

Allopatry – The occurrence or origin of species or populations in different geographical regions.

Apogamy – The production in plant of a sporophtyte directly from a gametophtyte cell without the production of a zygote deirved by fusion of gametes.

Apomict – An organism that reproduces by asexual means; it may be faculatives or an obligate apomict.

Apomixis – Asexual reproduction, including vegtative propagatio.

Apospory – The condition found in some higer plants in which a diploid embryosac is formed directly from a somatic cell of the nucellus of chalaza and an embryo is then formed without fertilization.

Autodeme – A group of individuals of a taxon composed of predominantly self-fertilizing individuals.

Autogamy – Self-fertilization; persistent autogamy results in an increase in homozygosity and the division of the population into a number of "pure lines".

Automimicry – Mimicry within a species.

Allasesthetic Selection – Disruptive selection via differential pollination.

B

Batesian Mimicry – The superficial resemblance of harmless organism (the mimic) to a sympatric and more plentiful organisms (the model).

Biosystemaics–The systematic study, usually including reproductive biology and chromosome studies, or the variation and evolution of a species of species complex.

Biotic Potential – The total reproductive potential of an individual or a population; it is directly related to the number of functional gametes, and seeds, produced.

Biotic Selection – Selection primarily by the biotic environment. See competitive exclusion and character displacement.

Bradytelic Evolution – Evolution at a relatively slow rate as compared with most other known or presumed evolutionary rates.

C

Canalisation – The propetry, possessed by developmental pathways, of achieving a standard phenotype despite genetic or environmental disturbance.

Catastrophic Selection – A harsh selective force (e.g., fire, severe drought) acting on an ecologically marginal population so there are only a few survivors and a small gene poll remaining; results in operation of the Founder Principle.

Character Displacement – The increase in the differential of a character of two competing species in an area of sympatry.

Cleistogamy – The flower self pollinaties, without opening in the bud.

Cline – A character gradient, often related to geographic distribution; a more or less uniform series of variants from one extreme to the other in a population or series of populations; a continuous variation pattern.

Clone – A group of individuals organisms produced by vegetative a sexual reproduction (rhizomes, stolons, budding, cuttings, etc.) and thus all of the same genotype.

Coadaptation – The adaptive response of two organsim to each other; e.g., flower structure and pollination.

Coenospecies – An ecolgical category approximating a Linnaean genus.

Commensalism – A relationship between individuals of two species in which one species benefits but does not harm the second, or host, species.

Comparium – A genetic term to designate a group of coenospecies which can hybridize either directly or through intermediates.

Competitive Exclusion – The effect of strong biotic selection which eliminates one of two competing sympatric/ organisms; environmental modification by one exclution of another; e. g., plant succession.

Convergence – A type of parallel evolution resulting in similar form or e.g., ecological equivalents; the morphological similarity of desert-dwelling plants of the Euphorbiaceae and Cactaceae.

Correlation Pleides – The tendency of two or more characters of an individual that are under the same selective force to have correlated patterns of variation.

Cross Pollination – Pollination between two different plants of the sanme species; outcross.

Cytodeme – A group of individuals of a taxon differing from others cytologically, e.g., in chromosome number.

D

Dimorphism – The occurrence of two morphological, or other, forms of a given organism; e.g, sexual dimorphism.

Diplospory – The phenomenon in some higer plants in which a diploid embryo-sac is formed directly from a megaspore mother-cell; an embryo is then formed without fertilization.

Directional Selection – Selection occuring when the environment is changing in a systematic fashion, leading to a regular directional change of the adaptive characteristics of a gamodeme.

Disjunct – A geographically isolated population or species outside of the range of other similar populations or species.

Disruptive Selection – Selection (usually biotic) which breaks up a weakly polymorphic population into a number of differently-adapted variants.

E

Ecodeme – A group of individuals of a taxon occurring in a specified kind of habitat.

Ecospecies – An ecological category indicating an infra-specific group of plants adapted to a particualr habitat.

Endemic – A popoulation or species with narrow physiological or other restrictions which limit it to a special habitat or a very restricted geographic range, or both.

Environmental Variance – That portion of the phenotypic variance caused by differences in the environments to which the individuals in a population have been exposed.

Epigenotype – The series of interrelated developmental pathyways through which the adult form is realized.

Euheterosis – The condition in which the heterozygotes in a population are at an adaptive or selective advantage over the homozygotes.

Evolution – A directional change in gene frequency, over time, in responses to selective factors of the environment; the continuous genetic adaptation of organisms to their environment.

F

Feed back – The influence of the result of a process upon the func-

tioning of the process; success in producing viable and fertile offspring.

Fittness – The survival value and reproductive capability of a given genotype relative to other genotypes in a population.

Founder Principle – The establishment of a new population in isolation so that its gene pool is not identical with that of the parent population; these differences are enhanced as different evolutionary pressures in the areas occupied by the two populations must therefore operate in different genetic environments; the result is increased divergence.

G

Gamodeme – A group of individuals of a specified taxon which, within the limits of the breeding system, can interbreed.

Gap – A discontinuity in varation.

Genecology – The study of intraspecific variation in plants in relation to environment.

Gene Flow–The distributional "movement" of genes within and among populations.

Gene Frequency – The percentage of an allele in a population; the number of loci at which a given allele is found within a population, divided by the total number of loci at which it could occur.

Gene Pool – The total genetic information possessed by a population.

Genetic Assimilation – The incorporation into the genotype, by a selective process, of characteristics appearing in ontogeney as a response to the environment.

Genetic Coadaptation – Adaptation of blocks or combinations of genes to each other in an interbreeding population.

Genetic Drift – Random fluctuation of gene frequency usually due to sampling error inherent in the genetic mechanism; present in all populations, its effects are most evident in very small populations.

Genetic Homeostasis – The tendency of populations under selection to maintain, or regress toward the adaptive mean rather than an adaptive peak.

Genetic Load – The average number of potential deaths per individual due to genetic causes, such as lethals and semilethals, in a population.

Genetic system – All the factors that affect genetic recombination in an organism.

Genotype – Refers either to the total genetic component of an organism or, in a different sense, to a specific allele, in a particular individual.

Genotypic Variation – Genetic variation; differences in genotypes within a population (or species) are a result of mutation and recombination.

Gynogenesis – Reproduction by parthenogenesis in which stimulation by a sperm is necessary for the development of the egg.

H

Haplophase – That part of the life cycle in which the gametic chromosome number is found in reproductive cells.

Hardy-Weinberg Law – The law stating in the absence of selction, mutation, migration and genetic drift, the frequency of autosomal genes at a given locus remains constant and that, after one genseration, the frequency of genotypes at the locus reaches equilibrium.

Heritability – The genetic variance divided by the phenotypic variance; resemblance between offspring and parent.

Hermaphorditic–Possessing the phenotype of both sexes in organisms with male-female differentiation.

Heteroblastic Change – The transition from a juvenile to an adult form accompanied by a more or less abrupt change in morphology.

Heterogametic – Producing gametes with differeing chromosome complements.

Heteromorphic – With many form; polymorphic.

Heterotrophic – Requiring organic carbon for nutrition.

Higher Categories or Ranks – Taxonomic categories or ranks above the level of genus in the established hierarchy of our taxonomic system.

Hologamodeme – A group of individuals of a taxon which are believed to inter-breed with a high level of freedom under a specified set of conditions. Theis term is preferred to 'biological' species' which relates to a taxonomic category.

Homomorphic – With a single form; monomorphic.

Horotelic Evolution – Rates of evolution falling within the distribution (asymetrical with a mode nearer the upper than the lower and) most commonly found when evolutionary rates are plotted in a frequency distribution.

Hybrid – An individual resulting from cross fertilization between organisms belonging to different taxa, usually different species.

Hybrid Zone – A zone of sympatry and hybridization between two species which may, despite hybridization within the zone, be a barrier to extensive gene flow or introgression between the species involved.

Hybridization – The formation of a zygote through the fusion of gametes from organis belonging to two different taxa.

I

Imprinting – The imposition of a stable behaviour pattern by exposure, during a particular period in development, to one of a restricted set of stimuli.

Inbreeder – A plant that is normally self-fertilized.

Induction – Determination of the developmental fate of one cell mass by another.

Introgression – Incorporation of genetic material form one population into that or another through hybridization followed by repeated backcrossings; usually restricted to populations regarded as distinct by taxonomists.

Introgressive   Hybridization – Genetic modification of one species by another by gene flow through fertile hybrids and backcrosses; introgression.

Isolating Mechanisms – Any mechansim or condition which prevents gene exchange between the individuals of two or more populations; isolation may be either prezygotic or postyzgotic and may be of several types.

M

Macroevolution – Evolutionary events viewed through the perspective of geologic time, such as the evolution of different and distinctive organisms.

Matroclinous   (Maternal)   Inheritance – The condition when a hybrid is closely similar to its seed parent.

Meiotic Drive – A higher probability of one allele at a locus being included in the gametes than that of others.

Mendelian Population – A reproductive group sharing a common gene pool.

Meristic Variation – Variation in numbers of parts or of organs.

Mertensian Mimic – When one deadly and one harmless organism both mimic moderately poisonous or distasteful model.

Mesopolyploid – A polyploid of intermediate biological or geological age. (See paleopolyploid, neopolyploid).

Microevolution – Evolutionary events usually viewed over a short period of time, such as changes in gene frequency within a population over a few generations.

Migration – The dispersal and establishment of organisms beyond their place of orgin; a periodic movement of individuals; both result in the transfer of genetic information among populations.

Mimicry–The superficial resemblance of one organism by another, presumably affecting the actions of predators.

Monomorphic – One form.

Morphogenesis – The process leading to the development of the characteristic mature form of an organism.

Mutualism– A type of symbiotic relationship between two organisms of different species which benefits both of the organisms involved.

N

Natural Selection – Nonrandom (differential) reproduction and survivial of geno-types without the conscious intervention of man.

Neopolyploid – A relatively recent polyplid, the parent (s) of which are still extant.

Niche – The status (way of life) of an organism in a community; e.g., its position in the food chain or relationships with the physcial substrate.

O

Ontogeny – The development of an individual.

Organizer – A portion of an embryo that determines the developmental fate of the cell masses with which it comes into contact.

Outbreeder – A plant that is normally cross fertilized; outcross.

Outcross – Pollination between two different plants of the same species; cross pollination.

Overdominance – The result of the

heterozygote being more extreme than either homozygote.

P

Paleopolyploid – An ancient polyploid the parent (s) of which are extinct.

Panmixis – Completely random mating within a population.

Parallel Evolution -the result of strong directional selection producing similar adaptations in related organisms.

Parallelism – Evolutionary convergence along similar pathways among closely related organisms.

Parasitism – When one organism obtains food or raw materials, or both from another living organism (the host) with no apparent benefit and often harm to the host; some organisms are obligate parasites and cannot live without their host, other organisms, facultative parasities, may be autotropic under some conditions.

Parthenogenesis – The development of an individual from an unfertilized gamete.

Penetrance – A measure of the proportion of individuals homozygous for a gene that shows its phenotypic effect.

Periodicity – The periodic occurrence of a given biological phenomenon, presumably as a result of natural selection.

Phenetic – Presumed natural (but not phyletic) relationship based on current total similarity; relationships indicated by the use of unweighted characters, as in numerical taxonomy.

Phenocopy–An environmentally produced specialized phenotype that resembles the phenotype of a different, more specialized, genotype; facultative dimorphism.

Phenology–Biological processes that are correlated with the seasons, e.g., flowering, mating, dormancy.

Phenotype – The actual expressed characters or observable form or physiology of an organism.

Phenotypic Variance – The total variance observed in a character.

Phenotypic Variation – Morphlogocial or physiological variation; the result of the action of different environments on one or more genotypes.

Phyletic Evolution – Pertaining to phylogeny, the evolutionary history of an organism; relationship through lineal descent from a common ancestral form.

Plasticity – The degree to which a given genotype may vary phenotypically under different environments; also, the total variation (both genotypic and phenotypic) of a species or population.

Polymorphism – The occurrence at a given stage of development, of two or more distinct variants of a species or population.

Polyploid – An individual with 3 or more (haploid) sets or chromosomes.

Polyploid complex – A series of inter-related polyploids, often with morphological similarities, involving one or more levels of polyploidy.

Polytopic – The occurrence of a species, or other taxonomic group, in two or more separate areas.

Population – A group of individuals of one kind in a given area at a given time; often used as a synonym for Mendelian population.

Population Structure – The sum of all the factors that govern the pattern in which gametes from various individuals unite with each other.

Postzygotic Isolation – The prevention of gene flow between hybridizing populations (or species) cused by the failure of the hybrids, or their derivatives, to be viable or reproductively competitive.

Preadaptation – The historical (evolutionary) or fotutious ability of an organism to survivie in a particular environment in which it finds itself.

Prezygotic Isolation – The prevention of gene flow between populations (or species) caused by geographical, ecological, seasonal, mechanical, ethological or other differences which variously act to prevent cross fertilization.

Primitive – An organism or character judged to be less changed from a presumed common ancestral state than another with which it is compared.

Pseudocopulation – A mode of pollination in some orchids in which structures of the flower resemble attempting copulation transfer pollen from one flowere to another.

Pseudogamy – The phenomenon found in some apomictic plants, where by pollination is necessary for seed development.

Pure Line – A lineage of individuals originating from a single homozygous ancestor.

Q

Quantum Evolution – Rapid evolutionary change resulting in what a taxonomist would regard as a new higher taxon.

R

Ramet – An individual belonging to a clone.

Random Sample – A subset of a population selected or obtained in such a way that all variants in the populations are equally likely to be included in the sample.

Ration-Cline – Clinal variation occurring in polymorphic species, in which successive populations show progressive change in the proportion of the variants.

Recapitualtion – The idea that in the course of development an individual form to adults of its presumptive ancestors; often stated as "ontogeny recapitulates phylogeny".

Recombination – The recombing of genetic material through crossing over and random assortment at meiosis and the fusion of genetically different gametes.

Recombination Index – A measure of the number of new gene combinations that can be produced in a given time.

Recombination System – The interaction of a series of genetically controlled processes which more or less determines the amount of recombination possible in a species; the system may be "open", "restricted", or "closed".

Replicate – To duplicate repeatedly.

Reproductive Cells – The gametes and their immediate predecessors from which they are produced by division.

Restitution Nucleus – A nucleus formed by the fusion of two daughter nuclei, usually in a cell which did not divide after its nucleus divided.

Reticulate Evolution – Evolution in which phyletic lines merge as well as well as diverge.

S

Sample – A subset of a population; a group of items picked by some procedure and from which one hopes to learn certain things about the populations.

Sampling Error – Variation due to random elements in the sampling process.

Selection – The action of environmental factors on each member of a population that results in the differential reproduction of genotypes.

Selection Coefficient – The measure of the disadavantage of a given genotype relative to other genotypes in the population.

Self-fertiles – Capable of self fertilization.

Sexual Reproduction– Reproduction involving the fusion of gametes produced by a prior meiotic process.

Specialized – Applying to an organism or character judged to be more changed from a presumed common ancestral state than another with which it is compared.

Speciation – The splitting process of evolution that results in the existence of different kinds of organisms that are classified as species by taxonomist.

Species – A category in the taxonomic system; a group of organisms judged by taxonomists (by diverse criteria) to be worthy of formal recognition as a distianct kind, category, or rank.

Spontaneous Generation – Abiogenesis.

Sporophyte – The diploid spore-producing phase of the life-cylce of a plant.

Stabilizing Selection – Selection in which genotypes closer to the mean for a character have an advantage over those at the extremes.

Stuuctural Hybridity – Heterozygosity for a chromosomal rearrangement.

Subspecies – The sequence of transient communities occurring in an area before the climax community is establishd.

Symbiosis – Living together; see mutualism, commensalism and parasitism.

Sympatric – Occurring in the same geographic area.

Sympatric Speciation – Speciation without geographic isolation; see divergent selection.

Sympatry – Populations (or species) occupyging the same geographical area.

Syngamodeme – A unit composed of all coenogamodemes which are connected by the ability of some of their members to form viable but sterile hybrids junder a specified set of conditions.

Systematic Pressure – One of the non-random evolutionary pressures: selection, mutation, or migration.

T

Tachytelic Evolution – Evolution at a relatively rapid rate, e.g., at a much faster rate in horotelic evolution.

Taxon – A unit of taxonomic classification such as family, genus, or species.

Territoriality – The defense of an area against organisms of the same or similar kind.

Topocline – A geographical variational trend which is not necessarily correlated with an ecological gradient.

Topodeme – A group of individuals of a taxon occuring in a specified geograpghical area.

Transgressive Variation – When the heterozygote (or individual segregating) is more phenotypically extreme than either homozygote (or parental form); overdominance.

V

Variant – A neutral term applied to any definable individual or group of individuals.

Variation Pattern – The type, form and extent of variation in one population of taxon as compared with that in another population or taxon.

Viability – The capability for living or continuing to develop; often, but incorrectly, used as synonymous with fitness.

Vicariads – Closely allied but allopatric taxa.

Vivipary – In plants, vegetative reproduction in which propagules replace flowers in the inflorescence; in animals, the production of living young rather than eggs.

X

Xenogamy – Outcrossing, outreeding.

TAXONOMY & BIOSYSTEMATICS

artificial – A system of classification based on one or a few important characters. The drawback with this system is that the plants closely resembling each other are often placed in widely separated groups, while those quite different from each ohter are placed in the same group.

authortiy – The name of the author of a name cited as such, after the name, usually in abbreviated form. Whether family, genus or species and gave it a name, the name of the person or persons written after the scientific name is known as the authority of the name. *Poa prantensis* was first named and described by Linnaeus, he becomes the authority for that name. Similarly, *Erythronitium grandiflorum* Pursh, shows that Pursh, first named the species. *Lomatium montanum* C & R, was first named and described by two men, Coulter and Rose.

apud (ap.) – Latin, meaning with in the work of; used in citing the work of an author contained within the publication of another author.

antonym – Award having the opposite meaning;

author – 1) The person who first published a name in such a way as to satisfy the criteria of availability or valid publication.
2) Ther person to whom is attrtibuted a work, a scientific name, a nomenclatural act or a nominal taxon.

asterisk (*) – A symbol used in taxonomy to denote categories of infraspecific rank; usually placed between the second and other third words of the trinomen.

associate type – Any of two or more type specimens listed in the original description of a taxon in the absence of a designated holotype; syntype.

available – To be taken into consideration for the purposes of scientific nomenclature; the equivalent of validly published of Bact, and Bot. 2) Bot. To be taken into consideration for the purposes of deciding the correct name of given taxon, i.e., legitimate and with its type falling within the range of variation of the taxon.

autonym – Automatically established name, applied to a nominate subordinate taxon.

auctorum non – (auct. non) – Latin, meaning of authors, not of; used to indicate a misapplied name.

admissible – Of a name, of a form that could permit it to be validly published; of the use of a name or epithet, in accordance with the provisons of the code.

artificial-classification – A classification structured for convenience, using easily observed phenotypic characters and not necessarily indicating phylogenetic relationships; key classification.

autographs – A document or text in the handwriting of the author, or a mechanical copy of such a document or text.

alpha taxonomy – That aspect of taxonomy concerned with the description and designation or species, typically on the basis of morphological characters; descriptive taxonomy.

artificial character – A character selected arbitrarily, without consideration of phylogenetic relationships.

adanasonian taxonomy – An early method of classification advoatiing the grouping of organisms on the basis of many equal-weighted attributes, a principle adopted in mdoern numerical taxonomy to avoid the subjectivity of classical taxonomic methodology.

analysis – A figure or figures, illustrative of a taxon, and showing the details necessary for identification.

autotype – The type, by original designation, of a taxon.

arbitrary – Used of a scientific name lacking formal derivation with regard to cynology (a name comprising an arbitrary combination of letters).

alternative names – Two names for the same taxon, of the same rank, published simultaneously by an author.

auetorum (auet.) – Latin, meaning of authors.

application – The use of a name to denote a taxon.

artifical key – An identification key based on convenient phenotypic characters and not indicating phylogenetic relationships.

aliorum – Latin, (al. alii) meaning others, of others.

affinis – Meaning akin to.

B

Botanical Gardens – 1) *Royal Botanic Gardens, Kew England.*

2) *National Botanic Gardens, Lucknow, India.*

3) *Llyod Botanic Gardens, Darjeeling, India.*

4) *Royal Botanic Garden, Shibpur, Calcutta, India.*

5) *Botanical Garden, Shahranpur, India.*

6) *Lalbagh Garden, Bangalore, India.*

binomen – The name of a species, consisting of the name of the genus in which it is classified followed by a word peculiar to the species; the equivalent of specific name of Bact. and Bot.

biotype – A given genotype expressed in one or more individuals. 2) (Bact.) A variant of a species, subspecies or serotype, which may be distinguished by the spossession of some special or usefully diagnostic physiological feature. The use of the term 'biovar' to replace 'biotype' is recommended by the ICNB.

basionym – The name, replaced by another making use of the same stem or epithet, as a result of a change in position and or rank of the taxon to which it refers.

brackets square brancketts – A pair of printed marks placed around a word or words. used to enclose a superspecies-group name, to indicate a misidentification in a synonym, and to enclose the name of the author of any nomenclatural act that was originally published anonymously but for which the author later became known.

bionumeric code – A numerical code in which separate categories of a hierarchy are assigned to different groups of digits.

bracketed key – A dichotomous key in which contrasting parts of a couplet are numbered and presented together, without intervening couplets, although the brackets joining each couplet are now omitted; bracket key; parallel key.

bigeneric – Resulting from the sexual cross of individuals assigned to two different genera.

C

changes in name when taxa are divided, fused or alterad – When a genus is divided into two or more genera, or a species is split into two or more species, the original generic or specific name must be retained for the new taxon containing the type; this applies also to infra-specific taxa.

When a section of a genus or species is transferrd to another genus or species without alternation in name, the original name must be retained whenever possible.

When rank of a genus of infra generic taxon is changed, the correct name or epithet is the earliest legitimate one available in the new rank.

When taxa of the same rank are united into one, of the oldest legitimate name must be used for the new combined taxon; if the names are of the same date, the author who first unites them has the right to choose one of the names and his choice must be followed by subsequent botanists.

category taxonomic – A concept to which taxa are assigned for the purpose of classification, a set of conventionlal categories constitutes the taxonomic hierarchy like *class, sub-class, order, family, genus, specice* etc.

circumscription – The limits set to a taxon by a gien author in terms of range of variation and or component adividuals, specimens, population or tax a.

compound – A word or name formed by the union of two or more words, concerning characters which are the same in the OTU's under comparison. (OTU = Operational Taxonomic Unit).

chemical taxonomy–Chemosystematics Primary metabolites parts of vital metabolic pathways and most of them are of universal occurrence. (e.g., organic acids, sugars and amino-acids, ii) Secondary metabolites. The compounds perform non-vital functions, and are therefore less widespread in plants. e.g., alkaloids phenolics, glucosino-lates, amino-acids, terpenoids, oils, waxes and carbohydrates.

Secondary plant products are lar-

gely waste substances, food stores, pigments, poisons, scents, structural nits or water repellents etc,. (iii) Semantides Information carrying molecules; DNA is primary semantide; RNA, a secondary semantide, and proteins are tertiary semantides phyloseny of plants based on chemical compounds is chenotaxonomy.

confer (cf. cfr) – Latin, meaning compare (with).

clone (cl) : (Bot. Cult) – A group of individuals formed by the vegetative (natural or artificial) or apomictic reproduction of a single original parent.

2) Bact. a population of bacterial cells derived from a single parent cell.

correctus: (correctsa, correctum, corr) – Latin, meaning corrected (by) used under to indicate a corrected orthography, the abbreviation corr, following the authority and prceding the name of the author who first effected the correction.

consortium – (Bact.) An aggregate or association of two or more organisms; names given to consoria are not regulated by the Bacteriological code, and have no standing in bacteriological nomenclature.

classification – 1) The process of establishing and delimiting taxa.

2) A system so produced.

combination nova – (comb. nov) Latin, meaning new combination; a combination made available (Zoo) or validly published combination (basionym,) from which the word peculiar to the taxon (epithet,) is transferred; used in citation to indicate a change in the position and or rank of a nominal taxon.

conservation – Procedure whereby the use of a name which would in the absence of such procedure contravene the provisions of a code is made possible.

centrotype – In numerical taxonomy,

that OTU (Operational Taxonomic Unit) closest to the geometric centre of the cluster of OTUs to which it belongs.

character index – A numerical value which quantifies the degree of difference between two taxa.

combination – The name of a taxon of below generic rank (Bot) or of specific or subspecific rank (Bact., Zoo), consisting of the name of the genus followed by one or more words peculiar to the taxon.

cultivar – 1) A taxon consisting of an assemblage of cultivated plants, maintained in cultivation, and retaining its distinguishing features when reproduced.
2) The corresponding taxonomic category. Abbreviated cv.

cultivar class – The taxon within which the use of the same cultivar name of two different cultivars would cause confusion.

continuity – The principle that continuity of usage of a particular name should take precedence over the priority of publication in determining which of two or more competing scientific names should be adopted.

commumicavit–(comm.) Latin meaning he (she or it) communicatied.

combined description – A description of a new species assigned to a monotypic genus that utilizes the same character states to diagnose both the species and the genus.

coenospecies – A group of ecospecies capable of limited genetic exchanbge by forming fertile hybrids.

commerical synonym – A name used for a cultivar in place of the internationally accepted name in countries where the latter is commericially inacceptable.

cluster analysis – A method that groups those variables within a set of variables that are highly correlated and that excludes from clusters those that are negatively correlated or correlated; used in numerical taxonomy as a procedure for arranging OTUs into homogeneous clusters based on their mutul similarities.

comparium – A biosystematic unit comprising one or more coenospecies capable of interbreeding and producing hybrids with partial fertility.

constitution – The constitution of the ICBN.

circa (ca.) – Latin, meaning about, approximately.

class – A rank within the hierarchy of taxonomic classfication; the principal category between phylum or division and order.

cladistic method – A method of classification employing phylogenetic hypotheses as the basis for classfication and using recently of common ancestry alone as the criterion for grouping taxa rather than date on phenetic similarity.

collective – (Bot. Cult. Orch) Of name or ephithets, used to denote hybrids or groups of hybrids.

commonality principle of – That the character state having the widest distribution among the taxa comprising a higer taxon is considered to be the most primitive.

classical – Pertaining to a name that is derived from Latin or ancient Greek.

article – A numbered section of the code, consisting of a rule or rules, which are mandatory, and sometimes also of examples, which are explanatory, and or supplementary recommendations.

cotype – A term now superseded, formerly used for syntype, isotype or paratype.

code – The system of rules and recommendations regulating nomenclature, published in the most recent edition of the International Code of Botanical Nomenclature.

code of ethics – A list of recommendations on the priority of certain taxonomic actions, set out in an appendix to the code.

conditional – Used with reference to the proposal of a scientific name which is made with definite reservations about its status.

chemotaxonomy – A recent trend rapidly expanding areas of plant taxonomy which utilizes chemical information to improve the classification of plants. (biochemical techniques and biochemcial characters).

citatus – (citata, citatum, cit) Latin, cited.

correctus (corr.) – Latin, meaning corrected (by) of a name, that by which a taxon should properly be known.

convivium – A biosystematic unit comprising those members of a *commiscuum* prevented from interbreeding by a geographical barrier.

checklist – A list of species arranged within a simple classification for convenience of reference.

confamilial – Belonging to the same family.

coenogamodeme – A biosystematic unit comprising all the hologamodemes which are capable of exchanging genes to some extent, but not with freedom, and which the capable of hybridizing with other coenogamodemes producing sterile hybrid offspring.

cyto-taxonomy – Classification based on a study of cell structrure, with special reference to chromosome structure.

cohort – A rank within a hierarchy of taxonomic clasification; a category comprising group of families.

correxit – (corr.) Lating, meaning he corrected.

coheironym – An unpublished scientific name; manuscript name.

common name – Colloquial or vernacular name.

cultivarietas – Latin, meaning cultivar.

cult – Internationla code of Nomenclature of Cultivated plants.

corrected original spelling – The corrected version of an incorrect original spelling.

conservandum(cons.)–Latin, meaning to be conserved.

D

date, publication, of – Of a work, the date of its becoming available to the general public or to relevant institutions.

2) Of a name, the date on which the criteria of availability (zoo) or valid publication (Bact, Bot) were first satisfied.

designation – 1) (Zoo) the act of an author in fixing, by an express statement, the type of the name of a taxon of the species-group or the genus-group.

2) (Bot) a statement of the features of a taxon which, in the opinion of the author, distinguish it from others; the equivalent of definition of (q.v.) Zoo.

DNA/RNA hybridization – A technique useful in chemo-taxonomy. The amount of association occurring between the DNA of one organism and a fraction of the RNA (usually ribosomal-RNA) of another is taken as a measure of relationship.

description generico specific – Latin, meaning combined description.

designation – The general name or formula of a taxon.

de novo – Latin, meaning arising a new.

data matrix – A tabulation of date, such as taxonomic characters, to show differences between categories (taxa) often in a machine readable form.

description – (descr.) Latin, meaning description; a statement of the attributes of a specimen or taxon.

determinavit – (det.) Latin, meaning he (she or it) identified, he (she or it) determined.

dichotomous Key – An identification key constructed as a sequence of alternative choices, each pair forming a characrter couplet; diagnostic key; sequential key.

E

emendatus – (emendata, emendatum , emend) Latin, altered (by); indicates a change in circumscription of a taxon without exclusion of the type of its name; the abbreviation emend. Follows the authority and proceedes the name of the author who affected the change.

errore typographico – (err. typogr.) Latin, meaning by typographical error.

exclusus – (*exclusa, exclusum, exxl.*) Latin, meaning excluded; used to indicate elements included in a taxon by a previous author or authors, but considerd not to belong to it by the writer and excluded from it by him.

epithet – (Bact., Bot) A word, other than a generic name or a term indicative of rank, forming part of a combination.

establish – 1. (Zoo) of a name, to make available.
2. Of a taxon, to describe and validly publish (or make available) a name for to erect.

elements – The constituent parts that fall within the limits of taxon, such as the species within a genus.

effective publication – Publication in accordance with the requirements of the Code.

excluso genere – (excal. gen.) Latin, meaning with the genus excluded.

ex – Latin, meaning from, according to, used to connect the names of two authors when the second author validly published a name proposed by, but not validly published by, the first author.

exclamation mark – In taxonomic publication, sometimes cited after a herbarium specimen to indicate that it has been examined.

evolutionary method – A method of classification employing hypotheti-

cal reconstructions of evolutionary history incorporation both cladistic date (on the sequence of branching events) and morphological divergence date; synthetic method.

euclidian distance – In numerical taxonomy, the distance bewteen OTUs calculated by extension of Pythagoras' theorem.

et – Latin, meaning and, used in nomenclature to connect the names of co-authors, often substitued by another person.

et. alic – (et. al.) Latin, meaning and others; used in an author citation to indicate that a publication or scientific name was written or published jointly by more than two authors.

epitheta specifica rejicienda–Latin, meaning rejected specific epithets.

excl. specim; (excl. spic.) – Latin, meaning exclusis speciminbus, with the specimens excluded.

exsiceatus -(exsic. exs.) Latin meaning dried; the code used for dried, preserved herbarium specimens.

epitheta hybrida – Latin, meaning specific epithets made up of parts of words from two or more different languages.

excluse speciei (excl. spec.) – Latin, meaning with the geus excluded.

exclusis varietatibus (excl. var.) – Latin, meaning with the varietics excluded.

emendavit – (emen,) Latin, meaning he emended.

ex. parte – (Latin, meaning in part.

excluded name – A scientific name that is not subject to the provisions of the Code Floras of India.
Flore of Upper Assam by Griffith (1836).
Naga Hill Flora by Masters (1844).
Banda District Flora by Edgoworh (1852).
Moulmein Flora by parish (1859).
Bihar and Lucknow Flora by Anderson (1863).
Burmese Flora by Kurz (1863).
Jhenlum Flora by Stewart (1881).

Darjeeling Flora by Gamble (1875-76).

Jeypore Flora by Beddome (1879).

Noth Western and Central India Flora by Stewart and Brandis (1874).

Rajuputana vegetation by King (1878) etc.

Flora of Madras Presidency by Gamble and Fischer.

Flora of Bombay Presidency by Cokke.

Flora of Upper Gangetic Plains by duthie.

Botany of Bihars Orissa by Haines.

Bengal Plants by Prian.

F

Form-group, (Bot) 1. A taxon of fossil plants of generic rank, which may be unassignable to a family, to which are referred apparentlty similar small isolated parts, e.g., leaf-fragments, roots, spores, seeds. 2. A taxon of imperfect fungi of generic rank.

The name of a form-genus can be used to refer only to the part or state represented by its type.

Form and spelling of names – According to the Code, authors have full liberty to coin their names. When the name is taken from the Latin Language, it retains the original gender and spelling. When the name is taken from the vernaculars, the author assigns a gender to it.

Names of genera and higher taxa are written with a capital initial letter, specific names should be written with a small initial letter.

formula – (Bot., Cult., Orch.) A designation of hybrid or hybrid group, formed by connecting the names of the partes by sing (X or +) or by combining them into one, forming a condensed formula preceded by a similar sign, e.g. *Crataegus* X *Mespilus*; X *Crataegomespilus*.

formula, condensed – A formula formed by combining parts of the names of two genera, and applied to intergeneric hybrids between them.

formulation – The way in which a name is formed.

fossil – An organism, or part of it, preserved by some natural means in the geological record, or an impression or petrification so preserved.

forma specialis – (f. sp.) Latin, special form; (Bact., Bot) a variant of parasitic or symbiotic species distinguished primarily by its adaptation or restriction to a particular host or hosts.

first reviser – The first author to cite simultaneously published synonyms, including different forms of the same name, or nomenclatural acts and to select one to take precedence over the other in the interest of nomeclatural stability.

fixation – The determination of a type species or type specimen, whether by designation, indication or subsequent fixation.

family name – The scienfic name of a taxon of family rank; with the ending *aceae*.

fancy name – A term given to a cultivar that is neither the proper scientific name nor its common vernacular name.

forma specialis – (f. sp.) Latin, meaning special form.

family – A rank within the hierarchy or taxonomic classificarion the principal category between order and genus.

form species – One of a group of species having similar morphological characters but which are not confined as having common ancestry.

fide – Latin, meaning according to on the authority of.

forma – The lowest category in the hierarchy or botanical classification; aplied to minor varations within a population.

form genus – A genus a taxon of generic rank comprising one or more form species.

figura – (fil. f.) Latin, meaning son.

formulation – Used in nomeclature to refer to the way in which a name is formed.

figure-type – An orginal figure of illustration of a specimen.

G

grex – (Cult., Orche) a hybrid taxon to which are referred all the progeny arising from any, each and every crossing of any two parent plants belong to different taxa that bear the same pair of specific, hybrid-specific or grex names.

group – 1. (Bact) Anm informal taxon, based upon antigenic analysis, consisting of an assemblage of serotype.
2. (Zoo) An assemblage of nomenclaturally coordinate categories; a name-group.
3. (Cult) An assemblage of similar cultivars whithin a species or interspecific hybrid.
4. (Vir) A viral taxon of uncertain rank, as yet not designated either as a family or as a genus.

genus – A rank in the hierarchy of taxonomic classification forming the principal category between family and species.

generitype – The type of the name of a genus.

gamma taxonomy – That aspect of taxonomy concerned with interspecific populations and with phylogenetic trends.

genoecodeme – A local interbreeding population occurring in a particular habitat (an ecodeme) that is characterized by genotypic features.

genolectotype – The primary type of the type species of a genus selected from a by the author in the original descripotion of the genus.

generic name – A scientific name of a taxon of genus rank; the first word of a binomen or trinonen.

genus et species nova – (gen. et. sp. nov.) Latin, meaning new genus and species.

gamodeme – An assemblage of individuals forming a relatively isolated naturally interbreeding poplation.

H

homonym – A name identical in orthography with another (or treated as such by the appropriate code) and based on a different type.

hortlanourum – (hert) Latin, meaning of gardeners; used in citation to denote a name of garden origin.

hierarchy, taxonomic – The framework formed by the conventional ordering of taxonomic categories into a series of consecutively subordinate levels or ranks.

hologamodeme – A local interbereeding population comprising all those individuals which are able to interbreed with a high level of freedom under a given set of conditions,; a unit capable of hybridizing with other hologamodemes to give hybrids showing some fertility.

holotype – the sole element used as the type by the author of a name or the one element designated or indicated by him as the type holotype.

hennigean systematics – See phylogenetic systematics.

heterotpe – A type which has been derived by combining the characters of two diferent species.

holoplesiotype – A pleasiotype of the same sex as the holotype.

herbarium (herb. hb.) – Latin, meaning a herbarium, a collection of preserved (usually dried) plant specimens.

1. Herbarium of Royal Botanic Gardens, Kew, Richomnd, Surrey, Great Britain -
Foundation 1841. Number of specimens about 6,000,000.

2. Museum National d' Histoire Naturelle, Laboratoire de Phanerogamic, Paris, France-
Foundatio 1635. Number of

Specimens about 5,000, 000.

3. Herbarium of the Department of Systematics and Plant Geography of the Botanical Institute of the Academy of Sciences of the U.S.S.R. Leningrad. Fondation 1714. Number of Speciments-Over 5,000, 000.

4. British Museum (Natural History), London, Great Britain-Fondation 1753. Number of Specimens about 4, 000, 000.

5. U.S. National Museum (Dept. of Botany) U.S. Smithsonian Institution, Washington, U.S.A. Foundation 1868. Number of Specimens 2,700,000.

6. Chicago Natural History Museum, Chicago, Illinois, U.S.A. Foundation 1893. Number of Specimens approximately 2,000,000.

7. Herbarium of the Royal Botanic Gardens, Edinburgh, Sctoland, Great Britain. Foundation 1761. Number of Specimens 1, 175, 000.

8. Museum Nation d' Historie, Natureile, Laboratoire de Cryptogamie, paries, France.F oundation 1635. Number of Specimens 1, 000, 000.

9. Herbarium of the Academy of Natural Sciences, Philadelphia. Foundation 1812. Number of Specimens 1, 000, 000.

10. Herbarium of the indian botanic Garden, Calcutta (India) Foundation 1787. Number of Specimens about 1, 000, 000.

11. Herbartium of the Botanical Institute of the Academy of Scien-ces of the Ukraniam S.S.R. (K. W.), Repin street, Kiev. U.S.S.R. foundation 1931. Number of specimens about 500, 000.

12. Herbarium of theDepartment of Botany, Faculty of Science, kyoto University, Sakyoku, Kansai, Kyoto, japan. foundation 1921. Number of Specimens 400, 000.

13. Herbarium of of the United States National Arboretum, Washington, U.S.A.

14. Herbarium of the Forest Research Institute and Colleges, Dehra Dun (India).

15. Madras Herbarium, Agricul-atural College and Research Institute, Coimbatore, South India. Foundation 1874. Number of Specimens about 2, 00, 000.

16. Herbarium of the Department of Botany, University of California, Los Angles, California, U.S.A. Foundation 1925. Number of Specimens 175, 000.

17. Forest Service Herbarium U.S. Department of Agriculture, Washington, U.S.A. Foundation 1910. Number of Specimens 110, 000.

18. Herbarium of the All-Uniai Institue of Plant Pathology of the Academy of Agriculture Sciences, Leningrad, U.S.S.R. Foundation 1907. Number of Specimens 1,00, 000.

19. Blatter Herbarium, St. Xavier's College, Fort, Bombay (India). Foundation – Nil. Number of Specimens about 100,00.

20) Botanisches Institute der universitat, Kiel, Germany. Foundation 1875. Number of Specimens about 90, 000.

21) Forest Herbarium (Assam), Shillong, Assam (India). Foundation 1913, Number of Specimens about 80, 000.

22) Herbarium of the Commonwealth Mycological Institute, Kew, Richmond Surrey, Great Britain. Foundation 1921. Number of Specimens 63, 347.

23) Heribarium of Gordon College, Rawalpindi (Pakistan). Foundation 1893. Number of Specimes 63, 347.

24) Herbarium of Punjab Univeristy, lahore (Pakistan). Foundation Nil. Number of Specimens 50, 000.

25) Herbarium of the National Botanical Gardens, Lucknow (India). Foundation 1948. Numbers os

Specimens about 40, 000.

26) Herbarium Cryptogamiae India Orientalis, Division of Mycology and plant Pathology, Indian Agricultural Research Institute New Delhi (India).
Foundation 1901. Number of Specimens about 20, 000.

27) Herbarium of the University of Alberta, Edomonton, Alberta, Candada.
Foundation 1916. Number of Specimens about 20, 000.

28) Herbarium of the Rangoon University, Rangoon, Burma.
Foundation 1947. Number of Specimens 15, 000.

29) Herbarium of the Division of Botany, indian Agricultural Research Institute, (I.A.R.I.) New Delhi (India). foundation Nil. Nimber of Specimens about 5,000.

30) Botanical Museum and Herbarium Osmaina University, Hyderabad (India). Founation 1933. Number of Specimens about 5, 000.

heterotypic synonyms - Synonyms based on different nomenclatural-types.

hierarchy - A series of consecutively subordinate categories forming a system of classification.

holoplastotype - A cast of holotype.

I

illegitimacy (Bot., Bact.) - A state of non-accordance with the rules that requires a name not ot be taken into consideration for the purposes of priority (except or the purposes of homonymy), When the correct name of a taxon is being decided.
2. (Bact., Bot., Cult., Vir.) A state of non-accordance with the arthicles of the Code.

in - Latin, meaning in used to connect the names of two persons, the second of which was the editor, or ovarall author, of a work in which the first was responsible for validly publishing, or making available, a name.

inadmissible - (Bot.) of a name, of a form that precludes its valid publication of the use of a name or epithet, contrary to the provisions of the Code.

invalid - 1. (Bact.) of names, not validly published, the equivalent of not validly published of Bot, and of unavailable of Zoo.
2. (Zoo) of names, available (q.v.) for a given taxon but not that by which it should properly be know; the equivalent of incorrect of Bact. and Bot.
3. Of taxa, not recognized as taxonomically distinct, at least at the rank in question.

IAPT - International Association of Plant Taxonomy. The body responsible for the publication of the ICBN and for the organization of committees dealing with the nomentelature of plants, at Botanical Congresses.

Indication - Information published before 1931 that allows a name to be treated as available in the absence of a description or definition.
2. Published information that determines the types species of a generic name in the absence of and original designation of the type species.

isotype - 1. A duplicate of a holotype from the single collection that contained the holotype.
2. A type described from two species of the same genus.
3. A form occuring in a variety of localites.

isoholotype - a plant specimen removed at a later date from the same plant form which the holotype was originally collected.

isoneotype-A duplicate of a neotype.

indirect reference - A clear statement that the description of diagnosis of a taxon for which a new name is given gas been effectively published in and earlier work.

icotype - A specimen used for identification, worked on by the orginal author, or collected from the type locality, but not one used for the published description.

indented key – A dichotomous key in which the first part of a contrasting couplet is followed by all subsequent couplets; each subordinate couplet being indented one step further to the right for clarity of presentation; yoked key.

incorrect subsequent spelling – Any change in the spelling of an available name other than a mandatory change or an emendation.

incidental mention -The use of a name by an author who demonstrates no intention of adopting or introducing it, but merely mentions it without comment on its suitability, such a mention does not fulfil the requirements of the Code.

inclusus (incl.) Latin, meaning include.

incertae cedis – inc. ced) Latin, meaning of uncertain seat (affinity); used of a category of uncertain taxonomic position.

in schedule (in sched.) Latin, meaning on a herbarium sheet or label.

ibidem – (ib.) Latin, Meaning the same, in the same place.

index – Kewensis.

iditoaxonmy – Taxonomic study of individuals, populations, species and higer taxz; traditional taxonomy.

ideotype – A specimen examined by the author of a species but collece collected from the type locality; idiotype.

interspecific – (Bot) of a hybrid, between species referred to the same genus.

ibidem – (ibid, ib) Latin, meaning in the same place, as in the same book or journal.

I.C.B.N – Internation code of Botanical Nomenclature. The internationally adopted set of rules governing botanical nomenclature.

inadmissible – Used of a name or epithet which cannot be used within the provisions of the Code.

onym – One of two ro more names based on the same nomenclatural type.

iconotype – A drawing or photgraph of a type specimen.

ineditus – (inedita, ineditum, ined) Latin, meaning unpublished.

IBPN – Internation Bureau of the International Code of Botanical Nomenclature, Taxon and *Regnum Vegetabile.*

genus (gen) Latin, meaning genus.

isoparatyp – A duplicate of a paratype.

in litteris – (in litt.) Latin, meaning in correspondence.

in correct original spelling – An original spelling that is correct under the provisions of the Code.

in situ – In the original location.

in adnotatione – (in adnot.) Latin, meaning in an annotatio.

isosyntype – A duplicate of a syntype.

in synonmis (in syn.) Latin, meaning in synonomy.

inguilinus (ing). Latin, meaning naturalized.

J

junior homonym – The later published or two homonyms; procupied name.

junior synonym – Any of two or more synonyms other than the earliest published later synonym; younger synonym.

K

kingdom – The highest categorty in the hierarchy of classification; the five kingdom classfication of living organisms proposed by whittaker and currently widely used, is Monera, Protista, Fungi, Plantae, and Animlia.

key classification – See artificial classification.

L

legitimacy – 1. (Bact., Bot) the state of accordance with the rules that requires a name to be taken into consideration for the purposes of priority when the correct name of a taxon is being decided.
2. (Bact., Bot., Cult., Vir.) the state of accordance with the articles of the Code.

loco citato – (loc.cit.) Latin, meaning in the place cited; used to avoid the repetition of a bibliographic reference already given.

legitimate name – Se nom. legit.

list approved–(Bact.) a list of names considered and approved by the International Committee on Systematic Bacteriology for use in bacterial nomenclature after 1 st Jan, 1980.

lapsus calani – Latin, meaning slip of names considered and approved by the Internatinonal Committee on Systematic Bacteriology for use in bacterial nomenclature after 1st Jan, 1980.

lapsus calani – latin, meaning slip of the pen; in nomenclature a misspelling made by the author.

legitimate name – See nomen legitimatum.

latus (lat.) Latin, meaning broad, wide.

lumping – The taxonomic practice of ignorin minor varation in the definition or recognition of species.

linnaean classfication – The syustem of hierarchical classfication and binomial nomenclature established by Linnaeus.

linnaean species – A broad concept of a species often comprising many varieties; later synonym junior synonym.

lat – Latin.

line precedence – The precedence of a name, for the purposes of priority, that occurs on as earlier line of the smae page than another name for the same taxon.

latinization – the treatment of a name not of Latin origin, according to the grammatical rules of Latin.

M

monotype – (Bact) a single strain used by an author as a holotype but not designated as such by him or her.

monotypic – Having only one immediatedly subordinate taxon.

monotype – 1. The state of being monotypic.
2. (Zoo) The situation in which a genus group taxon is established with only one originally included species.

monstrosity – (Bot.) a plant or specimen exhibiting an abnormal structural condition. an abnormality.

morphotype – (Bact.) a variant of a species distinguished by the presence of some special or unusual morphological feature. The use of the term 'morphovar' to replace 'morphotype' is recommended by the ICNB.

misapplication – The use of a name to denote a taxon whithin the range of variation of which its type does not fall.

misidentification – The assignment of a specimen, population or taxon to a taxon to which it is generally (or later) considered it does not belong.

mihi – (m) Latin, meaning to me, dative singular of do, 1. Used after a name to indicate the writer's responsibility for its proposal.

mutatis characteribus – (mut. char) Latin, meaning with the characters changed (by); used in the same way as named.

molecular systematics – The study of organisms and their interrelation-ships using biochemical characters, and employing technique such as electrophoresis, chromatography, serology and DNA hybridization; the study of evolutionary relationship using comparative molecular data.

macrotaxonomy – That branch of taxonomy concerned with classification of supraspecific taxa; beta taxonomy.

multiple-entry key – An identification key, often utilising some form of punch card system. Which allows the operator freedom to select any character in any sequence from a given list; multiaccess key.

metatype – 1. A specimen from the type locality determined by the species.
2. A spcimen subsequently determined by the author after comparison with the type.

manuscript – A text (MS) or texts (MSS) type written or hand written, not reproduced as multiple copies; typically that copy submitted for publications.

manuscript name – An unpublished scientific name.

merogamodeme – A part of a relatively isolated, naturally interbreeding population (gamodeme); a phenodeme of intrapopulation variants.

metanymous homonyms – Homonyms based on different nomenclatural types but which are considered to belong to a single taxon.

monothetic key – A dichotomous key in which each couplet has single contrasting statements requiring a simple answer.

mondelphous homonyms – Devalidated names taken up by different authors independently but validated with different nomenclatural types.

masculus – (masc.) Latin, meaning male.

mean character difference – (MCD) A measure of phenetic resemblance, calculated from the absolute values of the differenees.

monobasic – Used of a genus established on, or comprising, a single species; monotypic.

mandatory change – An alternation in spelling of the ending of a family-group name or epithet as required by the rules of nomenclature.

metonym – A taxonomic synonym; a name given to a taxon that already has a valid name.

monothetic taxon – A taxon specified by a unique combination of diagnositc characters which comprise both necessary and sufficient criteria for identification.

misapplied name – A name given to a taxon which excludes the nomenclatural type of the name.

monotaxic – Belonging to the same taxonomic group.

microtaxonomy – That branch of taxonomy conerned with the classification of species varieties and population.

mutatis characteribus (mut. char) Latin, meaning with the characters changed by equivalent to amend.

metatopotype – A metatype from the type locality.

museum (mus) – Latin, meaning museum.

monomial – Uninomial name.

merotype -a part of the organism furnishing the type specimen.

monotype – The establishment of a genus or subgenus upon a single species; or of a species upon a single speciomen.

maintenance – The continued use of, or support form a correct name in nomenclature; retention.

manhattan distance – The taxonomic distance between two OTUs plotted not as a straight libne but as the moves of a rock on a chess board; city block distance.

morphotype – A specimen selected to represent a given intrapopultion variant (morph).

material – The sample available for study.

misiquotation – A references to the wrong paper.

monograph – A comprehensive published work on a single group or subject.

Doctrine of Signatures – A herbalists doctrine according to which plant parts resembling portions of the human body must have been so created for the purpose of furnishing remedies for the ailments of those portions.

N

nomenclature, official– publications on of viruses.

1. Fenner. F (1976) The classification and Nomenclature of viruses. J. Gen. Virol. *31:* 463-470 (1976).

2. *Intervirology*. Basle.

Rules of Nomenclature :

1. Generic names one-and-half foot long, those of difficult pronunciation or repulsive names must be avoided.

2. A perfectly name of a plant is given a generic and a specific name.

3. The specific name must distinguish a plant form all its relatives.

4. Size does not distinguish species.

5. The original place of a plant does not give specific differences.

6. Colour in one and the same species plays some strange tricks, therefore, it is no good as a specific difference.

7. A generic name must be applied to each plant species.

8. The specific name should always follow the generic name.

9. The specific, name the shorther it is, the better it is.

nomenclature, official publications on – 1. Internation Code of Nomenclature of Baeteria. (1975) American Society for microbiology, Washington, U. S. A.

2. International Journal of systamatic Bacteriology, Iowa.

nomen legitimum – (non. legit) : Latin, meaning legitimate name:

nomenclature – 1. The giving of names to taxa.

2. The system of names so produced, or any part thereof.

nomen confusum – (nom. conf.)

Latin, meaning confused name, a name based on a type consisting of discordant elements form which it is impossible to select a satisfactory lectotype; in Bact., specifically a name, the type of which was an impure or mixed culture.

nomen rejiciendum – (nom. rejic) Latin meaning a name to be rejected a rejected name, ie., a name the use of which has been officially rejected, usually in favour of another (conserved) name; under Bact. and Zoo. names listed as officially rejected are to be permanently rejected; under Bot., rejected earlier homonyms and nomenclatural synonyms are to be permanently rejecte.

nomenclature – Of plants official publications on 1. International Code of Botanical Nomenclature (1972). International Association for Plant Taxonomy, Utrecht, Netherlands.

3. Taxon Utraecht.

neotypology – A method of classification involving the use of numerical taxonomy within typology as a means of assessing ranges of variation about the type.

neutral term – A taxonomic term of convenience having no nomenclatural significance or hierarchical rank, such as complex, group.

nomen perplexum – Latin, meaning perplexing name; (Bact) a name the application of which is known but which causes uncertainty in bacteriology.

nominate – (Zoo) containing the type of the name of the higher taxon to which it is subordinate.

nomen oblitum – (nom. oblit) Latin, meaning forgotten name.

nomen superfluum – (nom. superf.) Latin, meaning supperfluous name (bot., Bact.) a name which when first validly published, was applied by its author to a taxon so circumscribed as to include the type of another name which the author ought to have adopted under the rules.

nomen – Latin, meaning a name.

nomen revictum – Latin, meaning revived name.

nec – Latin, meaning and not (of) nor (of).

nomen illegitimum – (nom.illegit) Latin, meaning illegitimate name.

nothomorph – (nm) 1. A taxonomic category subordinate to the collective (hybrid) category equivalent to species and the equivalent of variety in the taxonomic hierarchy. 2. Any hybrid variant derived from the same parent species, forming a taxon of this category.

nomen dubium – (nom. dub.) Latin, meaning dubious name, a name of uncertain application, either becuase, through lack of the original type and sufficient information about it,. it is impossible to ascertain to which taxon its, type should be referred.

nomen conservandum – (nom. conserv.) Latin, meaning a name to be conserved; a conserved name.

natural system – In this type of system of classification all the important characters are taken into account and the plants are classified according to their related characters. It reflects the situation as it exists under natural conditions.

non-statutory – (Cult) established by voluntary agreement between organizations.

nomen hybridum – (Latin, meaning hybrid name. (Bact.) a name formed by combining words dervied from different languages.

neallotype – A newly designated type specimen selected in the absence of extant type material, of opposite sex to the neotype.

nomen non rite publicatum – (non.non rite public.) Latin, meaning name not properly published; used in citation to indicate a name that has been effectively but not validly published.

neotype – An element designated or selected subsequently to serve as the type of a name when all the original type elements are destroyed or missing or believed so to be.

nomen ambiguum – (nom. ambig) Latin, meaning ambiguous name, a name used in different senses i.e. applied by different authors to different taxa so that its has become a long persistent source of error.

nomen novum – (nom . nov) Latin, meaning new name; a name expressly proposed and published to replace an earlier name that cannot be used for some reason, e.g., if it is a later (junior) homonym.

nomen nudum – (nom. nud.) Latin, meaning naked name, a name published without such associate descriptive matter as is required by the appropriate Code to satisfy the criteria of availability (zoo) or valid publication (Bact., Bot.).

natural key – An identification key constructed from a natural classification and indicating the supposed evolutionary relation-ships of the group within the branching sequences of the key.

natural classification – A hierarchical classification based on hypothetical phylogentic relationships such that the members of each category in the classification share a single common ancestor.

nobis – (nobin.) Latin, meaning to us, used after a name to indicate the author's responsibility for its proposal.

nomen – (n) Latin, meaning name.

nomen triviale – Latin, meaning trivial name, the specific name (zoo) or specific epithet (Bact., Bot.)

nominal – (Zoo) of a taxon bearing a given name, in the sense of the type.

neocotype – A replacement syntype designated in the absence of the original type or type species; neosyntype.

nomen approbatum – Latin, meaning approved name. (Bact.)

neoparatype – A figured specimen used in addition to the neotype.

neoholotype – A new type specimen selected in the absence of the original type; neotype.

neotype – A newly designated type specimen selected in the absence of extant type material (holotype, paratype or syntypes).

nomen novum – (n. n.) latin, meaning new name.

novus – (n) Latin, meaning new.

non – Latin, meaning not (of).

numero – (no.) Latin, meaning number.

nee – Latin, meaning not (of); nor (of).

non visus – (n. v.) Latin, meaning not seen.

nominifer – A type.

O

organ-genus – A cateroy formerly employed in Bot, for genera of fossil plants, the characters of which were derived principally from a single organ, e.g., feructification, stem, and which were assignable to a family; it is now included in the category form-genus.

obligate synonym – See synonym, homotypic.

opinion – (Bact., Zoo) a decision of the International Commission Zoological Nomenclature, or of the Judical Commission of the International Committee on Systematic Bacteriology, on any particular case referred to it.

origin of angiosperms, monophyletic:

opero citato – (op.cit.) Lation, meaning in the cited; used to avoid the repetition of part of a bibliograhic reference alrady given.

oldest – First validly published (Bact., Bot) of first made available (Zoo).

orch – Handbook on Orchid Nomenclature and Registration.

orthographic variants – different spellings of the same name.

orthography – Spelling.

P

position – The jplace of a taxon in a system of classification indicated by the higher taxon to which it is referred.

Publication

1. The process of distributing graphic material in a way that statisfies the criteria of publication of the appropriate code.

2. Any item so published under 1.

3. In Zoo. Also a condition to be satisfied in order that a name may become aailable; the equivalent of offective publication of Bot. and Bact., See publicatio, effective.

publication effective, 16 – (bact., bot) the process of distributing graphic material of a kind and in a way that satisfies the criteria of effective publication of the appropriate Code.

paralectotype – (Zoo) a remaining syntype after a lectotype has been desinated from, amongst syntypes; not a type in the strict nomenclature sense.

publication, valid – (Bact., Bot, Cult.) the publication of names in accordance with the criteria of valid publication of the appropriate code; names that have not undergone valid publication are treated as non-existent for the purposes of nomenclature.

phylogenetic system – Classification of plants according to their evolutionary and genetic relationships. It enables us to find out the ancestors or derivatives of any taxon.

phagotype – (Bact) a variant of a species distinguished its sensitivity to a particular bacteriophage or by a distinctivae pattern of sensitivity to a set of specific bacteriophages, The use of the term 'phagovar' to replace 'phagotype' is recommended by the ICNB.

pro parte – (p.p.) : Latin, meaning in part; used in citatios to show that only a part of a taxon as circumscribed by a previous authior is being referred to be the writer.

protologue – The whole of the verbal and illustrative matter associated with a name at its place of first valid publication.

pro hybrida – (pro, hybr.) Latin, meaning as hybrid; used to indicate that a name of a taxon regarded as a species was originally published as a name of a hybrid.

phase – (Bact.) in Enterobacteriaceae, a well-defined stage of a naturally occurring alternating variation.

S

sensu stricto – (sens. str. s.s.) – Latin, in the strict sense, in the narrow sense, i.e., of a traxon, in the sense of the type of its name, or in the sense of its citcumscription by its original describer; or in the sense of its nominate subordinate taxon (in the case of a taxon with 2 or more subordinate taxa); or with the exclusion of similar taxa sometimes united with it.

sunfossifl – (Bot.) fossil but geologically young, soft in textue, organic in composition and usually found in a sort deposit, such as peat; the equivalent of fossil for nomenclature purposes.

status novus – (stat. nov.) Latin, meaning new status; 1. used in citation to indicate that a taxon has been altered in rank but retains in its name the epithet from its name in the former rank. 2. Used in citation to indicate that a taxon has been changed in status, from specific to hybrid or vice versa.

subspecific epithet – (Bact., Bot.) The third word (other thanb a word indicative of rank) in the name of a subspecies; the equivalent of subspecific name of Zoo.

subspecific name – (Zoo) the third word in the name of a subspecies; the equivalent of subspecific epithet of Bact. and Bot.

subsequent monotypy – The assumption of type status by the first species placed in a genus that was originally estabilshed without any included species.

starting-point dates – The date of publication of a work previous to which no name is consider to have been made available (Zoo. Code) or validly published (Bot. & Bact. Codes) Different groups of organisms have different starting points, depending upon which systematic work is considered to have laid the foundation of the modern nomenclature of the group concerned.

state – 1. (Bact.) One of certain variants which arise in cultures of many bacteria and which bring about a change in the gross appearance of a culture.
2. (Bot.) A phase in a life-cycle.

supplementary type – A decribed or figured specimen used to provide supplementary information about a previously described species.

subsequent spelling – Any modified spelling of a name other than the original spelling.

statutory – (cult.) established by a legal enactment or process of a country, or by legal tteaty between countries.

synonym – The existence of twop or more names spplied to the same taxon.
2. The relationship between any two such names.
3. The names considered to apply to a given taxon other than the name by which it should properly be knwon.
4. A list of such names.

synonytmum – (Byn.) Latin, meaning synonym.
1. In the abbreviated form, used befre a name to indicate it is

asynonym of the name properly to be used for the taxon concerned.

2. One of two or more names applied to the same taxon.

3. Commonly, a name applied to a given taxon other the one which it should properly be knowon; a name placed in and forming part of the synonymy thereof.

strain – (Bact.) the descendats of a single isolation in pure culture, sometimes showing marked differences in economic significance from other strains or isolations; analogous to clone (q.v.) of Bot and Cult.

starting-point – A published work and the date therof, before wich no name is considered to have been amde available (Zoo) or validly published (Bact., Bot., Cult.)

superfloous – (Bact., Bot) applied by its author to a taxon so circumscription by him as include the type of another name which could and should have been used as its correct name see also non. Superful.

synonym, commercial – (Cult.) a name that may properly be used for a cultiver in place of the internationally corrct oen in countries Where the latter is commercially unaccepthable.

stateus novus (stat. nov.) Latin, meaning new status, new rank; used to indicate that a taxon has changed either in rank, without a chagne of name or epithet, or in its specific or hhybrid status.

status conidialis – (stat. conid.) Latin, meaning the conidial date; an indication, used in synonmies, that a name refers to the conidial state of a pleomorphic fungus.

starting point – The date of a published work that for nomouclatural purposes is considered to be the first available or validly published for a particular group.

speciation –

syntype – Any of two or more elements used as types by the author

of a name, Whether or not designated as such by him.

system – The result of a given classification of a number of living orgainisms, usually one dealing with a taxon of high rank ᶢ , Kigndom, Phylum, division a   em of classification.

splitting – The tgaxonomic practice of subdividing pecies on the basis of more or less minor differences.

subsequent fixation – The fixation of the type of a taxon is a work published after the extablishment of the taxon; fixationby subsequent designation substitute name see nomen novum.

syntype – Any of two or more type specimens listed in the original description of a taxon when a holotype was not designated.

synonym, heterotypic – A synonym based on a different type (from that or another).

subsequent desingation – The designation of the type of a taxon in a work published subsequent to the establishment of the taxon.

specific epithet – (Bact., Bot.) The second word. in the name of a species, the equivalent of specific name of Zoo.

synonymum novum – (syn. nov.) Latin, meanign the imperfert state, an indication, used in synonomies, that a name refers only to the imperfect state of a pleomorphic fungus.

suppressed – (Zoo) made effectively unavailable (rejected) for the purposes prescribed, by actio of the International commission on Zoological Nomenclature under its Plenary Powers. See *non. rejic.*

supression – (Zoo) An act of the International commission on Zoological Nomenclature, using its Plenary powers, by which a name is made effectivel;y unavailable (rejected) for such purposes as the commission may prescribed (homonymum) and or priority). See *nom.*

rejic.

status pycnidialis – (Stat. pycid.) Latin, meaning the pycnidial states, an indication used in synonmes, that a name refers only the pycnidial state of a plemorphic fungus.

serotype – (Bact.) a varient of a species or subspecies distinguished from other such variants oif the same species or subspecies on the basis of its antigénic structure. The use of the term 'serovar' to replace 'serotype' is recommended by the ICNB.

sigla – (vir) Names made up from a few, generally initial, letters (e.g., Reovirus from respiratiory enteric orphan viruses); such names may be employed as the names of viruses or viral taxa, provides that they are meaningful to workers in the fields and are recommended by international virus study grounds.

strato type – The original or subsequently designated type of a named stratigraphic uniit or boundary, used as the standard for identification and definition; further gualifies by the prefixes, holo-para-neo-lecto-and hypo-as used in biological taxonomy.

syngamodeme – A biosystematic unit comprising all the coenogamodems, that are linked by the ability of some members to form viable but sterile hybrids; members are not capable of hybridizing with any other syngamodemes.

suppress – (Zoo) of a name, to make effectively unavailable (reject), for the purpose prescribed; only the International *rejic.* supression.

sensu lato (sens. lat.) Latin, meaning in the broad sense; i.e., of a taxon, including all its subordinate taxa and or other taxa sometimes considered as distinct.

sine – Latin, meaning without.

subforma – A category subordinate to forma; subform.

synonym, junior – (Zoo) the later published of two synonyms.

synonym taxonomic–See synonym, heterotypic.

suffix – The appended termination of a compound word.

synisonym – One of two or more names having the same name bearing epithet (basionym).

subfamily – A subdivision of a family; a taxon of the rank of a subfamily.

sphalmate – (sphalm) Latin, meaning in error, by mistake.

systematics – The scientific study of the variation of living organisms.

subordinate taxon – A taxon of lower rank than that with which it is compared.

SSI – Site of Special Scientific Interest. syn-prefix meaning joined (in space or time)

synonym, senior – (Zoo) the earlier published or two synonyms.

supra citato (sypra. cit.) – Latin, meaning cited above.

scheda – (sched) – Latin, meaning label (or a specimen).

status – (st.) Latin, meaning rank.

species nova – (n. sp.) Latin, meaning new species.

secundum – (sec.) Latin, meaning according to.

ssp – Abbreviation of subspecies.

saltem – Latin, meaning at least.

T

Type Method – *(typification)*
A modification recently introduced in the Code. The application of name of taxonomic groups is determined by means of nomenclatural types. This means that when a species is described as new, the author must indicate which is the type of specimen on which the new species is based. In other words the author of a species must designate a certain specimen or an acceptable substitute for a specimen as the type of that specimen. This becomes, then, the nomenclatural type of application of the name given to it. The nomenclatural type

of a genus therefore becomes the species on which the generic name was based. The nomenclatural type of a family is the genus on which the family name was based while the nomenclatural type of an order is the family on which the ordinal name was based. It follows therefore that if the type of a name is excluded from a taxon for any reason, the name of the taxon must be changed.

Holotype – This is the one specimen or element chosen of designated by the author as the nomenclatural type which automatically fixes the application of the name.

Isotype – This is duplicate of the type or holotype from the collection, with the same locality, date and number as the holotype. If several branches of a tree are collected at the same time, one specimen may be chosen as the type, the rest are isotypes. Similarly, if a number of small herbaceous plants are collected at the same time, all belonging to the same new spoecies one may be selected as the holotype, and the rest are isotypes. If they are mounted on the same herbarium sheet, they may be all part of the holotype.

Paratype – This is a specimen referred to or cited with the original description or publication of the taxon other than the holotype or isotype. A paratype usually is a specimen from a collection other then the type, but on which the descriprition of the new taxon has been based. Thus there can be only one sheet of holotype while there can be a number of isotypes and paratypes. Earlier workers often referred these paratypes as co-types.

Syntype – When an author describes a new species and cites two or more specimens, these become the syntypes, if none of them individually has been designated the type or one of two or more speci-

ments designated as types simultaneously in the origianl publi- cation.

Lectotype – It is a specimen or other element selected by a competent worker from the original material studied by the author to serve as a substitute for the holotype, if the latter was not designated in the original publication or so long as it is missing. When two or more specimens have been designated as types by the author, they become syntypes, and one must be chosen as a lectotype.

Neotype – It is a specimen selected to serve as a sustitute for the holotype when all the material on which the name was based is missing. When a selection has been once finally made for a neotype it must be followd by subsequent botanists. This seems to be the case with many of the new taxa described by Roxburgh (1814), for which no actual specimen was either designated or preserved.

The fixing of the type specimen for a species or of the type in general for taxa below the rank of order is considered so impoprtant that publication on or after January 1, 1958, of the name of a new taxon of recent plants of the rank of order or below is valid only when the nomenclatural type is indicated (Santapau, 1959).

trinominal – 1. consisting of three words, such as the name of a subspecies.
2. Making use nof names consisting of three words.

type specimen – A designated specimen or individual (Holotype) lectotype, neotype or one of a series or specimens (syntype) that is the type of a species or subspecies.

translatio nova – (trans. nova.) Latin, meaning new transfer, used to indicate that a taxon has been changed in position either by horizontal transfer to another genus or vertical transfer to a different taxonomic rank.

type concept – A method of nomenclature that attaches a name permanently to a type specimen, a name which is retained by the type throughout all subsequent taxonomic acts.

typus – (typ.) Latin, meaning type, the single element of a taxon to which its name is permanently attached, and on which the descriptive matter that satisfies the conditions of availability or valid publication is based.

typotype – The type of a type, e.g., if the type of name, studied by an author, is a description or illustration previously published by an earlier author, then the element on which the earlier author;s description or illustration was based, and which, as such, the later author did not study, is the typotype of the author's name.

typus conservandus (typ. cons.) – Latin, meaning type to be conserved used of the type of a conserved generic name which is different from that which would have to be adopted according to the Code but whose retention has been authorized by an International Botanical Congress.

trinomen – The scientific name of a subspecies. Comprising a generic name, a specific epithet and a subspecific epithet.

translato nova – (trans. nova) Latin, meaning new transfer; used in citation to indicate that a taxon has been altered in position but retains in its name the epithet from its name in the former position.

taxonomic distance – An expression of the relationship between individuals or taxa in terms of multidimensional space, where each dimension represents a character, based on quantitative estimates of dissimilarity.

taxon vagum – (tax. vag.) Latin, meanig uncertain taxon; occasionally used in names to indicate a taxon of uncertain rank.

taximetrics – The measurements of phenetic similarities between organisms and the incorporation of these data into systems of classification; numerical taxonomy.

taxonomy – The study of the principles and practice of classification. 2. Loosely, systematics.

termination – The end or ending of a name, described by the gender of a noun to which it is attached.

U

unavailable work – A work published prior to the starting point of the group or one that does not confirm to the provisions of the Code, or that has been annulled by the ICBN.

unused name – An available semior synonym that is not known to have been used as a valid name during the preceding 50 years.

uninomial – 1. Consisting of one word, such as the name of a genus, family or kingdom. 2. Making use of names consisting of one word.

usage – The sense in which a name has been applied, whether correctly or incorrectly occording to the Code.

uninomen – A scientific name comprising a single word; used for taxon at and above the rank of genus.

V

Valid Publication Conditions of –
1. Publication must be effective.
2. There must be a description of the new taxon, or a reference to a previously and effectively published description.
3. A combination is not validly published unless the author actually makes it. This is one of the changes recently introduced in the Code.
4. Publication must be done in Latin or with reference to a previously published Latin description.

5. When all the conditions for valid publication are not fulfilled at one and the same time, publication becomes valid when the last condition has been fulfilled. Thus, a publication is invalied if a botanist forgets to indicate the type of his new taxa, but it becomes valid when the type of the taxon has been indicated. The date of such publication will be when the type was indicated.

CELL BIOLOGY

A

Acetyl coenzyme A – An intermediate in energy-transferring reactions in metabolism, part of initial step of TCA cycle reactions.

Actin – A protein found in thin filaments of striated muscle, microfilaments, and many nonmuscle cells.

Actinin – A protein (M.W. – 95, 000) found in the Z-line of muscle fibers.

Actinomycin D – Antibiotic that inhibits the elongation of RNA chains.

Activation energy – Energy required by a system to allow a chemical reaction to proceed.

Active site – Region of the enzyme that binds and alters the substrate molecule.

Active transport – An energy-requiring movement of molecules across a membrane.

Actomyosin – A complex of two proteins, actin and myosin; the basic contractile element in muscle.

Adenosine triphosphatase – ATPase the enzyme that hydrolyzes ATP to form ADP and inorganic phosphate.

Adenosine triphosphate – ATP; a nucleoside triphosphate; a high-energy intermediate in energy-transferring metabolism.

Aerobes – Cells that live in and utilize oxygen.

Affinity chromatography – A technique for separation of molecules Molecules are attached to an insoluble (e.g., sepharose) matrix. Only those molecules that show affinity to the bound molecule (e.g., an antibody for its antigen) are retained. These trapped molecules can be subsequently eluted.

Allele – One of the alternative forms of a gene.

Allosteric effectors – Small molecules, usually metabolites, that bind to allosteric proteins at a site other than the active site so as to cause a change in protein shape.

Allosteric enzymes – Enzymes whose activity is modulated by the binding of allosteric effectors at sites other than the active site.

Aminoacyl adenylate – In protein synthesis, an activated compound that is an intermediate in the formation of a covalent bond between an amino acid and its RNA adaptor.

Aminoacyl synthetase – Any one of at least 20 different enzymes that catalyze (1) the reaction of a specific amino acid with ATP to form aminoacyl-AMP (activated amino acids) and pyrophosphate and (2) the transfer of the activated amino acid to tRNA forming aminoacyl-tRNA and free AMP.

Anaerobes – Cells that can live without oxygen.

Anaphase – Stage of mitosis or meiosis in which the chromosomes move toward opposite ends of the spindle.

Anaplerotic – Reactions that replenish intermediates depleted by other metabolic pathways.

Aneuploidy – Chromosome number that is not an exact multiple of the haploid number.

Angstrom (A) – A unit of length usually for describing molecular dimensions; equal to 10^{-8} cm.

Anticodon – The three-base group on a tRNA molecule that recognizes and pairs with a three-base codon of mRNA.

Antigen – A substance, usually a protein, that upon injection into a vertebrate is capable of stimulating the production of neutralizing antibodies.

Apoenzyme – The protein component of an enzyme; apoenzyme + coenzyme = holoenzyme.

Aster – The region at the poles of a dividing cell, composed of microtubules, a clear zone, and a pair of centrioles.

ATP – See adenosine triphosphate.

ATPase – See adenosine triphosphatase.

Attenuator region – A region of DNA within an operon at which most RNA polymerase molecules stop transcription. Receipt of a specific antitermination factor will cause transcription to proceed.

Autophagic vacuoles – Membrane-lined vacuoles containing morphologically recognizable cytoplasmic components. They include autolysosomes (which are secondary lysosomes) and autophagosomes (which are vesicles sequestering cytoplasmic organelles).

Autophagy – The process of sequestration of intracellular components in vacuoles.

Autoradiaogrphy – Determination of the location and geometry of radioactive components introduced into cells by means of exposure of photographic emulsions placed in contact with the cells.

Autotrophic cells – Cell that can synthesize macromolecules from simple nutrient molecules, such as carbon dioxide, ammonia, and water.

B

Bacteriophage – A virus requiring a bacterial host for its replication.

Basal body – An organelle located at the base of cilia and belived to be involved in the organization of cilary microtubules.

B.galactosidase – An enzyme catalyzing the hydrolysis of lactose into glucose and galactose.

Bivalent – A synapsed pair of homologus chromosomes.

C

Calorie – A unit of energy; the amount of heat required to raise the temperature of 1.0g of water from $14.5°$ to $15.1°C$.

Carcinogen – An agent that induces cancer.

Carrier – A transport protein within the membrane that binds temporarily with another molecule being transported across the membrane.

Catabolite repression – Decreased synthesis of specific enzymes in bacteria grow on glucose or other good catabolite source. Caused by low levels of cyclic AMP in such cells.

Catalyst – An agent that increases the rate of chemical reaction without altering the equilibrium point of that reaction.

C_4 cycle – A CO_2 – reducing pathway in photosynthesis; also known as the Hatch-Slack pathway.

Cell culture – A population of cells grown in vitro.

Cell cycle – The sequence of events in cell division cells including the G_1, S, G_2, and M periods.

Cell division – Formation of two daughter cells from a parent cell by enclosure of the two nuclei in separate cell compartments.

Cell-free extract – A fluid containing most of the suspended organelles and soluble molecules of a cell, made by breaking cell and removing the remaining whole cells.

Cellular affinity – Tendency of cells to adhere specifically to cells of the same type. The property is lost in cancer cells.

Cell wall - Rigid or semirigid structure enclosing the protoplast of most plant and procaryotic cells.

Centriole - Microtubule-containing organelle located at the spindle poles in dividing cells or forming the basal portion of a cilium or flagellum.

Centromere - The primary constriction of the chromosome to which the spindle fibers attach and which is required for chromosome movement to the poles at anaphase.

Chiasma - Site of DNA exchange between two chromatids of a bivlent.

Chlorophyll–Light-capturing pigment, located in chloroplast thylakoids or in procaryotic cells.

Chloroplasts - Membranous structures containing chlorophyll which are present in the cytoplasm of cucaryotic photosynthetic cells; site of photosynthesis.

Chromatid - One-half of a replicated chromosome, joined to the other chromatid at the centromere region.

Chromatin - The nuclear material easil;y stained for light microscopy. Seen as dense masses in transmission electron photomicrographs.

Chromatin fiber - The elongated deoxyribonucleoprotein molecule of the chromosome.

Chromomere - A beadlike or knobby region of chromosomes often seen in early stages of meiosis.

Chromosome - The gene-containing structure in the nucleus or nucleoid.

Cilium -(pl. cisternae) A flattened, membrane-bordered channel.

Clone - A group of cells that have descended from a single cell by mitosis.

Codon - A sequence of three nucleotides that code for an amino acid or chain termination.

Coenzyme - A small organic molecule associated with the protein portion of a holoenzyme, weakly bound at the active site of the enzyme, and required for enzyme activity.

coenzyme A - A small organic molecule that participates in energy-transfer reations, usually as a carrier of activated metabolites (e.g., acetate).

Colchicine - An alkaloid that binds tubulin on a molar basis and thereby causes breakdown of microtubules and dissolves the spindles also used as a polyploidizing agent.

Colinearity - The spatial correlation between codons in DNA and amino acids in the polypeptide translated from the DNA.

Colony - A group of continguous cells, usually derived from single cell, growing on a solid surrace.

Complement - A series of blood serum proteins which, when activated, lyse foreign cells.

Complementary base-pairing - Specific hydrogen bond inter-actions between a particular purine and a particular pyrimidine in nucleic acids: for example, guanine and cytosine, adenine and thymine, or adenine and uracil.

Concanavalin A - A lectin.

Constitutive enzyme - An enzyme synthesized at a constant rate.

Copolymer - A polymeric molecule containing more than one kind of monomer unit.

Corepressor - A metabolite that combines with repressor protein and blocks transcription of messenger RNA.

Coupled reactions - Two chemical reactions that have a common intermediate through which energy can be transferred form one reaction to the other.

Coupling factor - F_1 factor; the headpiece of the mitochondrial inner membrance subunit that has ATPase activity.

Covalent bond - Interaction between atoms by sharing electrons.

Cristae - Foldings of the mitochondrial inner membrane and the site of enzymes of oxidative phosphorylation and electron transport.

Crossing over – Exchange of homologous chromosome segments leading to recombination of linked genes.

Cyclin adenosine monophosphate – Cycli AMP; adenosine monophosphate with phosphate group bonded between 3'and 5' carbon atoms to form cyclic molecule; this nucleotide is active in regulating numerous reactions in cells.

Cycloheximide – An inhibitor of protein biosynthesis.

Cytochrome oxidase – Cytochrome - a; the terminal enzyme of aerobic respiration that transfers electrons to oxygen.

Cytochromes – Electron-transport intermediates containing heme or related prosthetic groups that undergo valency changes of the iron atom.

Cytogenetics – The study of biological systems using the combined methods of cytology and genetics.

Cytoplasm – The protoplasmic contents of the cell, exclusive of the nucleus.

Cytosol – The unstructued portion of the cytoplasm in which the organelles are suspended; the cytoplasmic fluid.

D

Dalton – Unit of molecular weight approximately equal to the weight of a hydrogen atom.

Deletion – Loss of part of a chromosome or DNA molecule from the genome.

Denaturation – change in the native configuration of a macromolecule resulting from heat treatment, extreme pH changes, chemical treatment, or other denaturing agents. It is usually accompanied by loss of biological activity.

Deoxyribonucleic acid (DNA) – The genetic material.

Diakinesis – Last of the stages of prophase in meiosis I.

Dictyosome – A stack of cisternae that forms part of the Golgi apparatus.

Diffusion – The ne overall movement of molecules in the direction of a lesser concentration.

Dimer – Structure resulting from association of two identical subunits.

Diploid – A cell or individual or species having two sets of homologous chromosomes in the nucleus of somatic cells.

Diplotene – A stage of prophase in meiosis I.

Disulfide bond – Covalent bond between two sulfur atoms in separate amino acids of a protein.

DNA polymerase I – Enzyme found to catalyze the formation of the 3'-5' phosphodiester bonds of DNA. It possesses 3' to 5' single-strand proofreading and 5'to 3' double-strand exonuclese activities, for use in DNA repair, its chief biological function.

Duplication – An extra copy of one or more genes in the chromosome complement.

Dynein – Potein component, in microtubule of the cilium of flagellum.

E

Effector – A regulatory metabolite that activates or inhibits an enzyme by binding to an allosteric site on the enzyme.

Electron carriers – Intermediates such as flavoproteins and cytochromes that reversibly gain or lose electrons.

Electron transport – The movement of electrons form substrates of oxygen catalyzed by the oxidative respiratory chain intermediates.

Electrophoresis – A method of separating macromolecules of particles according to their charge, size, and shape as they migrate throgh a gel or other medium in an electrical field.

Endergonic reaction – A chemical reaction with a positive standard free energy change; an energy-consuming reaction.

Endocytosis – Intake of solutes or particles by enclosure in a portion of plasma membrane bringing these materials into the cell.

Endoplasmic reticulum– ER; folded membrance system distributed within the cytoplasm of eucaryotic cells; frequently has attached ribosomes (rough ER).

Endothermic process – A process that absorbs heat.

End-product repression – A control mechanism in which the synthesis of an enzyme required to a metabolic pathway is inhibited by the final product of that metabolic pathway, thereby stopphing further pathway reactions.

Entropy – The randomness or disorder of a system.

Enzymes – The protein catalysts of biological systems.

Eurcaryotic cells – Cells having nuclear membranes and membrane-surrounded organelles.

Euchromation – Noncondensed, active chromosomes or chromosome regions of the interphase nucleus.

Exergonic reaction – A reaction with a negative standard free energy change; an energy releasing reaction.

Excited state – The energy-enhanced state of an atom or molecule existing after an electorn has been moved form its normal stable orbital to an outer orbital having a higer energy level.

Exocytosis – A mode of transport of substances out of the cell by enclosure is a vesicle, fusion with the plasma membrane, and subsequent expulsion to the outside.

Exothermic process – A process in which heat is evolved.

F

Facilitated diffusion– Assisted transport of molecules across the membrane along a conceentration gradient.

Fatty acid – Long hydrocarbon chain components of many lipids.

Feedback (end-product) inhibition – Inhibition of the first enzyme in a metabolic pathway by the end product of that pathway.

Fermentation – Oxidation of carbohydrate in non-oxygen-requring pathways such as glycolysis.

First law of thermodynamics – Energy can be neither created nor destroyed; statement of the principle of the conservation of energy.

Flagellum (pl. flagella) Elongated organelle produced by a centriole; ultrastructurally smilar to a cilium but usually longer than a cilium.

Flavin adenine dinucleotide – FAD; an celectron carrier molecule that acts in energy tranfer reaction as a coenzyme; the reduced form of the redox couple is $FADH_2$.

Fluid mosaic membrane – Model of cell membranes that postulates the distribution of proteins in a phospholipid bilayer and permits movements of particles within the membrane.

Fluorescent antibody technique – Detection of selected antigens in cells by stainng with a specific antibody conjugated with a fluorescent dye.

Formamide – A small organic molecule used in double-helical DNA denaturation. Formamide combines with the free HN_2 groups of adenine and prevents the formation of A-T base pairs.

Formylmethionyl-RNA – fmet-tRNA the initaial aminoacyl-transfer RNA complex that reacts with the small ribosome subunit at the beginning of polypeptide chain synthesis.

Free energy – A component of the total energy of a system that can do work under conditions of constant temperature and pressure.

Freeze-fracture – Procedure for preparing materials for electron microscopy by rapid freezing and fracturing of the tissue; the exposed fracture faces are used to create a replica that if observed and photographed in the electron microscope the fracture faces may or may not be further sublimed before the replica is made.

Furrowing – A cell division mechanism that involves a pinching -in, or cleavage, to form two daughter cells from the parent cell.

G

Galactosidase, B (beta) See B (beta) galactosidase.

Gap junction – Nexus; protions of the plasma membranes of adjacent cells that contain space between the two membranes that permits cell-to cell communication.

Gene – A portion of a chromosome that codes for RNA for a specific product.

Generation time – The time necessary for growing cells to double their numbers or mass.

Genetic information – The information contained in a sequence of nucleotide bases in DNA or RNA molecule.

Genetic map – The arrangement of genes on a chromosome as deduced from genetic recombination experiments.

Genome – The genes associatesed with a haploid set of chromosomes.

Genotype – The genetic constitution of an organism (as contrasted with its physical appearance or phenotype.)

Gluconeogenesis – Synthesis of carbohydrates from noncarbohydrate precursors such as fats or proteins.

Glycolipids – Lipids that contain polar, hydrophilic carbohydrate groups.

Glycolysis – The process of glucose catabolism.

Glycoprotein – A conjugated protein containing one or more sugar residues.

Glyoxylate cycle – An anaplerotic pathway replenishing intermediary metabolites.

Golgi apparatus – A region of the cytoplasm that functions in processing and packaging components for secretion form the cell.

Granulocytes – Leukocytes with distinct cytoplasmic granules. Includes eosinophils, basophils, and nectrophils.

Group-transfer reactions–Reactions (excluding oxidations of reductions) in which molecules exchange function groups.

Growth curve – The change in the number of cells of protoplasmic mass in a growing culture as a function of time.

Growth factor – A specific substance that must be present in the growth medium to permit a cell to multiply.

H

Hairpin loops – Regions of double helix formed by the pairing of two contiguous complementary structures of bases on the same single DNA or RNA strand.

Haploid – Cell or individual having one copy of each chromosome.

Haptens – Small nonantigenic molecules that are capable of stimulating specific antibody synthesis when chemically coupled to a larger molecule.

Heavy isotope – Form of atoms containing greater than the common number of neutrons and thus more dense that the commonly observed isotope (e.g. ^{15}N. ^{13}C).

Hela cells – An established line of human cervical carcinoma (cancer) cells derived from *Helen Lane.*

Helix – A spiral structure with a repeating pattern described by two simultaneous operations rotation and translation. It is the natural conformation of many regular biological polymers.

Heme – An iron-containing porphyrin

that serves as a prosthetic group in hemoglobins and in enzymes such as catalase and cytochromes.

Hemoglobin – Protein carrier of oxygen found in red blood cells; composed of two pairs of identical polypeptide chains and an iron-containing heme group.

Heterochromatin–Highly compacted chromatin regions of chromosomes during interphase.

Heterotrophic cells – Cells that require complex nutrient molecules such as glucose, amino acids, etc, from which to obtain enegy and to build their own macromolecules.

High-energy phosphate compound – A phosphorylated compound having a highly negative standard free energy of hydrolysis.

Histone – A protein component of the chromosome having a high content of the basic amino acids arginine and lysine.

Holoenzyme – The complete form of an enzyme.

Homologous – Having the same or similar gene content.

Homologous chromosomes – Chromosomes that pair during meiosis, have the same morphology, and contain genes governing the same characteristics.

Hormone – A chemical substance synthesized in one organ that, in small amounts, modulates biochemical functions in the cells of another tissue of organ.

Hydrogen bond – An electrostatic force between one electron negative atom and a hydrogen atom covalently linked to a second electron-negative atom.

Hydrolysis – The cleavage of a molecule into two or more molecules by the addition of a water molecule.

Hydrophilic – Molecules or parts of molecules that readily associate with water; usually containing polar groups with each other in aqueous solution.

I

Idiotype – The binding specificity of an immunoglobin for a specific antigen.

Immunoglobins – Y-shaped protein molecules that bind to and neutralize antigens.

Inducible enzymes– Enzymes whose rate of production can be increased by the presence of inducers in the cell.

Intermediary metabolism – The chemical reactions in a cell that transform food molecules into molecules needed as a source of energy and as precursors for cell growth.

Interphase – The state of the eucaryotic nucleus when it is not engaged in mitosis or meiosis, consists of G_1, S, and G_2 periods in cycling cells.

Inversion -Structural rearrangement of part of a chromosome so that genes within that part end up in inverse order.

In vivo (Latin; "in glass") Experiments done on isolated cells, tissues, or cell-free extracts rather than in situ, in place within the organism.

in vivo – (Latin; "in life") Experiments done on or with intact living organisms.

Ion pumps – systems that actively transport molecules across a membrane by expelling one substance out of the cell and thereby helping to drive many kinds of molecules into the cell along an energy gradient.

Isomers– Alternative molecular forms of a chemical compound.

Isopycnic density gradient centrifugation – A method used to separate macromolecules and cell components on the basis of differences that cause them to come to rest at equilibrium in a region of the gradient that has density of solute corresponding to their own buoyant density in the solute.

Isotopes – Alternative nuclear froms of an atom, all having the same atomic number (proton number) but different atomic weights (neutron number varies.)

Isozymes – Alternative molecular forms of an enzyme.

K

Karyotype – A photograph or diagram of a compete complement of chromosomes from a cell or individual arranged from the longest to the shortest chromosome.

Kinetochore – That which attaches laterally to the chromosomal centromere and is the site of cromosomal tubule attachemnt.

Krebs cycle – Most common pathway for oxidative metabolism of pyruvic acid, which is an end -product of glucose fermentation; also known as the citric acid cycle or the tricarboxylic acid cycle.

L

Label (radioactive) – A radioactive atom, introduced into a molecule to faciliate observation of its metabolic transformations.

Lampbrush chromosome – Giant diplotene chromosomes found in an ooocyte nucleus, with loops projecting in pairs from most chromomeres. Loops are sites of active gene expression.

Lectins – Cell-agglutining proteins. Most lectins are isolated from plant seeds.

Leptotene – The first of the prophase I stages in meiosis, before chromosome synapsis begins.

Ligase – Enzyme that joins together the parts of single strands of DNA between the 5' end of one strand and the 3' end of another.

Lipid – Class of organic compounds that are poorly soluble or insoluble in water but solube in nonaqueous

(organic) solvents such as ether.

Lipid bilayer – An early model for the structure of cell membranes based upon the hydrophobic interactions between phospholipids. The polar head groups face outwardly, while the hydrophobic tails are clustered in the interior.

Lysis – The bursting of a cell by the destruction of its cell membrane.

Lysogenic bacterium – A bacterium that contains a prophage.

Lysogenic viruses – Viruses that can become prophages.

Lysosomes – Intracellular granules that contain a large variety of hydrolytic enzymes; these fuse with ingested food vacuoles and break down their contents.

Lysozymes – Enzymes that degrade the polysaccharides found in the cell walls of certain bacteria.

Lytic infection – Viral infection leading to lysis of cell.

Lytic viruses – Viruses whose proliferation within the host cell leads to the cell's lysis.

M

Macromolecules – Molecules having molecular weights in the range of a few thousand to hundreds of millions of molecular weight units (Daltons).

Macrophage–Large, phagocytic white blood cell.

Matrix – The essentially unstructured substance of a cell or organelle consisting of a suspension of molecules and particles in a watery medium.

Meiosis – The reduction division of the nucleus in sexual organisms that produces daughter nuclei having half the number of chromosomes as the original nucleus.

Melting – The separation of the two strands of duplex DNA to form single strands by disruption of hydrogen bonds between the duplex strands.

Meromyosi, heavy – Portion of the

myosin molecule with ATPase activiy and Ca^2 binding propeties produced by trypsin digestion of myosin.

Mesosome – An extensively infolded portion of the procaryotic plasma membrane that functions in respiration and cell division.

Messenger RNA (mRNA) – The complementary copy of DNA that is made during transcription and that codes for protein during translation.

Metabolic pathway – A set of consecutive cellular enzymatic reactions that converts one molecule to another.

Metaphase – The stage of mitosis or meiosis when chromosomes are aligned along the equatorial plane of the spindle.

Microbody – A membrane-bounded cytoplasmic organelle with varied enzyme content and functions; may contain catalase and the enzymes of the glyoxylate cycle.

Microfilaments – Long, intracellular fibers that contain polymerized actin and that are thought to function in maintenance of cell structure and movement.

Micron (μ) – A unit of length convenient for describing cellular dimensions; it is equal to $\frac{1}{1000}$ of mm.

Microsome – A unbranched cylindrical assembly of protofilaments involved in cell movement phenomena; spindle fibers, ciliary subfibers, and centriole subfibers are microtubules.

Microvilli – Finger-like projections of plasma membranes of cells.

Mitochondria–Membrane surrounded organelles of aerobic cells that contain respiratory enzyme systems.

Mitosis – The division of the nucleus that produces two daughter nuclei exactly like the original parental nucleus; somatic nuclear division.

Monolayer – A single layer of cells, molecules, or other particles.

Monomer – The basic subunit from which, by repetition of a single reaction, polymers are made. For example, amino acids (monomers) yield polypeptides (polymers).

Multienzyme system – A group of enzymes active in the sequential steps of a metabolic pathway and in physical proximity to one another.

Mutagens – Physical or chemical agents, such as radiation, heat, or alkylating or deaminating agents, that can induce mutations.

Mutation – A change in the gene structure of a chromosome.

Myoblasts – Precursor cells that aggregate to form the multinucleated striated muscle cell.

Myofibril – Parallel units of a muscle fiber composed of bundles of myofilaments.

Myofilament – Individual thick (myosin) and thin (actin) filaments of the myofibril.

Myosin – Protein molecules, each composed of two coiled subunits (M.W. ~ 220, 000), that can aggregate to form a thick filament, globular at each end.

N

NAD, NADP – Nicotinamide adenine dinucleotide and nicotinamide adenine dinucleotide phosphate, carriers of electrons in many enzymatic oxidation-reduction reactions.

Negative control – Prevention of biological activity throguh a specific molecule; an example is inhibition of mRNA initiation by binding of specific repressor to specific sites along a DNA molecule.

Neutral fats – Glycerides; fatty acid esters of glycerol; a major storage form of fats.

Nuclear envelope – The double membrane surrounding the eukaryotic nucleus.

Nucleic acid – Polymer of nucleotides

in an unbranched chain; DNA and RNA.

Nucleoid – A region of segregated DNA in prokaryotic cells not separated from the cytoplasm by a membrane.

Nucleolar organizing region (NOR) – The specific part of the nucleolar organizing chromosome containing rRNA genes.

Nucleolus– Spherical structure found in nucleus of eucaryotic cells. Involved in rRNA synthesis and ribosome formation.

Nucleoplasm – The unstructured matrix portion of the nucleus in which the chromosomes and nucleoli are suspended.

Nucleoside – Molecule containing a nitrogenous base linked to a pentose sugar.

Nucleosome (Nu particles) – Spherical (100 A diameter) masses seen along partially dissociated chromatin.

Nucleotide – A nucleoside phosphate, a nitrogenous base linked to a pentose sugar linked to phosphate.

Nucleus – The major membrane-borderd compartment of the eucaryotic cell containing the chromosomes and nucleoli.

O

Open system – A system that exchanges matter as well as energy with its surroundings.

Operator – A specific nucleotide sequence in the operon that binds repressor and exerts control over transcription of adjacent structural gene(s).

Operon – A cluster of associated genes and recognition sites that participate in regulating and specifying amino acid polymerization into polypeptides; includes regulatory gene, promotor site, operator site, and structural gene(s).

Organelle – A discrete structural differentiation of the cell containing particular enzymes and performing particular functions for the whole cell, e.g., mitochondria, ribosomes, etc.

Oxidant – An oxidizing agent that loses electrons, or hydrogens, to a reducing agent or reductant.

Oxidation – The loss of electrons from an atom, in, or a compound.

Oxidative phosphorylation – The enzymatic phosphorylation of ADP to ATP that is coupled to electron transport along the respiratory chain to oxygen.

P

Pachytene – A stage of prophase I of meiosis characterized by synapsis of homologous chromosmes.

Peptide bond – A covalent bond between two amino acids in which the alpho-carboxyl group of the other.

Permease – A type of carrier protein situated in the plasma membrane and involved in transport of specific substrate molecules across that membrane.

Peroxisomes – Intracellular organelles that contain a fine granular matrix and often crystal-like cores. They contain enzymes involved in hydrogen peroxide metabolism including catalase. They may be important in during degradation, photorespiration, and the glyoxylate cycle.

pH – Measure of hydrogen ion concentration in aqueous solutions.

Phagocytosis – A form of endocytosis in which large amounts of particulate material, even whole cells, are taken up into large vesicles.

Phase-contrast microscope – An instrument that translates differences in the phase of transmitted or reflected light into gradations of contrast.

Phenotype – The observable properties of an organism; produced by the interaction of genotype and the environment.

Phosphodiester linkage– A covalent linkage involving esterfication to phosphoric acid.

Photophosphorylation – Process of fromation of ATP from ADP and inorganic phosphate in the ligh reactions of photosynthesis; occurs by a cyclic or noncyclic pathway involving photosystems I and II.

Photorespiration– Uptake of oxygen and release of carbon dioxide by photosynthetic cells or whole plants in the light.

Photosynthesis – The enzymatic conversion of light energy into chemical energy by forming carbohydrates and oxygen from CO_2 and H_2O in green plant cells.

Photosynthetic phosphorylation – The enzymatic formation of ATP from ADP in green plants coupled to light-dependent transport of electrons form excited chlorophyll.

Photosystem I (PS I) – A Photochemical reaction system in photosynthesis; coupled with photosystem II.

Photosystem II (PS II) – A photochemical reaction system in photosynthesis; coupled to photosystem I.

Phycobilin – An accessory photosynthetic pigment present in red and blue-green algae.

Pinocytosis – Endocytosis of soluble materials into small vesicles.

Plaque – Round, clear areas in a confluent sheet of cells; results from the killing or lysis or cluster of cells by several cycles of virus growth.

Plasmalemma – Plamsa membrane of the cell.

Plasmids – Cytoplasmic, autonomously replicating chromosomal elements found in bacteria.

Plasmodesmata – Cytoplasmic, autonomously replicating chromosomal elements found in bacteria.

Plasmodesmate – Cytoplasmic channels through the cell walls connecting the protoplasts to adjacent plant cells.

Plastid – Eucaryotic organelle that stores pigments or carbohydrates.

Polyacrylamide gel electrophoresis – A method of molecular separation of the differential migration of molecules, usualy proteins or polynucleotides, through a polya-crylamide matrix upon application of an electrical potential.

Polymer – An association of monomer units into a large molecule.

Polymerase -Enzyme catalyzing the synthesis of DNA or RNA from nucleoside triphosphate precursor.

Polynucleotide – A linear sequence of nucleotides in which the sugar of one nucleotide is linked through a phosphate group to the sugar on the adjacent nucleotide.

Polynucleotide ligase– Enzyme that covalently links DNA backbone chains.

Polynucleotide phosphorylase – A bacterial enzyme that catalyzes the polymerization of ribonucleoside diphosphates to yield phosphate and RNA.

Polypeptide – A long, unbranched polymer of amino acids.

Polyploid – Cell or individual having an excess or one or more whole complements of chromosomes.

Polysome – Polyribosome ; an aggregation of ribsomes, connected by a strand of messenger RNA.

Polytene chromosome – Giant chromosome composed of many fibrils (up to 2000) arising form successive rounds of chromatid duplication. Pairing of many identical chromomeres gives rise to characteristic banding pattern.

Pore – An opening in a membrane or other stucture; often referring to the nuclear pore complex of the nuclear envelope.

Positive control – control by a regulatory protein required for gene expression.

Primary constriction – Location of

centromere of chromosome.

Primary protein structure – The number of polypeptide chains in a protein, the sequence of amino acids within them, and the location of interchain and intrachain disulfide bridges.

Primer – A structure that severs as a growing point for polymerization.

Procaryote – Simple unicelluar organism, such as bacterium or bluegeen alga, with no nuclear membrane.

Procentriole – An immature plastid.

Prosthetic group – Coenzymes that are bound their enzymes.

Protamines – A class of proteins rich in the basic amino acid arginine. They are found complexed to the DNA in sperm in many invertebrates and fish.

Protist – Unicellular eucaryotic organisms such as protozoa, euglenoids, and algae.

Protoplasm – The living material of the cell.

Protoplast – The living structure of the cell, made of protoplasm, contained within but including the plasma membrane.,

Provirus – The state of a virus in which it is integrated into the genome of a host cell and is transmitted from one cell generation to another.

Puff – A region of expanded chromosome undergoing active transcription, usually observed in giant chromosomes.

Pulse chase experiment – A radioactively labeled compound is added to living cells or a cell extract (pulse), and a short time later, an excess of unlabeled compound is added. Samples are then taken at periods after the pulse to follow the course of the label as a compound is metabolized (chase).

Purine – Parent compound of the nitrogen-containing bases adenine and guanine.

Puromycin – Antibiotic that inhibits polypeptide synthesis by competing with aminoacyl tTNAs for the A binding site.

Pyrimidine – Parent compound of the nitrogen-containing bases cytosine, thymine, uracil.

Q

Quantum – The energy of a photon.

Quaternary structure – The manner in which the separtate polypeptide chains of a protein are held together and oriented with respect to one another in space.

R

Radioactive isotope – Isotope with an unstable nucleus that stabilizes itself by emitting ionizing radiations; important as tracers in biology.

Rannealing – Renaturation; specifically, the restoration of duplex DNA regions through complementary base pairing of single stranded DNA molecules.

Redox couple – compounds that occur in both the oxidized and reduced forms and that are participants in oxidation-reduction reactions, such as NAD^+ -NADH.

Reductant – A reducing agent that accepts electrons or hydrogens in oxidation-reduction reactions.

Reduction – Reactions involving gain of electrons or hydrogens.

Regulation – The modulation of metabolism or gene action through control mechanisms.

Regulatory genes – Genes whose primary function is to control the rate of synthesis of the products of other genes.

Release factor – Specific macromolecule involved in the reading of the "stop" singnal during protein synthesis.

Renaturation – The return of a protein or nucleic acid from a denatured and nonfunctioning state to its "native" functioning configuration.

Repetitive DNA – Repeated sequences of nucleotides that may occur in great numbers of reiterated copies in a chromosome complement.

Replicating fork – Y-shaped region of chromosome that acts as growing point in DNA replication.

Replicating forms (RF) – The structure of a nucleic acid during its replication; most frequently used to refer to double-helical intermediates in the replication of single-stranded DNA and RNA viruses.

Repressible enzyme – Enzyme synthesized in the absence of its substrate which then represses further synthesis of the enzyme.

Repressor – A protein product of the regulator gene of the operon that binds to the operator site and prevents transcription of structural genes.

Residual bodies – Secondary lysosomes containing undigested residues. membrane fragments, and whorls.

Respiration – The oxidative breakdown and release of energy from molecules by reaction with oxygen in aerobic cells.

Restrictiopn enzymes–Components of the restriction-modification cellular defence system against foreign nucleic acids. These enzymes cleave unmodified, double-stranded DNA at specific sequences that exhibit two-fold symmetry about a point.

Reticulocyte – Immature red blood cell still capable of limited hemoglobin synthesis.

Reverse transcriptase – An enzyme coded by certain RNA viruses that is able to make complementary single-standed DNA chains from RNA templates and then to convert thse DNA chains to double-helical form.

Ribonucleic acid (RNA) – Nucleic acids that function in transcription and translation of DNA.

Ribosomal DNA (rDNA) – The genes at the nucleolar organizing region that code for ribosomal RNA.

Ribosomal RNA – (rRNA) ribonucleic acids that are part of the ribosome structure and that function in protiein synthesis.

Ribosomes – Small cellular particles made up of rRNA and protein. Ribosomes are the site of protein synthesis; in eucaryotic cells, they are often attached to the endoplasmic reticulum.

RNA (ribonucleic acid) A polymer of ribonucleotides.

RNA polymerase – Enzyme that catalyzes the formation of RNA from ribonucleoside triphosphates, using DNA as a template.

RoughER (RER) – Portion of the endoplasmic reticulum bearing ribosomes.

S

Sarcolemma– The plasma membrane of a muscle cell or fiber.

Sarcomere – The contractile unit of muscle fiber, extending form one Z-line to an adjacent Z-line.

Sarcoplam – The cytoplasm of a cell or fiber.

Sarcoplasmic reticulum– Endoplasmic reticulum of a muscle cell or fiber.

Scanning electron microscope (SEM) – Electron-microscopic technique that permits observation of the surface structure (with a three-dimensional effect) rather than just thin sections.

Secondary constriction – Any pinched-in site along a chromosome other than the primary constriction at the centromere.

Secondary structure -Structure of a polypeptide chain describing the location, extent, and types of helices (as well as non-helical regions).

Second law of thermodynamics – The principle that physical and chemical change proceeds in a direction such that the entropy of the universe increases.

Secretion – Release of cellular products into the extracellular space.

Sedimentation coefficent – A quantitative measure of the rate of sedimentation of a given substance through water at 20(145)C in a unit centrifugal field. expressed in *Svedberg* units, S.

Semi-conservative replication – The usual mode of duplex DNA synthesis resulting in daugther duplex molecules that contain one parental strand and one newly formed stand.

Serum protein – Protein found in the serum (cell-free) component of blood, includes globulins, immunoglobulins, albumin, clotting factor, and enzymes.

Sex chromosme – Any chromosome involved in sex determination; such as the X and Y chromosomes.

Sliding filament mechanism – A model used to explain the structural basis of cell movements, such as muscle contraction and bending of cilia and flagella.

Smooth ER (SER) Portion of the endoplasmic reticulum devoid of ribosomes.

S period – Interval during the cell cycle in which DNA replcation occurs.

Spindle– Aggregation of microtubules during nuclear division that functions in the alignment and movement of chromosmomes at anaphase.

Spindle fiber–A microtubule in mitotically or meiotically dividing cells that extends from one pole to an attachment in the centromere region of a chromosome or that extends form pole to pole.

Spontaneous process – A process accompanied by a decrease in free energy.

Standard electrode potential – E_o ; the oxidation-reduction potential of a substance relative to a hydrogen electrode; expressed in volts.

Standard free-energy change – DG_o; a thermodynamic constant representing the difference between the standard free energy of the products of a reaction; energy-requiring reactions have a negative DG_o .

Standard state – Most stable form of a pure substance at 1.0 atmosphere pressure and 25°C (298K). For reactions occurring in solution, the standard state of a solute is a 1.0 M solution.

Steady state – A nonequilibrium state of an open system through which matter is flowing and in which all components remain in constant concentration.

Stereoisomers – Molecules that have the same structural formula but different spatial arrangement of dissimilar groups bonded to a common atom. Steroisomers have differences in their crystal structures and differ in the direction in which they rotate polarized light; they also differ in their ability to be used in an enzyme-catalyzed reaction.

Steric (stereochemical) – Relating the arrangement in space of the atoms in molecules.

Steroids – Compounds that are derivatives of a tetracyclic structure composed of a cyclopentane ring fused to a substituted phenanthrene nucleus.

Streptomycin – An antibiotic isolated from *Streptomyces griseus* (a soil bacterium) that binds specifically to bacterial 30 s ribosomal subunits, thereby blocking protein biosynthesis.

Stroma – Unstructured matrix of the chlooroplast that bathes the grana and stroma thylakoids.

Substrate – Specific compound acted upon in the active site of an enzyme.

Supercoils -Twisted forms taken by covalently close, circular, double-stranded DNA molecules when purification has removed the protein components of the chromosome, thereby slightly changing the pitch of the double helix.

Suppressor gene – A gene that can reverse the phenotypic effect of a variety of other genes.

Svedberg unit – The unit of sedimentiation equal to 10^{-13} seconds. The number of s units of a molecule or particle in a given centrifugal field is related to the weight, shape, and density of the molecule or particle.

Synapsis – Specific pairing of homologous chromosomes, typically during zygotene of prophase I in meiosis.

Synaptinemal complex – A complex structural component situated between a pair of synapsed homologous chromosomes during pachytene of meiosis I.

System – An isolated collection of matter and energy. All other matter and energy in the universe apart from the system is said to be outside the system or its surroundings.

T

Telophase – Stage of nuclear division when the nucleus reestablishes its interphase structure.

Temperature-sensitive mutation – Mutation yielding a protein that is functional at low (high) temperature but that is inactivated by temperature elevation (lowering).

Template – A macromolecular pattern that can be used for the synthesis of another, complementary macromolecule.

Tertiary structure–The three-dimensional folding of a polypeptide chain into a complex structural form, brought about by interactions among side chains of amino acids.

Thermodynamics – The branch of physical science that deals with exchanges of the energy in collections inherent in matter.

Thylakoid – A closed membrance sac that may be disk-shaped in grana or may be greatly elongated in a chloroplast; light-requiring reactions of photosynthesis take place here.

T^m – Midpoint melting temperature; temperature at the midpoint of transition of a preparation of uplex DNA molecules to single strands during melting.

Transcription – Process by which the base sequence of DNA is copied into a complementary RNA molecule.

Transfer RNA (tRNA) – The RNA molecule that carries an amino acid to a specific codon in messenger RNA during translation.

Transformation – The genetic modification induced by the incorporation into a cell of DNA form another source.

Translation– Process by which amino acids are assembled into a polypeptide on the ribosome, under the direction of the base sequence transcribed from DNA into messenger RNA.

Translocation – A structural rearrangement involving parts of or entire non-homologous chromosomes.

Tritium – ^3H; a radioactive isotope of hydrogen; extremely important in tracer studies.

Tropomyosin – A muscle protein that associates with actin to form long, thin fibers; plays a role in the regulation of muscle contraction.

T-system – Invaginations of the sarcolemma in muscle fibers of striated muscle, producing a system of transverse tubular infoldings.

Tubulin – Globular protein subunits (M.W. 55, 00 and 57, 000) whose regular helical packing forms the hollow, cylindrical microtubules.

U

Ultracentrifuge – Centrifuge capable

of rotor speeds up to 75, 000 rpm and able to rapidly sediment tiny particles and macromolecules.

Ultraviolet light – Electromagnetic radiation having a wavelength shorter than that of visible light (3900-2000 A). Causes DNA base-pair mutations and chromosome breaks.

Uncoupling agent – A substance (example, 2,4-dinitrophenol) that can uncouple phosphorylation of ADP from electrom transport; the energy is therefore released as heat.

Unit membrane – Membrance showing a railroad-track or dark-light-dark pattern of electron density in the electron microscope; a model of membrane structure proposing that a phospholipid bilayer is coated on its outer and inner surfaces by proteins.

V

Vacuole – A membrane-enclosed sac in the cell cytoplasm filled with molecules and particles a watery medium, frequent in plant cells.

Van der Waals force – A weak, attractive force between atoms; particularly important in hydrophobic bonding of amino acids in proteins.

Vesicle – A small, spherical, membrane-bordered element.

Viruses – Infectious, disease-causing particles that require a host cell for replication and that contain either DNA or RNA as their genetic material.

W

Weak bonds – Forces between atoms that are weaker than the forces involved in a covalent bond such as ionic bonds, hydrogen bonds, and Van der Waals force.

Wobble – Ability of third base in tTNA anticodon (5' end) to hydrogen bond with any two or three bases at 3' end of codon. Thus, a single tRNA species can recongnize several different codons.

X

X-ray crystallography – The use of X-ray scattering y crystals to determine the three-dimensional structure of molecules, especially proteins and nucleic acids.

Z

Zygote – The product of fusion of two gametes, the cell from which a new individual develops in each sexual generation.

Zygotene – Stage during prophase of meiosis I in which homologous chromosomes undergo synapsis.

Zymogen – A digestive enzyme precursor lacking catalytic activity in this form, for example, pepsinogen (converted to active pepsin).

GENETICS

A

A priori probability – Probability determined by the nature or geometry of an event or a situation.

A (aminoacyl) site – The site on the ribo-some occupied by an aminoacyl-tRNA just prior to peptide bond formation.

Acentic fragment – A chromosome piece without a centromere.

Acrocentric – A chromosome whose centromee lies very near one end.

Active site – The part of an enzyme where the actual enzymatic function is performed.

Adaptive value – *See* fitness.

Additive model – A mechanism of quantitative inheritance in which alleles at different loci either add a fixed amount to the phenotype or add nothing.

Adenine – *See* purines.

Adjacent -1 segregation – A separation of centromeres during meiosis in a reciprocal translocation heterozygote such that unbalanced gametes are produced.

Adjacent-2 segregation – Sparation of centromeres during meiosis in a reciprocal translocation heterozygote such that homologous centromeres are pulled to the same pole.

Affected – Individuals in a pedigree that exhibit the specific phenotype under study.

Allele – Alternative form of a gene.

Allopatric speciation – Speciation in which the evolution of reproductive-isolating mechanisms occurs during physical separation of the populations.

Allopolyploidy – Polyploidy produced by the hybridization of two species.

Allosteric protein – A protein whose shape in changed when it binds a particular molecule. In the new shape the protein's ability to react to a a second molecule is altered.

Allotype – Mutant of the nonvariant parts of immunoglobulin genes that follows the rules of simple Mendelian inheritance.

Alternate segregation – A separation of centromeres during meiosis in a reciprocal translocation heterozygote such that balanced gametes are produced.

Aminoacyl-tRNA synthetases – Enzymes that attach amino acids to their proper tRNAs.

Amphidiploid – An organism produced by hybridization of two species followed by somatic doubling. It is an allotetraploid that appears as a normal diploid.

Anaphase – The stage of mitosis and meiosis where sister chromatids or homologous centromeres are separated by spindle fibers.

Aneuploids – Individuals or cells exhibiting aneuploidy.

Aneuploidy – The condition of a cell or of an organism that has additions or deletions of whole chromosomes from the expected, balanced number of sets.

Angiosperms – Plants whose seeds are enclosed within an ovary. Flowering plants.

Anticodon – The complementary sequence to a codon.

Antigen – A foreign substance capable of triggering an immune response in an organism.

Antimutator-mutations – Mutations of DNA polymerase that decrease the overall mutation rate of a cell or of an organism.

Antiparallel strands – Strands, as in DNA, that run in opposite directions.

Ascospores – Haploid spores found in the asci of Ascomycete fungi.

Ascus – The sac in Ascomycete fungi that holds the ascospores.

Assignment Test – A test that determines whether a locus belongs on a specific chromosome by the observation of the concordance of the locus and the chromosome in hybrid cell lines.

Assortative mating – The mating of individuals with similar phenotypes.

Attenuator Region – A control region at the promoter end of repressible amino acid operons that exerts tanscriptional control dependent on the translation of a small leader peptide gene.

Autogamy – Nuclear reorganization in a single *Paramecium* cell similar to the changes that occur during conjugation.

Autopolyploidy – Polyploidy in which all the chromosomes come from the same species.

Autoradiography – A technique in which radioactive molecules make their location known by exposing pho-tographic plates.

Autosomal set – A combination consisting or one chromosome from each homolgous pair in a diploid species. The nonsex chromosomes.

Autotrophs – Organisms that can utilize carbon dioxide as a carbon source.

Autozygosity – Homozygosity in which the two alleles are identical by descent.

Auxotrophs – Strains of organisms that have specific nutritional requirements.

B

Bacillus – A rod-shaped bacterium.

Backmutation – The proces that causes reversion. A changes in a nucleotide parir in a mutant, gene that restores the original sequence and hence the original phenotpe.

Backcross – the cross of an individual with one of its parents or an organism with the same genotype as a parent.

Bacterial lawn – A continuous cover of bacteria on the surface of the growth medium.

Balanced lethal system – An arrangement of recessive lethals that maintains a heterozygous chromosome arrangement. Homozygotes for any lethal-bearing chromosome perish.

Balbiani rigns – The larger polytene chromosomal puffs. Generally synonymous with *puffs*. See chromosome puffs.'

Barr body – Heterochromatic body found in the nuclei of normal females but absent in the nuclei of normal males.

Binary fission – Simple cell division in single -celled organisms.

Binomial expansion – The terms generated when a binomial raised to a particular power is multiplied out.

Binomial theorem–The theorem that gives the terms of the expansion of a binomial raised to a particular power.

Bivalents – Structures, formed during prophase of meiosis I, consisting of the synapsed homologus chromosomes.

Bottleneck – A marked reduction in size of a population that potentially leads to genetic drift.

Breakage and reunion hypothesis – A model that suggests that breakage of homologous chromatids and their rejoining account for recombination.

Bubbles – Nucleic acid configuration relating to replication in eukaryotic chromo-somes or the shape of heteroduplex DNA at the site of a deletion or insertion.

Buoyant density of DNA – A measure of the density or lightness of DNA determined by the equilibrium point reached by the DNA in a density gradient.

C

Cancer – An informal term for a diverse class of diseases marked by abnormal cell proliferation.

Cancer-family syndromes – Pedigree patterns in which unusually large numbers of blood relatives develop certain kinds of cancers. A sequence of methyl groups added to the 5' end of eukaryotic mRNA. A protein that when bound with cyclic AMP can bind to sites on sugar-metabolizing operons to enhance transcription of these operons. Repression of certain sugar-metabolizing operons in favor of glucose utilization when glucose is present in the environment of the cell.

Catalyst – A substance that increases the rate of a chemical reaction without itself being permanently changed.

Cell cycle – The cycle of cell growth, replication of the genetic material, and nuclear and cytoplasmic division.

Central dogma – The original postulate that information can only be transferred from DNA to DNA, from DNA to RNA, and from RNA tro protein

Centric frangment – A chromosome piece with a centromere.

Centrioles – Cylindrical organelles, found in eukaryotes (except in higher plants), that organize the formation of the spindle.

Centromeres markers – Loci located near their centromeres.

Centromeres – Constrictions in eukaryotic chromosomes in which the kinetochores lie.

Centromeric fission – Creation of two chromosomes from one by splitting the centromere.

Chargaff's rule – Chargaff's discovery that, in the base composition of DNA, the quantity of adenine equalled the quantity of guanine equalled the quantity of cytosine (equal purine to pyrimidine content).

Charon phages – Phage lambda derivatives used as vehicles in recombinant DNA work.

Chiasmata – X-shaped configurations seen in tetrads during the latter stages of pro-phase I of meiosis. They represent physical crossovers.(Singualr: *chiasma*).

Chimaeric plasmid – Hybrid, or genetically mixed, plasmids used in recombinant DNA work.

Chimeras – *See* mosaics.

Chi-square distribution – The sampling distribution of the chi-square statistic. A family of curves depending on degrees of freedom.

Chloroplast – The organelle that carries out photosynthesis and starch grain formation in plants.

Chromatids – Two identical units (sister chromatids) held together at the centromere that, at prophase of nuclear divisions, make up each chromosome. When the centromeres divide and the chromatids separate, each chromatid is then a chromosome.

Chromatin – The nucleoprotein malerial of the eukaryotic chromousome.

Chromomeres – Dark regions in eukaryotic chromosomes at meiosis or mitosis.

Chromosome Theory – The theory that chromosomes are linear sequences of genes.

Circular DNA – The form of the genetic material in viruses and cells. A circle of DNA in prokaryotes; a DNA or an RNA molecule in viruses; a linear nucleoprotein complex in eukaryotes.

Chromosome puffs – Diffuse, uncoiled regions in polytene chromosomes where transcription is actively taking place.

Cis – Meaning "on the near side of" and referring to geometric configurations of atoms, or mutants usually on the same chromosome.

Cis-trans complementation test – A mating test to determine whether two mutants on opposite chromosomes will complement each other. A test for allelism.

Cistron – Term coined by Benzer for the smallest unit that exhibits the cis-trans position effect. Synonymouys with *gene or locus*.

CIB method – A technique devised by Muller to rapidly screen fruit flies for recessive Z chromosome lethals. The *CIB* chromosome carries a recessive lethal (L), a dominat marker (B) and an inversion (crosover suppressor, C).

Clone – A group of cells arising form a single ancestor.

Coccus – A spherical bacterium.
The situation in which the phenotypes independently produced by each allele as a homozygote are visible together as the phenotype in the heterozygote. The sequences of three RNA nucleotides that specify either an amino acid or terminuation of translation.

Coefficient of coincidence – The percent=age of observed double crosovers divided by the percentage expected.

Colicinogenic factors – *See* col plasmids.

Col plasmids – Plasmids that produce antibiotics (colicinogens) used by the host to kill other strains of bacteria.

Common ancestry – The state of two individuals when they are blood relatives. When two parents have common ancestry their offspring will be inbred.

Competence factor – A surface protein that binds extracellular DNA and enables the cell to be transformed.

Complementarity – The correspondence of DNA bases in the double helix such that adenine in one strand is opposite theymine in the other srtrand and cytosine in one strand is opposite guanine in the other. This explains chargaff's rule.

Complementation – The production of the wild-type phenotype by a cell or an organism that contains two mutant genes. If complemenation occurs, the mutants are non-allelic.

Complete linkage – The state in which two loci are so close together that alleles of these loci are virtually never separated by crossing over.

Complete linkage – The state in which two loci are so close together that alleles of these loci are virtually never separated by crossing over.

Complete medium – A medium that is enriched to contain all of the growth requirements of a strain of organisms.

Component of fitness – A particular variable in the life cycle of an organism upon which selection acts.

Concordance – The amount of similarity in phenotype among individuals.

Conditional-lethal mutant – A mutant that is lethal under another condition.

Confidence limits – A statistical term for a pair of numbers that predict, with a particular probability level, the region in which a particular parameter lies.

Conjugation – A process whereby two cells come in contact and exchange genetic material. In prokaryotes the transfer is a one-way process.

Consanguineous – Mating between blood relatives.

Conservative replication – A postulated mode of DNA replction where an intact double helix would act as a template for a new double helix.

Constitutive heterochromatin – Hetero-chromatin that surrounds the centromere. *See* satellite DNA.

Constitutive mutant – A mutant that is no longer under regulatory control but instead produces a fixed quantity of gene product.

Continuous replication – In DNA, uninterrupted replication allowed in the 5' to 3' direction by a 3' to 5' template.

Continous variation–Variation measured on a continuum or distribution rather than in discrete categories (e.g., height in humans).

Copy-choice hypothesis – An incorrect hypothesis that stated that recombination resulted from the swithching of the DNA replicating enzyme from one homologue to the other.

Corepressor – The metabolite that when bound to the repressor (of a repressible operon) forms a functonal unit that can bind to its operation and block transscription.

Correlation coefficient – A statistic that gives a measure of how closely two variables are related. The simultaneous transduction of two or more genes.

Cot values (cot1/2) – The product of C_0, the original concentration of denatiured, single-stranded DNA and t, time in seconds, giving a useful index of renaturation. Cot1/2 values are the midpoint values on cot curves -cot values plotted against concentration of remaining single-stranded DNA - and estimate the lenthe of unique DNA in the sample.

Coupling – Allele arrangement in which mutants are on the same chromosome and wild-type alleles on the homologue.

Covariance – A statistical value measuring the simultaneous deviations of x and y variables from their means.

Criss-eross pattern of inheritance – The phenotypic patern of inheritance shown by traits controlled by X-linked recessive alleles in a diploid XY species.

Critical chi-square – A chi-square for a given degree of freedom and proobability level to which an experimental chi-square is to be compared.

Crossbreed – Fertilization between separate individuals.

Crossing over – A process in which homologous chromosomes exchange parts by a breakage-and -reunion process.

Cryptic coloraton – Coloration that allows an organism to match its background and hence become less vulnerable to predation. A form of AMP used frequently as a second messenger in eukaryotic hormone nets and in catabolkite repression in prokaryotes.

Cytokinesis – The division of a cell into two daughter cells.

Cytoplasmic inheritance – Extra-chromo-somalk inheritance controlled by nonnuclear genomes. *See* pyrimidines.

D

Dauermodification – The persistance for several gnerations of an environmentally induced trait. A code in which several coc'e words have the same meaning. The genetic code is degenerate because there are many instances in which different condons specify the same amino acid.

Degrees of freedom – An estimate of the number of independent categries in a particular statistical test or experiment.

Deletion Chromosome – A chromosome with part deleted.

Denatured – Loss of natural configuration (of a molecule) through heat or other treatment. Denatured DNA is single stranded. A method of separating molecular moieties dependent upon their differential sedimentation in a centrifugal gradient.

Depauperate fauna – A fauna, especially common on islands, lacking many species found in similar habitats.

Dereprssed – The condition of an operon that is transcribing because repressor control has been lifirted.

Development – The process of orderly change that an individual goes through in the formation of struction of structure.

Diakinesis – The final stage of prophase I of meiosis when chaismate terminalize.

Dicentrie Chromosome – A chromosome with two centromers. An organism heterozygous at two loci.

Dimerization – The chemical union of two identical molecules.

Diploid – Thestate of having each chromosome in two copies per nucleus or cell.

Diplotene – The stage of prophase I of meiosis in which chromatids appear to repel each other.

Directional selection – A type of selection that removes individuals from one ednd of a phenotypic distribution and thus causes a shift in th dustribution.

Disassortative mating – The mating of individuals with dissimilar phenotypes.

Discontinuous replication – In DNA, only interrupted replication allowed backward in 5' to 3' segments by a 5' to 3' template strand.

Discontinuous variation – Variation that falls into discrete categories.

Discrete generations – Generations that have no overlapping reproduction. All reproduction takes place between individuals in the same generation.

Dispersive replication – A postulated mode of DNA replication combining aspects of conservative and semiconservative replication.

Disruptive selection – A type of selection that removes individuals

form the center of a phenotypic distribution and thus causess the distribution to become bimodal.

DNA-DNA hybridization – A technique in which when DNA from the same or different sources is heated and then cooled, double-helix configurations will reform at homologous regions. The technique is useful for determining sequence similarities and degrees of repetitiveness among DNAs.

DNA ligase – An enzyme that closes nicks or discontinuties in one strand of double-stranded DNA by creating an ester bond betweenadjacent 3'OH and a 5'PO$_4$ ends on the same strand.

DNA polymerase – One of several classes of enzymes that polymerize DNA nucleotides by using single - strnded DNA as a template and that require a double-helical primer.

DNA-RNA hybridization – A technique in which, when a mixture of DNA and RNA is heated and then coooled, RNA can bybridize (form a double helix) WITH DNA that has a complementary mucleotide sequence. A trait that expresses itself even when heterozygous.

Dosage compensation – The mechanism used in species with sex chromosomes to ensure that one sex does bnot suffer due to the different number of sex-linked alleles in the two sexes.

Double helix – The structure of DNA that is made of two helixes rotating about the same axis.

Double reduction – The condition in polyploids in which a heterozygous individual produces homozygous gametes.

Doublexes – An allele that converts fruit fly males and females into developmental intersexes.

Dyad – A centomere with two chromatids attached. The separation of molecular moieties by using electric current.

E

Elongation factors (EF-Ts, EF-Tu, EF-G) – Proteins necessary for the proper elongation and translocationprocesses during translation at the ribosome in prokaryotes.

Emperical probability – Probability determined by observing a large number of relevant cases.

Endogenote – Bacterial host chromosome.

Endomitosis – Chromosomal replication without nuclear or cellular division that results in cells with many copies of each chromosome.

Endonucleases – Enzymes that make nicks internally in the backbone of a polynu-cleotide. They hydrolyze internalo phosphodiester bonds.

Enriched medium – See complete medium.

Enzyme – Protein catalyst.

Episomes – Genetic particles that can either exist independently in a cell or can become intergated into the host chromo-some.

Epistasis – The masking of the action of alleles of one gene by allele combinations of another gene.

Equational division – The second meiotic division that is equational because it does not reduce chromosome numbers.

Euploidy – The condition of a cell or organism that thas one of more complete sets of chromosomes.

Evolution – In Darwinian terms, a gradual change in phenotypic frequencies that results in a population of individuals better adapted to survive.

Evolutionary rates – The rate of divergence between taxa, measurable as amino acid substitutions per million years.

Excision repair – A process whereby cells repair certain kinds of mutations by the removal of the mutated DNA starand and replacement using the good strand as a template.

Exconjugant – Each of the two cells that separates after conjugation has taken place.

Exogenote – DNA that a bacterial cell has incorporated through one of its sexual processes.

Exon – A region of a gene that has intervening sequences (introns) and that is actually translated. Enzymes that digest nuclecotides from the ends of polynucleotide molecules. They hydrolyze terminal phosphodiester bonds.

Experimental design – A branch of statistics that attempts to outline the way in which experiments should be carried out so that the date gathered will have statistical value.

Exploitation competition – A form of competition that, revoles around the superior ability to gather resources, rather than an active interaction among organisms for resources.

Expressivity – The degree of expression of a genbetically controlled trait.

Eyes – Referring to the configuration of replicating DNA in eukaryotic chromosomes.

F

F-pili – Sex pili, Hair-like projections of an F^+ or Hfr bacterium involved in anchorage during conjugation and presumabley through which DNA passes.

Factorial – The product of all integers from the specified number down to 1.

Fecundity selection – The forces acting to cause one genotype to be more fertile than another genotype.

Feedback inhibition – A postranslational control mechanism in which the end product of an enzymatic pathway inhibits the activity of first enzyme of this pathway.

Fertility factor – The plasmid that

allows a prokaryote to engage in conjugation with and pass DNA into an F- cell.

Filial generation – Offspring generation. F^1 is the first offspring, or filial, generation; F^2 is the second; and so on.

Fimbriae – Fringed. Referring to the surfacr of fbacteria with pili (hair-like projections).

First-division segregation (FDS) – The allele arrangement is spores of Ascomycetes with ordered spores that indicates no recombination between a locus and its centromere.

Fitness – The relative reproductive success of a genotype as measured by survival, fecundity, or other life history parameters.

Fluctuation test – An experiment by Luria and Delbruck that compared the variance in number of mutations between small cultures and subsamples of a large culture to the expense of the individual possessing the trait.

Fokker-Planck equation – An equation tht describes diffusion processes and that is used by population geneticists to describe random genetic drift.

Founder effect – Genetic drift observed in a population founded by a small, nonrepre-sentative sample of a larger population.

Frameshift, – A shift in the codon reading frame. *See* frameshift mutation.

Frameshift mutation – An addition or deletion of nucleotides that causes the codon reading frame to shift.

Frequency-dependent selection – Selection whereby a genotype is at an advantage when rare and at a disadvantage when common.

Functional alleles – Mutants that fail to complement each other in a *cis- rtans* complementation test.

Fundamental number – The number of chromosome arms in a somatic cell of a particular species.

G

G-bands – Eukaryotic chromosomal bands produced by treatment with Giemsa stain.

β-galactosidase – The enzyme that splits lactose into glucose and galactose (coded by a gene in the lac operon).

β-galactoside permease – An enzyme involved in concentrating lactose in the cell (coded by a gene in the *lac* operon).

Gametic selection – The forces acting to cause differential reproductive success of one allele ovar another in a heterozygote.

Gametophyte – The stage of a plant life cycle that produces gametes (by mitosis). Alternates with a diploid, sporophytge generation. INherited determinant of the phenotype. *See* cistron, locus.

Gene conversion – In Ascomycete fungi, where a 2:2 ratio of alleles is expected after meiosis and a 3:1 ratio is sometimes observed. The mechanism of this gene conversion is explained by the Holliday model of recombination.

Gene pool – All of the alleles available among the reproductive members of a population form which gametes can be drawn.

Genetic code – The linear sequences of nucleotides that specify the amino acids during the process or translation at the ribosome.

Genetic fine stucture – The structure of the gene in relation to the number and size of the smallest units of recombination and mutation.

Genetic load, – The relative decrease in the mean fitness of a population due to the presence of genotypes that have less than the highest fitness.

Genetic polymorphism – The occurrence together in the same population of more than one allele at the same locus, with the least frequent allele occurring more frequently than can be accounted for by mutation.

Genic balance theory – The theory of bridges that stated that the sex of a fruit fly is determined by the relative number of X-chromosomes and antosomes.

Genome – The genetic complement of a prokaryote or virus. A haploid cell or gamete of a eukaryotic species.

Genotype – The genes that an organism possesses.

Ger-line theory – A theory to account for the high degree of antibody variability. The germ-line theory suggests that every B lymphocyte has all the genes for every type of immunoglobulin but only transcribes one. *See* somatic-mutation theory.

Giemsa stain – A complex of stains specific for the phosphate groups of DNA.

Gray crescent – A cortical region of the egg of frogs and some salamanders that forms just after fertiliztion on the side opposite sperm penetration.

Group selection – Selection for traits that would be beneficial to a population at the .

Guanine – *See* purines.

Gynandromorphs – Mosaic individuals having simultaneous aspects of both the male and female phenotype.

H

H-Y antigen – The Histocompatibility Y-antigen, a protein found on the cell surfaces of male mammals. The state havifng one copy of each chromosopme per nucleus or cell.

Hemizygous – The condition of loci on the Z chromosome of the heterogametic sex of a diploid species.

Heritability – A measure of the degree to which the variance in the distribution of a phenotype is due to genetic causes.

Heterochromatin – Chromatin that

remains tightly coiled (and darkly staining) throughout the cell cycle.

Hetero duplex analy SIS – Analysis in which, if double-helix DNA is formed by strands from different sources, loopps and bubbles identfy regions where the two DNAs differ. This heterogeneous DNA is referred to as *hetero-duplex*. Elkectronmicroscpic observation of this DNA is a useful tool in recombinant DNA work. '

Hetero duplex DNA – See hybrid DNA.

Hetero geneous nuclear m RNA (hn RNA) – The original RNA transcripts found in eukaryotic nucei proir ot post tarns scriptional modifications.

Heterokaryon – A cell that contains two or more nuclei from different origins.

Heteromophic chromosomes – Chromosomes of which the members of a homologous pair are not morphologically identical (e.g., the sex chromosomes).

Heterotrophs – Organisms that require an organic form of carbon as a carbon source.

Heterozygote – A diploid or polyploid with different alleles at a particular locus.

Heterozygote advantage – A selection model in which heterozygotes have the highest, fitness.

Heterozygote DNA – *See* hybrid DNA.

Hfr – High frequency of recombination. A strain of bacteria that has incorporated an F factor into its chromosome and can then transfer the chromosome during conjugation.

Histones – Arginine and lysine-rich basic proteins making up a substantial portion of eukaryotic nucleoprotein.

Holoenzyme – The complete enzyme. Usually refers to RNA polymerase when indicating the core enzyme plus the sigma factor.

Homogametic – The sex with homomorphic sex chromosomes and which therefore only produces one kind of gamete in regard to the sex chromosomes.

Homologous chromosomes – Morphologically identical members of a homologous pair of chromosomes. A diploid or polyploid with identical alleles at a locus.

Hybid DNA – DNA whose two strands have different origins.

Hybrid – Offspring of unlike parents.

Hybrid screening – Radioisotope technique used to determine whether a hybrid plasmid contains a particular gene or DNA region.

Hypostatic gene – hA gene whose expression is masked by an epistatic gene.

Hypotheses, testing of – Statistical methods for determining the probabiloity that a date set firs a particular hypothesis about it.

I

Idiogram – A photograph or diagram of the chromosomes of a cell arrangeed in an orderly fashion.

Idiotypic variation – Variation in the variable parts of immuno-globulin genes.

Immunity – The ability of an organism to resist infection.

Immunoglobulins – Specific proteins produced by derivatives of B lymphocytes that protect an organism from antigens.

Inbreeding – The mating of genetically related individuals.

Inbreeding coefficient, F – The probability of autozygosity.

Inbreeding depression – A depression of vigor or yield due to inbreeding.

Incestuous – A mating between blood relatives who are more closely related than the law of the land allows.

Incomplete dominance – The situation in which both alleles of the heterozygote influence the phenotype.

Independent assortment, rule of – Mendel's second rule describing the independent segregation of alleles of different loci.

Inducibell system – A system, in which a coordinated group of enzymes is involved in a catabolic pathway, is inducible if the metabolite on which it works causes transcription of the genes controlling these eenzymes. These systems are primarily prokaryotic operons.

Induction – Regarding temperate phage, the process of causing a prophage to become virulent.

Industrial melanism – The darkening of moths during the recent period of industrializtion in mnay countries.

Initiation eodon – The mRNA sequence AUG, which specifis methionine, the first amino acid used in the translation process.

Initiation complex – The initiation complex of translation consisting of the 30S ribosome subunit, mRNA, N-formyl methionine tRNA, and three initiation factors.

Initiation factors (IF1, IF2, IF3) – Proteins required for the proper initintion of translation.

Insertion sequences (IS) – Regions of homology between host chromosomes and plasmids that allow the latter to synapse with the formar and become inserted into the host chromosome by a crossover.

Inside marker – The middle locus of three linked loci.

Intercalary heterochromatin – Heterochromation, other than centromeric hetero-chromatin, dispersed through eukaryotic chromosomes.

Intergenic suppression – A mutation at a second locus that apparently restores the wild-type phenotype to a mutant at a first locus.

Interphase – The metabolically active, nondividing stage of the cell cycle.

Interrupted mating – A mapping technique that disrupts bacterial conjugation after specified time intervals.

Intersex – An organism with external sexual characteristics that have attributes of both sexer.

Intervening sequences – Sequences of DNA within a gene that are transcribed but later removed prior to translation. *See* intron.

Intragenic suppression – A second change within a mutant gene that results in an apparent restoration of the original phenotype.

Intron – A length of DNA that makes up an intervening sequence.

Inversion – The replacement of an internal section of a chromosome in the reverse orientation.

In vitro – Biological or chemical work done in the test tube (literally, "in glass") rather than in living systems.

Iojap – A locus in corn that produces variegation.

Ionizing radiation – Radiation, such as X rays, that causes atoms to release electrons and become ions.

Isochromosome– A chromosome with two genetically and morphologically identical arms.

K

Kappa particles – The bacteria-like particles that give a *Paramecium* the "killer" phenotype.

Karyotype – The chromosome complement of a cell.

Kinetochores – The chromosomal attachment points for the spindle fibers, located within the centromeres.

L

Lampbrush chromosomes–Chromosomes of amphibian oocytes having loops suggestive of a lampbrush.

Leader – The length of mRNA from the 5' end to the initiation codon

(AUG).

Leader peptide gene – A small gene within the attenuator control region or repressible amino acid operons, Translation of the gene tests the content of the amino acid whose operon is being regulated.

Leader transcript – The mRNA transcried by the attenuator region of repressible amino acid operons. The transcript is capable of several alternate stem-loop structures dependent on the translation of a short leader peptide gene.

Leptotene – The first stage of prophase I of meiosis where chromosomes become distinct.

Level of significane – The probability value used to separate agreement or disagreement with the null hypothesis.

Linkage – The association of genes to chromosomes and the association of different loci on the same chromosome.

Linkage groups – Associations of loci on the same chromosome. In a species there are as many linkage groups as there are homologous pairs of chromosomes.

Locus – The position of a gene on a chromosome. Used synonymously with *gene*. (Plural: *loci*.)

Lyon hypothesis – The hypothesis that suggests that the Barr body is an inactivated X chromosome.

Lysate – The contents released from a lysed cell.

Lysis – The breaking open of a cell by the destruction of its wall or membrane.

Lysogenic – The state of a bacterial cell that has an integrated phage in its chromosome. One percent recombination be tween two loci.

M

Mapping – The study of the position of genes on chromosomes.

Mapping function – The mathematical relationship between measured map distance and actual recombination frequency.

Marker – A locus whose phenotype provides information about a chromosome or chromosome segement during genetic analysis.

Mutation rate – The proportion of mutants per cell division in bacteria or single-celled organisms or the proportion of mutants per gamete in higher organisms.

Mutator mutations – Mutations of DNA polymerase that increase the overall mutation rate of a cell or of an organism.

Muton – A tgerm coined by Benzer for the smallest mutable site within a cistron.

Maternal inheritance – Extrachromosomal inheritance controlled by non-DNA cytoplasmic substances.

Mean – The arithmetic mean, or the sum of the data values divided by the sample size.

Mean fitness of the population, - The sum of the fitnesses of the genotypes of a population weighted by their proportions; hence, a weighted mean fitness.

Meiosis – The nuclear process that results, in diploid eukaryotes, in gametes or spores with only one member of each original homologous pair of chromosmes per nucleus.

Merozygote – A partially diploid bacterial cell arising from one of the sexual processes.

Metacentric Chromosome – A chromosome with a centrally located centromere.

Meta female – A fruit fly with an X/A ratio greater than 1.0.

Metagon – An RNA necessary for the maintenance of mu particles in *Paramecium.*

Metamale – A fruit fly with a X/A ratio below 0.5.

Metaphase – The stage of mitosis or meiosis in which spindle fibers are attached to kinetochores and the chromosomes are positioned in the center of the cell.

Metaphase plate – The plane of the equator of the spindle into which chromosomes are manipulated at metaphase.

Microtubules – Hollow cylinders made of the protein tubulin and making up, among other things, the spindle.

Mimicry – A phenomenon in which an individual of one species gains an advantage by looking like individuals of a different species.

Minimal medium – A culture medium for microorgansms that contains the minimal necessities for growth of the wild type.

Missense mutations – Mutations that chnage a codon for an amino acid to a codon for a different amino acid.

Mitochondrion – The eukaryotic cellular organelle in which the Krebs cycle and electron transport reactions take place.

Mitosis – The nuclear division producing two daughter nuclei identical to the original nucleus.

Monohybrids – Offspring of parents that differ in only one characteristic. Usually implies heterozygosity at a single locus under study.

Monosomic – A diploid cell missing a single chromosome.

mRNA – Messenger RNA. The basic function of the nucleotide sequence of mRNA is to determine the amino acid sequence in proteins.

Mu particles – Bacteria-like particles found in the cytoplasm of *Paramecium* that cause the mate-killer phenotype.

Multihybrid – An organism heterozygous at numerous loci.

Multinomial expansion– The terms generated when a multinomial raised to a particualr power is multiplied out.

Mutants – Alternate alleles to the wild type. The phenotypes produced by alternate alleles.

Mutation – The process by which a

gene or chromosome changes structrually and the end result of this process.

Mutational load – Genetic load, caused by mutation, that brings deleterious alleles into a population.

N

Natural selection – A process whereby one genotype leaves more offspring than another genotype.

Nearest-neighbor analysis – A technique of transferring radioactive atoms between adjacent nucleotides in DNA that demonstrated that the two strands of DNA run in opposite directions.

Negative interference – The phenomenon whereby a crossover in a particular region enhances the occurrence of other apparent crossovers in the same region of the chromosome.

Neo-Darwinism – The merger of classical Darwinian evolution with population genetics.

Neutral gene hypothesis – The hypothesis that suggests that most genetic variation in natural populations is not maintained by selection.

NF – *See* fundamental number.

Nickase – An enzyme that nicks one strand of double-stranded DNA during DNa replication presumably to allow torsion to be released.

Nondisjunction – The failure of a pair of homologous chromosomes to separate properly during meiosis.

Nonhistone proteins – The proteins remaining in chromatin after the histones are removed. The scaffold structure is made of nonhistone proteins.

Nonparental ditype (NPD) – A spore arrangement in Ascomycetes that indicates a four-strand, double crossover between two linked loci.

Nonparentals – *See* recombinants.

Nonrecombinants – In mapping studies, offspring that have alleles arranged as in the original parents.

Nonsense codon – One of the mRNA sequences (UAA, UAG, UGA) that signals the termination of translation.

Nonsense mutations – Mutations that change a codon for an amino acid to a nonsense codon.

Normal distributions – Any of a family of bell-shaped curves defined on the basis of the mean and standard deviation.

Nuclear transplantation – The technique of placing a nucleus from one source into an enucleated cell.

Nuclease – One of several classes of enzymes that degrade nucleic acid. *See* endonucleases and exonucleases.

Nucleolus – The globular, nuclear organelle formed at the nucleolus organizer.

Nucleolus organizer – The chromosomal location of the ribosomal RNA genes around which the nucleolus forms.

Nucleoprotein – The substance of eukaryotic chromosomes consisting of proteins and nucleic acids.

Nucleosomes – Arrangements of DNA and histones forming regular spherical

Nucleotide – Subunits that polymerize into nucleic acid (DNA or RNA). Each nucleotide consists of a nitrogenous base, a sugar, and one or more phosphates.

Null hypothesis – The statistical hypothesis that states that there are no differences between observed and expected data.

Nullisomic – A diploid cell missing both copies of the same chromosome.

Nutritional-requirement mutants – *See* auxotrophs.

O

Okazaki fragments – Segments of newly replicated DNA produced during discontinuous DNA replica-tions.

Oogenesis – The process of ovum formation in female animals.

Operator – A DNA sequence that is recognized by a repressor protein or repressor-corepressor complex. When the operator is complexed with the repressor, transcription is prevented.

Operon – A sequence of adjacent genes all under the transcriptional control of the same operator.

Operon – The inducible operon including three loci involved in the uptake and brakdown of lactose.

Outbreeding – The mating of genetically unrelated individuals.

Outside marker – Loci on either side of another locus or specified region.

P

P (peptidyl) site – The site on the ribosome occupied by the peptidyl-tRNA just prior to peptide bond formation.

P₁ – Parental generation.

Pachytene – The stage of prophase I of meiosis where chromatids are first distinctly visible.

Paracentric inversion – An inversion that does not include the centromere.

Paramecin – A toxin liberated by "killer" *Paramecium*.

Parameters – Measurements of attributes of a population; denoted by Greek letters.

Parental ditype (PD) – A spore arrangement in Ascomycetes that indicates no recombination between two linked loci.

Parentals – *See* nonrecombinats.

Partial dominance – *See* incomplete dominance.

Pascal's triangle – A triangular array made up of the coefficients of the binomial expansion.

Passenger DNA – DNA incorporated into a plasmid to form a hybrid plasmid.

Path diagram – A modified pedigree showing only the direct line of descent form common ancestors.

Pedigree – A repesentation of the ancestry of an individual or family. A family tree.

Penetrance – The normal appearance in the phenotype of genetically controlled traits.

Peptidyl inversion – The enzyme responsible for peptide bond formation during translation at the ribosome.

Pericentric inversion – An inversion that includes the centromere.

Permissive temperature – A temperature at which temperature-sensitive mutants are normal.

Petite mutations – Mutations of yeast that produce small, an aerobic-like colonies.

Phages – *See* bacteriophages.

Phenocopy – A phenotype that is not genetically controlled but that looks like a genetically controlled phenotype.

Phenotype – The observable attributes of an organism.

Phosphodiester bond – Diester bond linking nucleotides together (between phosphoric acid and sugars) to form the nucleotide polymers DNA and RNA.

Photoreactivation – The process whereby dimerized pyrimidines (usually thymine dimers) are restored by an enzyme requiring light energy (deoxyribodipyrimidine photolyase).

Pili (fimbriae) – Hair-like projections on the surface of bacteria (Latin for "hair").

Plaques – Clear area on a bacterial lawn caused by cell lysis due to viral attack.

Plasmid – A genetic particle that can exist independently in a cell's cytoplasm without the ability to integrate into the host chromosome.

Plastid – A chloroplast prior to the development of chlorophyll.

Point mutatins – Mutations that are single changes in the nucleotide sequence and that consist of a replacement, addition, or deletion of a base pair.

Poky mutations – Mutations in *Neurospora* that produce a petite phenotype.

Polar bodies – The small cells (that eventually disintegrate) that are the by-products of meiosis in female animals. One functional ovum and three polar bodies result from meiosis of each primary oocyte.

Polarity – Referring either to an effect seen in only one direction from a point of origin or to the fact that linear moieties (such as a single strand of DNA) have ends that differ from each other. Polarity means directionality.

Polarity gene – A gene in mitochondrial DNA with alleles that are preferentially found in daughter mitochondria after recombination between mitochondria.

Poly-A-tail – A sequence of adenosine nucleotides added to the 3'end of eukaryotic mRNAs.

Polygenic inheritance – *See* quantitative inheritance.

Poly nucleotide phosphorylase – An enzyme that can polymerize diphosphate nucleotides without theneed for a primer. The in vivo function is probably inits reverse role, as an RNA exonuclease.

Polyploids – Organisms with whole chromosome sets greater than two.

Polysome – The configuration of several ribosomes simultaneously translating the same mRNA. Polyribo-some.

Polytene chromosome – Large chromosome consisting of many chromatids formed by rounds of endomitosis followed by synapsis.

Position effect – An alteration of phenotype caused by the relative arangement of the genetic material.

Positive interference – When the occurrence of one crossover reduces the probability that a second crossover will occur in the same region.

Postrrplicative repair – A DNA repair system initated when DNA polymerase bypasses a damage area. Uses enzymes in the *rec* system.

Posttranscriptional modification – The changes in eukaryotic mRNA made after transcription has been completed. These changes include additions of caps and tails and removal or introns.

Preemptor stem – A configuration of leader transcript that does not terminate transcription in attenuator-controlled amino acid operons.

Pribnow box – Relatively invariant sequence of 7 nucleotides in DNA that signal the start of transcription.

Primer – In DNA replication, a length of double-stranded DNA that continues as a single-stranded template leaving a 3'-OH end.

Probability – The expectation of the occurrence of a particular event.

Probability theory – The conceptual framework concerned with quantification of probabilities. *See* probability.

Product rule – The rule that states that the probability of the occurrence of independent events in the product of their separate probabilities.

Progeny testing – Breeding of offspring to determine their parents', genotypes.

Promoter – A region of DNA that signals the initiation of transcription to RNA polymerase.

Proofread – Technically, to read for the purpose of detecting errors for later corection. DNA polymerase has 3' to 5' exonuclease activity, which it uses during polymerization to romove nucleotides it it has recently added.

Prophage – A temperate phage integrated into a host chromosome.

Prophase – The initial stage of

mitosis or meiosis in which chromosomes become visible and the spindle apparatus forms.

Proplastid – Mutant plastids that od not grow and develop into chloroplasts.

Propositus (proposita) – The person through whom a particular pedigree was discovered.

Prototrophs – Strains of organisms that can survive on the minimal medium.

Pseudoalleles – Alleles that are functionally but not structurally allelic.

Pseudodominance – The phenomenon in which a recessive allele shows itself in the phenotype when only one copy of the allele is present as in hemizygous alleles or alleles opposite deletions.

Punnett square – A diagrammatic representation of a particular cross used to determine the progeny of this cross named after R.C. Punnett.

Purines – Nitrogenous bases of which thymine is present in DNA, uracil in RNA, and cytosine in both.

Q

Quantitative inheritance – The mechanism of genetic control of continuous variation. *See* continuous variation.

Quaternary structure – Of a protein, the association of polypeptide subunits to form the final protein.

R

R plasmids – Plasmids that carry genes that control resistance to various drugs.

RAM mutants – Referring to ribosomal ambiguity (RAM). Ribosomal mutants that allow incorrect tRNAs to be incorporated into the translation proces.

Random genetic drift – Changes in allelic frequency due to sampling error.

Random mating – The mating of individuals in a population such that the union of individuals with the trait under study occurs according to the product rule of probability.

Random strand analysis – Mapping studies in organisms that do not retain all the products of meiosis in a recoverable form.

Realized heritability – Heritability determined by response to selection.

Recessive – A trait that does not express itself in the heterozygous condition.

Reciprocal cross – Testing of the role of parental sex on a phenotype by repeating a particular cross withthe phenotype of each sex reversed as compared to the original cross.

Reciprocal translocation – A chromosomal configuration in which the ends of two non homologous chromosomes are broken off and become attached to the non-homologues.

Reciprocity – In relation to recombination, the conservation of the total amount of genetic material while allowing changes in the arrangement of alleles.

Recombinants – In mapping studies, off-spring with allelic arrangements made up of combinations of the original parental arrangements.

Recombination – The nonparental arrangement of alleles in progeny that can result from either independent asortment or crossing over.

Recon – A term coined by Benzer of the smallest recombinable unit within a cistron.

Rec system – Several loci controlling genes (recA, recB, recC, and others) involved in postreplicative DNA repair.

Reductional division – The first meiotic division that is reductional because it reduces the number of chromosomes and centromeres to half the original per daughter cell.

Regression to the mean – A pheno-menon of polygenmic traits in which the offspring of extremes tend toward the population mean.

Regulator gene – A gene primarily involved in control fo the produc-tion of another gene's product.

Release factors (RF-1, RF-2, RF-3) – Proteins responsible for proper termination of translation and release of the newly synthesized polypetide when a nonsense condon appears in the A site of the ribosome.

Renner complexes – Specific game-tic chromosome combinations in *Oenothera*.

Repetitive DNA – DNA containing copies of the same nucleotide se-quence.

Replica plating – A technique to rapidly transfer microorganism colonies to num-erous petri plates with different media.

Replication – The process of copying.

Replicons – A replicating genetic unit including the site for the initiation of replication.

Repressible system – A system in which a coordinated group of en-zymes is involved in a synthetic pathway (anabolic) if excess quan-tities of the end product of the pathway lead to the termination of transcription of the genes for the enzymes. These systems are porimarily prokaryotic operons.

Repressor – The protein product of a regulator gene that acts to control transcription of inducible and repressible operons.

Reproductive isolating mecha-nisms – Environmental, behavioral, mechanical, and physiological bar-riers that prevent two individuals of different populations from producing viable progeny.

Reproductive success – The unit of natural selection that is measured as the relative production of off-spring by a particular genotype.

Repulsion – Allele arrangement in which each homologue has mutant and wild-type alleles.

Resistance transfer factor – A plas-mid that confers on its host the simultaneous resistance to several antibiotics.

Restriction endonucteases – En-donu-cleases that recongnize cer-tain DNA sequences and cleave that DNA. Thought to protect cells from viral infection; useful in recombinat DNA work.

Restrictive temperature – A tem-perature at which temperature-sensitive mutants display the mutant phenotype.

Reverse transcriptase – An enzyme that can synthesize single-stranded DNA by using RNA as a template.

Reversion – The return of a mutant to the wild type through the pro-cess of a second mutational event.

Rho – A protein that is involved in the termination of transcription and release of the transcript at the terminator sequence.

Ribosomes – Organelles at which translation takes place. Made up of two subunits consisting of RNA and proteins.

RNA phages – Phages whose genetic material is RNA. They are the simplest phages known.

RNA polymerase – The enzyme that polymerizes RNA by using DNA as a template. (Also known as *transcriptase* or RNA *transcrip-tase.*)

RNA replicase – A polymerase en-zyme that catalyzes the self-replication of singl-stranded RNA.

Robertsonian fusion – Fusion of two acrocentric chromosomes at the cenbtromere.

Rolling circle replication – A model of DNA replication that accounts for a circular DNA molecule producidng linear danghter double helixes.

rRNA – Ribosomal RNA. RNA com-ponents of the subunits of the ribosomes.

Satellite DNA – Highly repetitive eukaryotic DNA primarily located around the centromeres. Satellite DNA usually has a different buoyant density than the rest of the cell's DNA.

S

Scaffold - The eukaryotic chromosome structure remaining when DNA and histones have been removed; made from nonhistone proteins.

Sereening technique – A technique to determine the genotype or phenotype of an organism.

Secondary oocytes–The cells formed by meiosis I in female animals.

Secondary spermatocytes– The products of the first meiotic division in male animals and which undergo the second meiotic division.

Secondary structure – Of a protein, the flat or helical configuration of the polypeptide backbone.

Second-divison segregation (SDS) – The allele arrangement in spores of Ascomycetes with ordered spores that indiates a cross-over between a locus and its centromere.

Segregation, rule of – Mendel's first principle describing how genes are passed from one generation to the next.

Segregational load – Genetic load caused when a population is segregating less fit homozygotes under heterozygote adantage.

Selection coefficients, s, t, – The sum of forces acting to prevent reproductive success of a genotype.

Selection mutation equilibrium – An equilibrium allele frequency resulting from the balance between selection removing an allele and mutaion recreating this allele.

Selective medium – A medium that is enriched with a particualr substance to allow the growth of particaular strains of organisms.

Self-fertilization – Fertilization in which the two gametes are from the same individual.

Semiconservative replication – The mode by which DNA replicates. *See* template.

Semisterility – Nonviabillity of a proportion of progeny.

Sex chromosomes – Heteromorphic chromosomes whose distribution in a zygote determines the sex of the organism.

Sex-controlled traits – Traits that appear more often in one sex than another but are neither sex linked, sex limited, or sex influenced.

Sexduction – A process whereby a bacterium gains access to and incorporates foreing DNA brought in by a modified F factor during conjugation.

Sex-influenced traits – Traits controlled by alleles that show a different dominance-recessiveness relationship depending on the sex of the heterozygote.

Sex-limited genes– Autosomal genes whose phenotypes are expressed in only one sex.

Sex linked – The inheritance pattern of loci located on the sex chromosomes (usually the X chromosome in XY species). Also refers to the loci themselves.

Sex-ratio phenotype – A trait in *Drosophila* where females produce mostly, if not only, daughters.

Sexual selection – The forces acting to cause one genothype to mate more frequently than another genotype.

Siblings (sibs) – Brothers and sisters.

Sigma factor – The protein that gives promoter-recognition speci-ficity to the RNA plymerase core enzyme.

Skew – A distortion of the shape of the normal distribution toward one side or the other.

Somatic doubling – A disruption of the mitotic process that produces a cell with twice the normal chromosome number.

Somatic-mutation theory – A

theory to account for the high degree of antibody variability. The somatic-mutation theory suggests that mutation of a basic immunoglobulin gene accounts for all of the different types of immunoglobulins produced by B lymphocytes. *See* germ-line theory.

Spacer DNA – Regions of nontranscribed DNa between transcribed segments, as in the numerous spacer regions in the nucleolus organizer.

Speciation – A process whereby, over time, one species evolves into a different species or where one species diverges to become two or more species.

Species – A group of organisms belonging to the same species because they are capable of interbreeding to produce fertile offspring.

Spindle – The microtubule apparatus that controls chromosome movement during mitosis and meiosis.

Spiral cleavage – The cleavage process in mollusks and some invertebrates where the spindle at mitosis is tipped in relation to the original egg axis.

Spirillum – A spiral bacterium.

Sporophyte – The stage of a plant life cycle that produces spores by meiosis and alaternates with the gametophyte stage.

Stabilizing selection – A type of selection that removes individuals from both ends of a phenotype distribution and thus maintains the same mean of the distribution.

Standard deviation – The square root of the variance.

Standard error of the mean – The standard deviation divided by the square root of the sample size. It is the standard deviation of a sample of means.

Statistics – Measurements of attributes of a sample from a population; denoted by Roman letters.

Stem-loop structure – Structures formed when nucleic acid loops back on itself to form complementary double helixes (stems) topped by the loops. Lollipop-shaped structures.

Stochastic – A proces with an indeterminate or random elecment as compared to a deterministic process that has no random element.

Stringent factor – A protein that catalyzes the formation of two unusual nucleotides during the stringent response under amino acid starvation.

Stringent response – A translational control mechanism of prokaryotes that represses tRNA and rRNA synthesis during amino acid starvation.

Structural allele – Mutant alleles that have changes at identical base pairs.

Stuctural genes – Nonregulatory genes.

Submetacentric chromosome – A chromosome whose centromere lies between the middle and the end, but closer to the middle.

Subtelocentric chromosome – A chromosome whose centromere lies between the middle and the end, but closer to the end.

Sum rule – The rule that states that the probability of the occurrence of one of several of a group of mutually exclusive events is the sum of the probabilites of the individual events.

Supergenes – Close physical association of several loci that usually control related aspects of the phenotype.

Suppressor gene – A gene that, when mutated, apparently restores the wild-type phenotype to a mutant of another locus.

Survival of the fittest – In evolutionary theory, survival of only those organisms best able to obtain and utilize resources (fittest). This phenomenon is the cornerstone of Darwin's theory.

Svedberg unit – A unit of sedimentation during centrifugation. Abbrebviated as S, as in 50S.

Swivalase – *See* nickase.

Sympatric speciation – Speciation in which the evolution of reproductive isolating mechanisms occurs within the range and habitat of the parent species. This specation is common in parasites.

Synapsis – The point-by-point pairing of homologous chromosomes during zygotene or in certain dipteran tissues prior to endomitosis.

Synaptinemal complex – A proteinaceous complex that mediates synapsis during zygotene and breaks donw shortly there-after.

Synteny test – A test that determines whether two loci belong to the same linkage group by observing concordance in hybrid cell lines.

Synthetic medium – A chemically defined substrate upon which microorganisms are grown.

T

Target theory – A theory that predicts response curves based on the number of events required to cause the phenomenon. Used to determine that point mutations are single events.

Tantomeric shift – Reversible shifts in proton position in a molecule. Bases in nucleic acids shift between keto and enol forms or between amino and imino forms.

Telocentric Chromosome – A chromosome whose centromere lies at one end.

Telophase – The terminal stage of mitosis or meiosis in which chromosomes uncoil, the spindle breaks down, and cytokinesis usually occurs.

Temperate phage – A phage that can enter into lysogency with its host. Mutants that are normal at a permissive temperature, but mutant at a restrictive temperature.

Template – A pattern serving as a mechanical guide. In DNA replication , each strand acts as a template for the synthesis of a new double helix.

Terminator sequence – A sequence in DNA that signals to RNA poloymerase the termination of transcription.

Terminator stem – A configuration of leader transcript that signals transcription termination in attenuatr-controlled amino acid operons.

Tertiary structure – Of a protein, the further folding beyond the secondary stucture as well as the formation of disulfide bridges between cysteines.

Testcross – The cross of an F_1 hybrid female with a male homozygous recessive organism.

Testing of hypotheses – The determination of whether to accept or reject a proposed hypothesis based on the likelihood that the hypothesis is correct.

Tetrads – The configuration made of four chromatids first seem in pachytene. There is one tetrad – bivalent -per homologous pair of chromosomes.

Tetranuclecotide hypothesis – Hypo-thesis, based on incorrect information, that DNA could not be the genetic material because its structure was too simple -repeating subunits containing one copy each of the four DNA nucleotides.

Tetraploids – organisms with four whole sets of chromosomes.

Tetratype (TT) – A spore arrangement in Ascomycetes tht indicates a single cross-over between two linked loci

Theta structure – An intermediate structure formed during the replication of a circular DNA molecule.

β-thiogalactoside acetyltransferase – An ezyme that is involved in lactose metabolism and encoded by a gene in the *lac* operon.

Three-point cross – A cross involving three loci.

Thymine – *See* pyrimidines.

Trailer – The length of mRNA from the nonsense codon the the 3' end (or, in prokaryotes, from a nonsense condon to the next initiation codon).

Trans – Meaning "across" and referring to geometric configurations of atoms or mutants usually on different homologous chromosomes.

Transcription – The process whereby RNA is synthesized form a DNA template.

Transduction – A process whereby a cell can gain access to and incorporate foreign DNA. The new DNA is brough in by a viral particle.

Transfer operon *(tra)* – Sequence of loci that impart the male (F-pili producing) phenotype on a acterium. The cell can then transfer its genes to antoher bacterium.

Transformation – A process whereby prokaryotes take up DNA from the environment and incorporate it into their genomes.

Transformer – An allele in fruit flies that converts chromosomal females into sterile males.

Transition mutation – A mutation in which a purine/pyrimidine base pair is replaced by a base pair in the same purine/pyrimidine relationship.

Translation – The process of protein synthesis wherein the primary structere of proteins is determined by the nucleotide sequence in RNA.

Translocease (EF-G) – Elongation factor necessary for proper translocation at the ribosome during the translation process.

Translocation – A chromosomal configuration in which part of a chromosome becomes attached to a differeent chromosome.

Transversion – A mutation in which a purine replaces a pyrimidine or

vice versa.

Trihybrid – An organism heterozygous at three loci.

Triploids – Organisms with three whole sets of chromosomes.

tRNA – Transfer RNA. Small RNA molecules that transfer amino acids to the ribosome for polymerization.

Two-point cross – A cross involving two loci.

Two-strand double crossovers – Double crossovers that occur in only two of the four chromoatids of a tetrad.

Type I error – In statistic, the rejecting of a true hypothesis.

Type II error – In statistics, the accepting of a false hypothesis.

Type-species concept – The concept that organisms that are morphologically similar belong to the same species.

U

Uninemie chromosome – A chromosome connsisting of one double helix of DNA.

Unique DNA – A length of DNA with no repetitive nucleotide sequences.

Unusual bases – Other bases, in addition to adenine, cytosine, guanine, and uracil, found primarily in tRNAs.

Uracil – *See* pyrimidines.

V

Variance – The average squared deviation about the mean of a set of data.

Variegation – Patchiness. A position effect caused when particular loci are contiguous with heterochromatin.

Vehicle plasmid – A plasmid containing a piece of passenger DNA forming a hybrid plasmid, used in recombination DNA work.

Virion – A virus particle.

Viroids – Bare RNA particles that are plant pathogens.

W

Wild type – The phenotype of a particular organism as first seen in nature.

Wobble – When the third position of an anticodon is not as closely constrained as the other positions (wobbles) and thus allow additional complementary base pairing.

X

X linked – *See sex* linked

X ray crystallography – A photographic technique, using X rays, to determine the atomic structure of molecules that have been crystallized.

Y

Y linked – Inheritance pattern of loci located on the Y chromosome. Also refers to the loci themselves.

Z

Zygotene – The stage of prophase I of meiosis in which synapsis occurs.

Zygotic inductions – When a prophage that is passed into an F cell during conjugation becomes virulent.

Zygotic selection – The forces acting to cause differential mortality of an organism at any stage in its life cycle (other than gametes).

BIO-TECHNOLOGY

A

Adventitious – Developing from unusual points of origin, such as shoots or root tissues, developing from callus or embyros, developing from sources other than zygotes.

Adventive embryony – Embryo formation and development resulting from asexual cells as occurs *in vivo* in certain members of the Rutaceae.

Androgenesis–Development of plants from male gametophytes.

Aneuploid – When the nucleus of a cell does not contain an exact multiple of the haploid number of chromosomes; one or more chromosomes being present in a greater or lesser number than the rest. The chromosomes may not show rearangements.

Antibiotics – Substances secreted by fungi or some bacteria which are capable of inhibiting the growth or killing various bacteria. They can be made naturaly using appropriate microrganisms, or synthetically in the laboratory.

Antibody – Specific protein produced by the immune system of higher animals and humans as part of the immune response to the presence of a specifric antigen.

Antigen – Substance or well defined part of a substance which is recognized and bound by a matching antibody.

Aseptic cuture – Raising cultures from a tissue or an organ after freeing it of bacteria, fungi, and other mieroorganisms.

Assay – Technique for measuring a biological process.

Attenuated – Weakened. Applied to vaccine formation , a pathogen used to induce antibody formation has been treated to render it incapable of causing a disease.

Autoimmune Disease – An immunological disorder in which the body produces antibodies against its own tissues.

Auxins – A class of growth hormones which cause cell elongation, apical dominance, root initiation, Indole acetic acid (IAA), naphthalene acetic acid (NAA), indolebutyric acid (IBA), and 2,4-dichlorophenoxyacetic acid (2, 4-D) are some of the auxins commonly used in tissue culture.

Axenic – Totally free form association with other organisms.

Axillary – Developing in the axil of leaves, e.g., axillary bud.

B

Bacillus subtilis – A bacterium used in genetic engineering experiments because of its importance in secreting proteins it makes.

Bacillus thuringiensis – A common soil bacterium which produces a protein toxic to insects.

Base – A component of the nucleotides which make up the DNA. Four different organic bases are involved in the structure of the DNA: adenine, guanine, cytosine and thymine. Their sequence is responsible for the genetic information in the DNA molecule.

Base Pair–A unit of the DNA double helix consisting of a base of each DNA strand loosely connected to each other. Only two pair combinations are possible: adenine-thymine and guanine-cytosine.

Batch Culture – A suspension culture in which cells grow in a finite volume of nutrient medium and follow a sigmoid pattern of growth.

Batch Processing– An industrial fermentation technology in which defined amounts of inorganic and living material are joined in a bioreactor. The desired product is selectively removed upon completion.

Biocatalyst–An enzyme that accelerates a biochemical reaction.

Biochips – Biological molecules that can replace semiconductors in electronic circuits.

Bioconversion – Capable of being broken down into simpler molecular components by living organisms such as bacteria or fungi.

Bioelectro catalysis – Incorporation of enzymes into fuctioning electrodes and their use to catalyze electrochemical and energy transfer reactions.

Biohazard – Term used by scientists to denote that genetically engineered organisms are being experimented on; intended as a warning to be cautions.

Biomass – The total mass of living matter in a given area in biotechnologically produced organismes or substances.

Bioprocess – Use cells or cellular components to produce a desired end product.

Bioreactor – Apparatus used for bioprocessing.

Biosynthesis – Formation of complex biological molecules from simpler ones by living organisms.

Biotechnology – Development of products by exploiting biological processes or substances. Production may be carried out by using intact original or modified organisms, e.g., bacteria or yeasts, or by using active cell components such as enzymes form organisms.

B Lymphocytes (B Cells) – Lymphocytes developed in bone marrow and involved in the production of antibody.

C

Callus – Cluster of undifferentiated plant cells that may be induced to form whole plants by suitable treatments.

Carcinogen – Agent that can induce cancer.

Cell culture – A technique for growing cells outside the body of an organism.

Cell Fusion – Formation of a hybrid cell by fusing two different cells.

Cell Line – Cells grown by culture methods having the same genetic makeup.

Chemostat – A growth vessel that maintains constant volume of cells or microbes by adding fresh nutrient medium and removing spent culture.

Clone – A group of individual organisms or cells all derived from a single progenitor by asexual reproduction and genetically identical to it.

Cloning – Multiplication method via clones.

Co-Metabolism – Process in which a substrate is modified but is not utilized for growth by an organism that is grown on or metabolizing another substrate.

Complementary DNA (cDNA)–DNA synthesized from a messenger RNA rather than from a DNA template. This type of DNA is used for cloing a gene or as a DNA probe to locate specific genes.

Continuous Processing – A bioprocessing method in which new material is added and the products removed continuously at a rate that maintains the volume at a constant level.

Cosmid – A vector for carrying large DNA fragments into host cells, made in the laboratory. It is formed from a plasmid by introducing "cos" (insertion) sites from lambda phage DNA at two exposed ends of the plasmid.

Cybrid – The viable cell resulting from the fusion of a cytoplast with a whole cell, thus creating cytoplasmic hybrid.

Cytokinins – A class of growth hormones which cause cell division, cell diferentiation, shoot differentiation, breaking apical dominance, etc. Some of the cytokinins commonly used in tissue culture are kinetin, benzylaminopurine (BPA), 2- isopentenyladenine (2-ip), and zeatin.

Cytoplast – The intact cytoplasm remaining following the enucleation of a cell.

Cytotoxic – Capable of killing cells.

D

Diagnostics – Agents used as a help to diagnose disease or disorders, i.e., identify a disease or disorder and distinguish one from another.

Differentiation – The development of different physiological and/or morphologial characteristics; the formation of different cell types, roots, shoots, embryos, or any other organ in the callus or cell culture.

DNA Probe – Usually a nucleic acid (or some other molecule) which has been labelled in some way and used to locate a particular base sequence or gene or a DNA molecule.

Downstream Processing – Stages in industrial processing following the bioconversion step (e.g., fermentation). Includes separation, purification and packaging of the product.

E

Electrophoresis – Separation of molecules based on their differing electrical charges.

Embryo – A very young plant developing inside the female gametophyte with or without fertilization.

Embryogenesis – The process of embryo initiation and development.

Embryoid – Non-zygotic embryo formed in culture.

Encapsulated Embryos – Plant embryos derived from somatic cells (somatic embryogenesis) artificially encapsulated together with nutrients and possibly also growth enhancers and pesticides. Intended for replacing seeds.

Endonuclease – Enzyme that breaks nucleic acids at specific internal sites, producing fragments of various lengths.

Epigenetic Variation – Phenotypic variability which has a non genetic basis.

Escherichia coli : (E.Coli) – A bacterium that normally inhabits the intestine of mammals. commonly used in genetic experiments.

Excise – A segment of DNA that is transcribed into mRNA and translated into protein.

Exon – A segment of DNA that is transcribed into mRNA and translated into protein.

Explant – Tissue taken from its origins site and transferred to an artifical medium for growth or maintenance.

F

Feedstock – Raw material used in chemical or biological processing.

Fermentation – An anaerobic process that releases energy from a sugar of other fermentable substance. Used to synthesize various products.

Friability – A term indicating the tendency for plant cells to separate from one antoher.

G

Gene – A segment of DNA carrying, in its base sequence, a very specific informatino. Some genes carry the information for the synthesis of proteins (structural genes), others carry information for regulatory fuctions (regulatory genes).

Gene Therapy – Therapy for congenital diseases involving the replacement of a deficient gene.

Genetic Engineering – A technology used to alter the genetic make up of living cells through direct interference with the genome in order to make them capable of producing substances or performing functions alien to the unmanipulated cells.

Genome – The total set of hereditary elements in a cell or organism.

Genomic Library – Fragments of cloned DNA from a single species of organism obtained by restriction enzyme digests; fragments are used to locate specific genes using the hybridizaton technique.

Genotyope – The genetic make-up of a individual as determined by the set of genes carried in the chromosomes.

Germplasm – The sum total of genetic variability available to a particular population of organisms.

H

Habituation – Ability of cells to grow in the absence of phytohormones after prolonged cultivation in the presence of hormones.

Heterokaryon – A cell possessing two or more genetically different nuclei, in a common cytoplasm usualy derived as a result of cell-to cell fusion.

Heteroploid – The term given to a cell culture when the cell comprising the culture possess nuclei containing chromosome numbers other than the diploid number. This is a term used only to describe a culture and is not used to describe individual cells. Thus, a heteroploid culture would be one which contains aneuploid cells.

High Fructose Corn Syrup (HFCS) – Sweeteners made of maize, rich in fructose.

Homokaryon – A cell possessing two or more genetically identical nuclei, in a common cytoplasm, derived as a result of cell-to-cell fusion.

Host-Vector system – Combination of DNA-transporting unit (vector) and DNA -receiving cell (host); used for introducing foreign DNA into a cell or organism.

Huymoral Response – An immune responese involving the production of antibodies by B lymphocytes.

Hybrid cell – The term used to describe the mononucleate cell which results from the fusion of two different cells, leading to the formation of a synkaryon.

Hybridoma – A hybrid cell that produces monoclonal antibodies in culture; formed by the fusion of a myeloma (cancer) cell with a normal antibody producifng lymphocyte.

I

Immune System – In vertebrates, the surveillance system that recognizes and acts against alien invaders or foreign cells.

Immunoassay – Technique for identifying substances by using antibodies that combinbe with them.

Immunodeficiency – Deficiency in the normal defence reaction of higher organisms against a foreign substance especially against disease causing agents.

Immunoglobulins – Circulating proteins comprising the antibodies.

Immunosuppressant – A substance which suppresses the immune system of an animal to lessen the body's rejection of, for example, transplanted organ.

Immunotoxin – A molecule that kills cells attached to an anthibody.

Induction – Intitiation of a plant structure, organ, or process i.i vitro.

Interferon – A chemical messenger of the immune system that inhibits viral replication and may have anticancer properties; three major types (alpha, beta and gamma) are known.

Interleukin – A class of lyumphokines important in the function of the immune system.

Intron – 'A part of eucaryotic gene that does not code for protein.

In vitro – Literally "in glass", now applied to any process carried out in sterile cultures.

In Vitro Propagation – Propagation of plants in a controlled, artificial environment, using plastic or glass culture vessels, aseptic techniques, and a defined growing mdium.

In vivo – Literally "in life", applied to any process occurring in a whole organism.

J

Juvenile – A phase in the sexual cycle of a plant characterized by differences in appearance from the adult and which lacks the ability to respond to flower-inducing stimuli.

K

Karyoplast – A cell nucleus surrounded by a narrow rim of cytoplasm and a plasma membrance which is obtained form the cell by enucleation.

L

Ligase – Enzyme used to join nucleic acid fragments together.

Lymphokine – A class of soluble proteins that play a role in the immune response whose mechanism is not yet understood.

Lymphoma – Cancer of the lymph tissue.

M

Macrophages – White blood cells that ingest dead cells and other debris in tissues and are involved in the production of interleukin; may also kill tumour cells when exposed to lymphokine "macrophage activating factor."

Mericloning – A popular term, not in scientific usage, referring to the in vitro vegatative propagation of orchids from excised shoot tip, axillary buds or floral organs.

Meristem Culture – In vitro culture of a generally shiny dome like structure measuring less than 0.1 mm in length when excised, most often excised from the shoot apex.

Micropropagation – This term is synonymous with in vitro propagation.

Monoclonal Antibodies – Highly specific antibodies derived from only one clone of a specific hybridoma cell and hence exactly of the same type. They recognize only one specific site of an antigen.

Morphogenesis – The porocess of growth and development of differentiated structures; the evolution of a structure from an undifferentiated to differentiated state.

Mutagenesis – A process by which changes in genetic constitution of a cell through alterations in its DNA are brought about.

Mutant – A phenotypic variant resulting from a changed or new gene.

Myeloma – A type of cancer cell used in the monoclonal antibody technique to form hybridomas.

O

Oligonucleotide – Short segments (up to about 10 nucleotides) of DNA or RNA.

Oncogene -Cancer-causing gene.

Organogenesis – The evolution, from dissociated cells, of structure which show natural organ form or function or both.

P

Parasexual Hybridization – Hybridization by non-sexual methods, e.g., by protoplast fusion.

Phagocyte – The name adopted by Tulecke et al., (1965) for an apparatus designed for the semi-continuous chemostat culture of plant cells.

Plantlet – A small rooted shoot or germinated embryo.

Plasmid – A small rooted shoot or germinated embryo.

Plasmid – A small bacterial DNA circle occurring in the bacterial cell separated from the bacterial chromosome. A plasmid carries additional genes not essential for bacterial growth. Plasmids can be replicated independently from the bacterial chromosome and can be passed on to another bacterial cell upon contact. They are efficient tools for genetic engineering.

Promoter – A DNA site where RNA polymerase initiates transcription.

Protoplast – A plant cell without its outer retaining cell wall.

Pseudodiploid – This describes the condition where the number of chromosomes in a cell is diploid but, as a result of chromosomal rearrangements, the karyotype is abnormal and linkage relationships may be disrupted.

Pseudogene – A silent gene; a copy of a gene that is not transcribed.

R

Radioimmunoassay – A technique that uses a radioactively labelled antibody to identify a molecule or measure a process.

Recombinant DNA (rDNA) – Segments of DNA from two differet organisms spliced together in the laboratory into a single molecule.

Regulatory Gene – A gene that acts to control the protein-synthesizing activity of other genes.

Replicon – Any genetic element that can reproduce independently, e.g., chromosome or a plasmid.

Restriction Enzyme – An enzyme that catalyzes the cleavage of DNA at a highly specific site. Each one recognizes a specific DNA sequence where it catalyzes cleaveage.

Retroviruses – A virus belonging to a group of extremely small viruses whose carrier of genetic information is not DNA but RNA. In these retroviruses, RNA is first reverse transcribed into DNA which is then inserted into the genome of the infected host cell, thus changing its genetic composition. This can be one of the reasons for cancerous diseases.

S

Somatic Cell Hybridization – The *in vitro* fusion of animal cells or plant protoplasts derived from somatic cells which differ genetically.

Somatic Embryogenesis – Induction via hormones of the development of an embryo out of a somatic cell cluster, in plants.

Subculture– With plant cultures, this is the process by which the tissue or explant is first subdivided, then transferred into fresh culture medium.

Super ovulation – The maturation and release of more than the usual number of ova in an animal induced by the additional application of hormones.

Splicing – A stage in RNA processing in which introns are removed and exons are joined to form a continuous coding sequence of RNA.

Suspension culture – A type of culture in which cells, or aggregates of cells, multiply while suspended in liquid medium.

Synchronous Culture – A culture in which the cell cycles (or a specific phase of the cycle) of a proportion of the cells (often a majority) are synchronous.

Synkaryon – A hybrid cell which results from the fusion of the nuclei it carries.

T

Tissue Culture – Commonly used as a blanket term to refer to all types of aseptic plant or animal cultures. Strictly speaking, however, this term should include only the culture of unorganized tissue or callus

Tisssue Plasminogen Activator – A protein involved in the process of dissolving blood clots. It activates the enzyme plasminogen which is important in dissolving blood clots.

Tlymphocytes (TCELLS) – Lymphocytes which recognize, engulf, and destroy specific foreign cells.

Totipotency – A cell characteristic in which the potenital for forming all the cell types in the adult organism is retained.

Transfection – The transfer, to cells in culture, of a gene(s) from another cell.

Transgenic Organism – A genetically manipulated organism containing in its genome one of more inserted genes of another species.

Transposon – A segment of DNA carrying one or more genes that can move from one DNA molecule to another to result in a change of the altered DNA.

Turbidostat – An open continuous culture into which fresh medium flows in response to an increase in the turbidity of the culture. A preselected biomass density is uniformly maintained by wash-out of excess cells.

V

Vaccine – A preparation containing an antigen made up of whole disease causing organisms (attenuated or parts of such organisms; used ot confer immunity against the antigen.

Variant – A culture exhibiting a stable phenotypic change whether genetic or epigenetic in origin.

Vector – Agent used to transfer DNA into a host cell, e.g., plasmid or bacteriophage.

Vegetative Propagation – Reproduction of plants using a non-sexual proces involving the culture of plant parts such as stem and leaf cuttings.

GENE CLONING

2 μm circle – A plasmid found in the yeast *Saccharomyces cerevisiae* and used as the basis for a series of cloning vectors.

Adaptor – A synthetic, double-stranded oligonucleotide used to attach sticky ends to a blunt-ended molecule.

Agrobacterium tumefaciens – The soil bacterium which, when containing the Ti plasmid, is able top form crown galls on a number of dicotyledonous plant species.

Autoradiography – A method of detecting radioactively labelled molecules through exposure of an X-ray sensitive photographic film.

Auxotroph – A mutant microorganism that will grow only if supplied with a nutrient not required by the wild-type.

Avidin – A protein that has a high affinity for biotin and is used in the detection system for biotinylated probes.

B

Bacteriophage or Phage – A virus whose host is a bacterium. Bacteriophge DNA molecules are often used as cloning vectors.

Baculovirus – A virus that has been used as a cloning vector for the production of recombinant protein in insect cells.

Batch culture – Growth of bacteria in a fixed volume of liquid medium in a closed vessel, with no additions or removals made during the period of incubation.

Biolistics – A means of introducing DNA into cells that involves bombardment with high-velocity microprojectiles coateed with DNA.

Biological containment – One of the prcautionary measures taken to prevent the replication of recombinant DNA molecules in microorganisms in the natural environment. Biological containment involves the use of vectors and host organisms that have been modified so that they will not survive outside of the laboaratory.

Biotechnology – The use of living organisms, often but not always microbes, in industrial processes.

Biotin – A molecule that can be incorporated into dUTP and used as a non-radioactive label for a DNA probe.

Blunt end or Floush end – An end of a DNA molecule at which both strands terminate at the same nucleotide position with no single-stranded extension.

Bovine papilloma virus (BPV) – A group of mammalian viruses, derivatives of which have been used as cloning vectors.

Broth culture – Growth of microorganisms in a liquid medium.

Buoyant density – The density possessed by a molecule or particle when suspended in an aqueous salt or sugar solution.

C

Capsid – The protein coat that encloses the DNA or RNA molecule of a bacterio-phage or virus.

Cassette – A DNA sequence consisting of promoter-ribosome binding site -unique restriction site-terminator carried by certain types of expression vector. A foreign gene inserted into the unique restriction site will be placed under control of the expression signals.

Cauliflower mosaic virus (CaMV)– The best studied of the caulimoviruses, used as a cloning vector for some species of higher plant.

Caulimoviruses – One of the two groups of DNA viruses to infect plants, the members of which have potential as cloning vectors for some species of higher plants.

Cell extract – A preparation consisting of a large number of broken cells and their released contents.

Cell-free translation system – A cell extract containing all the components required for protein synthesis (i.e. ribosomal subunits, tRNAs, amino acids, enzymes and cofactors) and able to translate added mRNA molecules.

Chimaera–A recombinant DNA molecule made up of DNA fragments from more than one organism, named after the mythological beast.

Chromosome – A self-replicating nuc-leic acid molecule carrying a number of genes.

Chromosome walking – A technique used to identify a series of overlapping restriction fragments, often to determine the relative positions of genes on large DNA molecules.

Cleared lysate – A cell extract that has been centrifuged to remove cell debris, subcellular particles and possibly chromosomal DNA.

Clone – A population of identical cells, generally those containing identical recombinant DNA molecules.

Compatibility – Refers to the ability of two different types of plasmid to coexist in th same cell.

Competent – Refers to a culture of bacteria that have been treated to enhance their ability to take up DNA molecules.

Complementary – Refers to the ability of two different types of plas-

mid to coexist in the same cell.

Competent – Refers to a culture of bacteria that have been treated to enhance their ability to take up DNA molecules.

Complementary– Refers to two polynucleotides that can base-pair to form a double-stranded molecule.

Complementary DNA (cDNA) cloning – A cloning technique involving conversion of purified mRNA to DNA before insertion into a vector.

Conformation – The spatial organization of a molecule. Linear and circular are two possible conformations of a polynucleotide.

Conjugation – The process whereby two bacteria exchange genetic material via a transient intercellular connection called a pilius.

Consensus sequence – A nucleotide sequence used to describe a large number of related though non-identical sequences. Each position of the consensus sequence represents the nucleotide most often found at that position in the real sequences.

Continuous culture – The culture of microorganisms in liquid medium under controlled conditions, with additions to and removals form the medium ovar a lenghty period of time.

Copy number – The number of molecules of a plasmid contained in a single cell.

Cos site – One of the cohesive, single-stranded extensions present at the ends of the DNA molecules of certain strains of l phage.

Cosmid – A cloning vector consisting of the l *cos* site inserted into a plasmid, used to clone DNA fragments up to 40 kb in size.

Covalently closed-circular (CCC) – A completely double-stranded circular DNA molecule, with no nicks or discontinuities, usually with a supercoiled conformation.

D

Defined medium– A bacterial growth

medium in which all the components are known.

Deletion analysis – The identification of control sequences for a gene by determining the effects on gene expression of specific deletions in the upstream region.

Denaturation – Of nucleic acid molecules; breakdown by chemical or physical means of the hydrogen bonds in base pairing.

Density-gradient centrifugation – Separation of molecules and particles on the basis of buoyant density, by centrifugation is a concentrated sucrose or caesium chloride solution.

Dideoxynucleotide– A modified nucleotide that lacks the 3'hydroxyl group and so prevents further chain elongation when incorporated into a growing polynucleotide.

Disarmed plasmid – A Ti plasmid that has had some or all of the T-DNA genes removed, so it is no longer able to promote cancerous growth of plant cells.

DNA sequencing – Determination of the order nucleotides in a DNA molecule.

Double digestion – Cleavage of a DNA molecule with two different restriction endonucleases, either concurrently or consecutively.

E

Electrophoresis–Separation of molecules on the basis of their net electric charge.

Electroporation – A method for increasing DNA uptake by protoplasts through prior exposure to a high voltage which results in the temporary formation of small pores in the cell membrance.

End-filling – Conversion of a sticky end to a blunt end by end by enzymatic synthesis of the complement to the single-stranded extension.

Endonuclease – An enzyme that breaks phosphodiester bonds

within a nucleic acid molecule.

Episome – A plasmid capable of integration into the host cell's chromosome.

Ethanol precipitation– Precipitation of nucleic acid molecules by ethanol plus salt, used primarily as a means of concentrating DNA.

Exonuclease – An enzyme that sequentially removes nucleotides from the ends of a nucleic acid molecule.

Expression vector – A cloning vector designed so that a foreign gene inserted into the vector will be expressed in the host organism.

F

Fermenter – A vessel used for the large-scale culture of microorganisms.

Footprinting – The identification of protein-binding site on a DNA molecule by determinging which phosphodiester bonds are protected from cleavage by DNase l.

G

Gel electrophoresis– Electrophoresis performed in a gel matrix so that molecules of similar electric charge can be separated on the basis of size.

Gel retardation – A technique that identifies a DNA fragment that has a bound protein by virtue of its decreased mobility during gel electrophoresis.

Geminivirus – One of the two groups of DNA viruses that infect plants, the members of which have potential as cloning vectors for some species of higher plants.

Gene – A segment of DNA that codes for an RNA and/or polypeptide molecule.

Gene cloning – Insertion of a fragment of DNA, carrying a gene, into a cloning vector, and subsequent propagation of the recombinant DNA molecule in a host organism.

Gene mapping–Determination of the relative positions of different genes on a DNA molecule.

Genetic engineering – The use of experimental techniques to produce DNA molecules containing new genes or new combinations of genes.

Genetics – The branch of biology devoted to the study of genes.

Genome – The complete set of genes of an organism.

Genomic library – A collection of clones sufficient in number to include all the genes of a particular organism.

H

Harvesting – The removal of microorganisms from a culture, usually by centrifugation.

Helper phage – A phage that is introduced into a host cell in conjunction with a related cloning vector, in order to provide enzymes required for replication of the cloning vector.

Heterologous probing – The use of a labelled nucleic acid molecule to identify related molecules by hybridization probing.

Homology – The degree of identity displayed by the nucleotide sequences of two related but not complementary polynucleotides. 85% homology means that 85 nucleotide positions out of 100 are identical in the two polynucleotides (e.g. AAAAA) to the end of a nucleic acid molecule, usually referring to the synthesis of single-stranded homopolymer extensions on the ends of a double-stranded DNA molecule.

Horseradish peroxidase–An enzyme that can be complexed to DNA and with is used in a non-radioactive procedure for DNA labelling.

Host-controlled restriction – A mechanism by which some bacteria prevent phage attach through the synthesis of a restriction endonuclease that cleaves the non-bacterial DNA.

Hybrid-arrest translation (HART) – A method used to identify the polypeptide coded by a cloned gene.

Hybrid-release translation (HRT)– A method used to identify the polypeptide coded by a cloned gene.

Hybridization probe – A labelled nucleic acid molecule that can be used to identify complementary or homologous molecules through the formation of stable base-paired hybrids.

I

Immunological screening – The use of an antibody to detect a polypeptide synthesized by a cloned gene.

Incompatibility group – Comprises a number of different types of plasmid, often related to each other, that are unable to coexist in the same cell.

Induction – (1) Of a gene; the switching on of an expression of a gene or group of genes in response to a chemical or other stimulus. (2) Of l, phage: excision of the integrated form of λ, and swith to the lytic mode of infection, in response to a chemical or other stimulus.

Insertional inactivaton – The cloning strategy whereby insertion of a new piece of DNA into a vector inactivates a gene carried by the vector.

Insertion vector – A λ vector constructed by deleting a segment of non-essential DNA.

In situ **hybridization** – A technique for gene mapping involving hybridization of a labelled sample of a cloned gene to a large DNA molecule, usually a chromosome.

In vitro **mutagenesis** – Any one of several techniques used to produce a specified mutation at a predetermined position in a DNA molecule.

In vitro **packaging** – Synthesis of infective λ particles form a preparation of a capsid proteins and a concatmer of DNA molecules separated by *cos* sites.

K

Klenow fragment – (of DNA polymerase 1) The enzyme that sythesizes a new DNA strand on an existing template, used primarily in DNA sequencing.

L

Labelling – The incorporation of a radioactive nucleotide into a nucleic acid molecule.

Lambda(λ) – A bacteriophage that infects *E. coli*, derivatives of which are extensively used as cloning vectors.

Ligase (DNA ligase) – An enzyme that repairs single-stranded discontinuities in double-stranded DNA molecules in the cell. Purified DNA ligase is used in gene cloning to join DNA molecules together.

Linker – A synthetic, double-stranded oligonucleotide used to attach sticky ends to a blunt-ended molecule.

Lysogen – A bacterium that harbours a prophage.

Lysogenic infection cycle – The pattern of phage infection that involves integration of the phage DNA into the host chromosome.

Lysozyme – An enzyme that weakens the cell walls of certain types of bacteria.

Lytic infection cycle – The pattern of infection displayed by a phage that replicates and lyses the host cell immediately after the initial infection. Integration of the phage DNA molecule into the bacterial chromosome does not occur.

M

M13 – A bacteriophage that infects *E. coli*, derivatives of which are extensively used as cloning vectors.

Microinjection – A method of introducing new DNA into a cell by injecting it directly into the nucleus.

Minimal medium – A defined medium that provides only the minimum number of different nutrients needed for growth of a particular bacterium.

Multicopy plasmid – A plasmid with a high copy number.

Multigene family – A number of identical or related genes present in the same organism, usually coding for a family of related polypeptides.

N

Nick – A single-strand break, involving the absence of one or more nucleotides, in a double-stranded DNA molecule.

Nick translation – The repair of a nick with DNA polymerase I, usually to introduce labelled nucleotides into a DNA molecule.

Northern transfer – A technique for transferring bands of RNA from an agarose gel to a nitrocellulose or similar membrane.

Nucleic acid hybridization – Formation of a double-stranded molecule by base pairing between complementary or homologous polynucleotides.

O

Oligonucleotide-directed mutagenesis – An *in vitro* mutagenesis technique that involves the use of a synthetic oligonucleotide to introduce the predetermined nucleotide alteration into the gene to be mutated.

Open-circular – The non-supercoiled conformation taken up by a circular double-stranded DNA molecule when one or both polynucleotides carry nicks.

Origin of replication – The specific position on a DNA molecule where DNA replication begins.

Orthogonal field alternation gel electrophoresis (OFAGE) – A gel electrophoresis technique that employs a pulsed electric field to achieve separation of very large molecules of DNA.

P

Partial digestion – Treatment of a DNA molecule with a restriction endonuclease under such conditions that only a fraction of all the recognition sites are cleaved.

Phenotype expression – A technique designed to maximize the transformation frequency obtained when using a plasmid vector.

Pilus – One of the structures present on the surface of a bacterium containing a conjugative plasmid, through which DNA transfer occurs during conjugation.

Plaque – A zone of clearing on a lawn of bacteria caused by lysis of the cells by infecting phage particles.

Plasmid – A usually circular piece of DNA, primarily independent of the host chromosome, often found in bacterial and some other types of cell.

Plasmid amplification – A method involving incubation with an inhibitor of protein synthesis aimed at increasing the copy number of certain types of plasmid in a bacterial culture.

Polyethylene glycol – A polymeric compound used to precipitate macromolecules and molecular aggregates.

Polylinker – A synthetic double-stranded oligounucleotide carrying a number of restriction sites.

Primer – A short single-stranded oligonucleotide which, when attached by base pairing to a single-stranded template molecule, acts as the start point for complementary strand synthesis directed by a DNA polymerase enzyme.

Promoter – The nucleotide sequence, upstream of a gene, that acts as a signal for RNA polymerase binding.

Prophage – The integrated form of the DNA molecule of a lysogenic phage.

Protease – An enzyme that degrades protein.

Protoplast – A cell from which the cell wall has been completely removed.

R

Radioactive marker – A radioactive atom used in the detection of a larger molecule in which it is incorporated.

Random priming – A method for DNA labelling that utilizes random DNA hexamers which will anneal to single-stranded DNA and act as primers for complementary-strand synthesis by a suitable enzyme.

Recombinant – A transformed cell that contains a recombinant DNA molecule.

Recombinant DNA molecule – A DNA molecule created in the testtube by ligating together pieces of DNA that are not normally contiguous.

Recombinant DNA technology – All the techniques involved in the construction, study and use of recombinant DNA molecules.

Recombinant protein – A polypeptide that is synthesized in a recombinant cell as the result of expression of a cloned gene.

Recombination – The exchange of DNA sequences between different molecules, occurring either naturally or as a result of DNA manipulation.

Relaxed – Refers to the non-supercoiled conformation of open-circular DNA.

Replacement vector – A vector designed so that insertion of new DNA is by replacement of part of the non-essential region of the DNA molecule.

Replica plating – A technique whereby the colonies on an agar plate are transferred *en masse* to a new plate, on which the colonies will grow in the same relative positions as before.

Replicative form of M13 – The double-stranded form of the M13 DNA molecule found within infected *E. coli* cells.

Reporter gene – A gene whose phenotype can be assayed in a transformed organism, and which is used in, for example, deletion analyses of regulatory regions.

Repression – The switching off of expression of a gene or a group of genes in response to a chemical or other stimulus.

Restriction analysis – Determination of the number and sizes of the DNA molecules only at a limited number of specific nucleotide sequences.

Restriction fragment length polymorphism (RFLP) – A mutation that results in a detectable change in the pattern of fragments obtained when a DNA molecule is cut with a restriction endonuclease.

Restriction map – A map showing the positions of different restriction sites in a DNA molecule.

RFLP linkage analysis – A technique that uses a closely linked RFLP as a marker for the presence of a particular allele in a DNA sample, usually as a means of screening individuals for defective genes responsible for genetic diseases.

Ribonuclease – An enzyme that degrades DNA.

Ribosome binding site – The short nucleotide sequence upstream of a gene, which after transcription forms the site on the mRNA molecule to which the ribosome binds.

S

Selectable marker – A gene carried by a vector and conferring a recognizable characteristic on a cell containing the vector or a recombinant DNA molecule derived from it.

Selection – a means of obtaining a clone containing a desired recombinant DNA molecule.

Shotgun cloning – A cloning strategy that involves the insertion of random fragments of a large DNA molecule into a vector, resulting in a large number of different recombinant DNA molecules.

Shuttle vector – A vector that can replicate in the cells of more than one organism (e.g. in *E. coli* and in yeast).

Simian virus 40 (SV40) – A mammalian virus used as the basis for a series of cloning vectors.

Southern transfer – A technique for transferring bands of DNA from an agarose gel to a nitrocellulose or similar membrane.

Sphaeroplast – A cell with a partially degraded cell wall.

Stem-loop – A hairpin structure, consisting of a base paired stem and a non-base paired loop, that may form in a polynucleotide.

Sticky end – An efficient promoter that can direct synthesis of RNA transcripts at a relatively fast rate.

Stuffer fragment – The part of a λ replacement vector that is removed during insertion of new DNA.

Supercoiled – The conformation of a covalently closed circular DNA molecule, which is coiled by torsional strain into the shape taken by a wound-up elastic band.

T

T-DNA – The portion of the Ti plasmid transferred to the plant DNA.

Temperature-sensitive mutation – A mutation that results in a gene product that is functional within a certain temperature range (e.g. at less than 30°C), but non-functional at different temperatures (e.g. above 30°C).

Template – A single-stranded polynucleotide (or region of a polynucleotide) able to direct synthesis of a complementary polynucleotide.

Terminator – The short nucleotide

sequence downstream of a gene that acts as a signal for termination of transcription.

5'-terminus – One of the two ends of a polynucleotide that which carries the phosphate group attached to the 5' position of the sugar.

3'-terminus – One of the two ends of a polynucleotide; that which carries the hydroxyl group attached to the 3' position of the sugar.

Ti plasmid – The large plasmid found in those *Agrobacterium tumefaciens* cells able to direct crown gall formation on certain species of plants.

Total cell DNA – Consists of all the DNA present in a single cell or group of cells.

Transcript analysis – Experiment aimed at determining which portions of DNA molecule are transcribed into RNA.

Transfection–The introduction of any DNA molecule into any living cell.

Transformation fequency – A measure of the proportion of cells in a population that are transformed in a single experiment.

U

Undefined medium – A growth medium in which not all the components have been identified.

UV absorbance spectroscopy – A method for measuring the concentration of a compound by determining the amount of ultraviolet radiation absorbed by a sample.

V

Vector – A DNA molecule, capable of replication in a host organism, into which a gene is inserted to construct a recombinant DNA molecule.

Vehicle – Often used as a substitute for the word 'vector', emphasizing that the vector transports the inserted gene through the cloning experiment.

W

Watson-Crick rules – The base pairing rules that underlie gene structure and expression. A pairs with T, G with C.

Western transfer – A technique for transferring bands of protein from an electro-phoresis gel to a membrane support.

Yeast artificial chromosome (YAC) – A cloning vector comprising the structural components of a yeast chromosome and able to clone very large pieces of DNA.

Yeast episomal plasmid (YEp) – A yeast vector carrying the 2 μm circle origin of replication.

Yeast integrative plasmid (YIp) – A yeast vector that relies on integration into the host chromosome for replication.

Yeast replicative plasmid (YRp) – A yeast vector that carries a chromosomal origin of replication

CYTOGENETICS

Allelism - The relationship between the characters which are alleles *allelomorph* two contrasting but closely parallel genetic characters e.g., smooth or wrinkled skin in peas; *dominant allele* one which determines the phenotypic expression in a heterozygous form; *Isolallele* one allele which so closely resemble another that the two can only be distinguished by special techniques; *Pseudoallele* a group of closely linked loci once thought to be a single lous, Pseudoalleles do not complement and recombine very rarely; *Recessive allele* which produces the phenotypic expression only when in the homozygous state; *Reciprocal translocation* are in which each chromosome receives a portion of the other.

allometry - A genetic change in the proportion of an existing character, such as an increase of depth of colour in a flower.

allopatric speciation - A morphological discontinuity arising from geographical fragmentation combined with the passage of time.

allochromic speciation- The productions of morphological discontinuity between species to be caused slowly by the passage of time.

allosome - A chromosome which is different form the rest, usually the sex chromosomes.

allosynapsis - Pairing at meiosis of chromosomes derived from different ancestors in an amphipolyploid.

allopolyploid - A polyploid in which replicated diploid sets of chromosomes come from genetically different strains.

allotetraploid - A tetraploid in which one diploid set has been believed from a genetically different parent.

allele - Alternate forms of genetic characters which occur at the same locus on the chromosome are said to be alleles, or allelic to each other see dominant allele, pseudoallele, recessive allele, allelism, allelomorph.

amber mutation - A suppressible genetic change which results in the terminators codon UAG in messenger RNA. The corresponding polypeptide chain is terminated at the site of amber oodon if no phenotypic or genotypic suppression occur.

allosteric - Of an enzyme whose activity is altered when its structure is distorated by an organic compound at a nonsubstrate site. Chromosome mutation any structural change involving the gain, loss or relocation of chromosome segments. Chromosome mutations arise spontaneously or are induced experimentally by chemical or physical mutagens. Zeta karyology the study of chromosome, the mapping of bands, chromosome puffs, nucleolar organizers and other landmarks on the basis of polytene chromosome analysis.

acquired characteristics - A theory that argues that traits that an organism acquires by accommodating to the environment are assimilated in to the genetic material and transmitted to the next generation.

acrocentric chromosome - A chromosome or chromatic which has the centromere near an end.

allele - One of two or more forms of a given gene.

aneuploidy - The loss or gain of one or more chromosome, as comapred to the basic chromosome complement.

anticodon - The triplet of tRNA nucleotides that is complementary to, and pairs with, a codon in the messenger RNA.

assortive mating - The nonrandom formation of mating pairs, can be positive or negative.

autosome - A chromosome other than the sex chromosome.

B

backcross - The cross of an F1 hybrid with one of the parental lines.

bands, chromosomes - Areas of light or dark staining produced by a variety of techniques.

barr hody - The sex chromation as seen in female somatic cells of animal.

base analog - A DNA or RNA base that resembles the normal base, but is different, so that it is incorporated in to the nucleic acid molecule in place of the normal base.

base pair - In DNA, A must pair with T, and G must pair with C.

bivalent - The figure produced by the pairing of two homologous chromosomes.

carrier – An individual who is heterozygous for a normal allele and for an abnormal allele that is not phenotypically expressed.

centric fusion - Fusion of the long arms of two aerocentric chromosomes at the centromere.

C

chromosome map - A representation of the linear arrangement of genes on chromosomes, as deduced from genetic and cytological observations.

chromosome theory of inheritance - The established theory that genes and chromosomes are linked and that chromosomes are the carrier of genetic information.

cistron - A gentic unit of function, usually equated with the term *gene.*

clone - A population of cells originally derived from a single cell by mitosis.

codominance - The expression of both of two different alleles in a heterozygote.

constitutive enzyme - An enzyme that is produced at a fixed rate, irrespective of need.

co-repressor - A molecule that combines with a repressor to inhibit the function of an operon.

cytogenetics - The study of the relationship between the appearance of chromosomes and the genotype/phenotype of the individual.

cytoplasmic inheritance - The inheritance of traits through the cytoplasm instead of through the nuclear chromosomes.

D

deletion - The loss of a chromosome segment.

DNA – Deoxyribo Nucleic Acid, the chemical basis of heredity.

diakinesis - A stage in meiosis in which the chromosomes are maximally condensed.

dihybrid cross - A cross between individuals differing with respect to two pairs of alleles.

discontinuous variation - Variation in a population that falls into two or more nonoverlapping classes.

E

endonuclease - An enzyme that breaks the bonds between adjacent nucleotides in the interior of a DNA or RNA chain.

enzyme - A protein molecule that catalyzes a specific chemical reaction.

epistasis - The masking by one gene of the expression of another, nonalletlic gene.

equilibrium - In population genetics, the state at which the forces that tend to change gene frequencies are counter balanced so that there is no net change in gene frequencies from one generation to the next.

euchromation - The chromosomal region that carries genes and characteristically stains lightly.

eugenics - A philosophy concerned with improving the genetic quality of a population.

euploid - Having the basic haploid complement of chromosomes or a multiple of the basic complement.

exon - A sequence of DNA that is translated into protein.

exonuclease - An enzyme that breaks bonds between adjoining nucleotides only when one of the nucleotides is a terminal one in a DNA or RNA chain.

F

F1 - The first-generation progeny of a cross.

feedback inhibition - Inhibition of an enzyme by a product of the pathway catalyzed by the enzyme.

fitness - A quantitative measure of an organism's ability to survive and transmit its genes to the next generation.

founder principle - A type of genetic drift, in which a small number of individuals from a larger population break off and start a new population. These individuals are not representative of the gene frequencies expressed by the larger parent population.

frameshift mutation - A shift in the read CDEFHPX translation caused by the addition or deletion of bases in a DNA molecule.

G

gene - The basic unit of inheritance that occupies a specific locus on the chromosome and has a specific function.

genetic code - Sequences of three nucleotides in DNA or RNA that specify amino acids when translated into polypeptides.

genetic - drift changes in allelic frequencies due to chance sampling of the parent population.

genetic marker - A single gene trait used to follow the transmission of chromosomes in a cross or mating.

genetic load - The decrease in fitness of a population due to detrimental alleles.

genome - The totality of genes in the haploid set; can also refer to the complete gene complement without regard to the ploidy.

genotype - The genetic constitution of an organism. Hardy-Weinberg law; A law stating the expected gene frequencies under conditions of random matting.

H

hn RNA - Heterogenous nuclear RNA; the in the nucleus that still contains inctrons.

heterokaryon - A cell having two or more genetically different nuclei.

heterosis - The vigour of hybrids, which is greater than that of the parental lines.

heterozygous - Pertaining to two different alleles at a specific locus in a diploid organism.

histone - A group of basic proteins associated with chromosome.

homologous chromosomes - Chromosomes that are identical in their content of gene loci.

homozygous - In a diploid organism, characterized by both alleles being the same at a specific locus.

hybrid cell - A cell derived from two different cultured cell lives that have fused.

I

incomplete dominance - A condition that exists in the phenotype of the heterozygote which is intermediate between the two parental extremes.

independent assortment - The distribution that occurs when a pair of alleles located on different chromo-some pairs live up and segregate in an independent fashion during meiosis.

inducer - A small molecule that causes a cell to produce larger amounts of the enzyme (s) needed to metabolize that molecule.

intron - A group of excess DNA bases that interrupts the sequence of amino acid-coding base at irregular intervals within the gene; the group is eliminated enzymatically from the DNA or mRNA transceripts, leaving the amino acid-coding bases in an uninterrupted sequence; intervening sequence.

in vitro - In the test tube; outside the living body.

in vivo - Within the living body.

isochromosome - A chromosome with two identical armes.

isozyme - An enzyme that can be distinguished by some property, usually electrical charge, but that acts enzymatically on the same substance.

K

karyotype - The chromosome complement.

L

leptotene - The early stages of prophase in meiosis I before the chromatids are visible as separate structures.

linkage - Association of loci on the same chromosome.

M

map unit - In a linkage map, one map unit is equal to 1% recombination between linked genes.

mRNA - Messenger RNA, the gene transcript that is translated into polypeptide.

Metacentric chromosome - A chromosome with a centromere in or near the middle.

leaky gene - A mutant allele with a lesser effect than that of the wild type; lymphomorphic allele, neomorphic allele, hypermorphic allele.

misssense mutation - A codon change that results in the substitution of one amino acid for another.

modifier gene - A gene that modifies the expression of another nonallelic gene.

monoclonal - Derived from the mitotic division of single cell.

monhybrid cross - A cross between individuals differing in a singe pair of alleles.

monosomy - The existence of only one copy of a particular chromosome instead of two.

mosaic - An individual with two or more genetically different cell lines derived from a single zygote.

multifactorial - Pertaining to a tract influence by variation at several genetic loci; polygenic.

multiple alleles - One of three or more alternative froms of an allelic series.

N

nondisjunction - The failure of chromosomes to separate properly in cell division.

nonsense mutation - A mutation that changes a codon specifying an amino acid into one that specifies no amino acid.

nucleic acid - A polymer composed of

a sequence of nucleotides; DNA and RNA.

nucleotide - One of the nucleic acid bases, along with a sugar and a phosphate group, that makes up a unit of DNA or RNA.

nullisomic - Lacking a particular chromosome.

O

operator - In the Jacob-Monod operon model, the site at which the repressor molecule binds on DNA, shutting of transcription.

operon - A unit of coordinately controlled genes under the control of an operator and regulatory genes.

P

pairing - The side-by-side alignment of homologous chromosomes during the prophase of meiosis I.

paracentric inversion - An inversion that does not included the centromere.

pharmacogenetics - The study of genetic variability, in response to, and metabolism of, drugs.

phenocopy - An environmentally induced phenotype that resembles an inherited tracit.

phenotype - The observed properties of an organism, produced by the interaction of the genotype with the environment.

pleiotropy - A condition in which a single gene has a wide range of effects.

point mutation - A mutation affecting a single nucleotide pair.

polymerase - An enzyme that catalyzes the formation of a polymer from its constituent building blocks.

polyploidy - The condition in which the chromosome number of an individual is three or more times the haploid chromosome number.

position effect - The change in the expression of a gene when its position is changed with respect to neighbouring genes.

promoter - In the operon, the region between the operator and the structural genes to which RNA polymerase binds.

pure line - A homozygous strain produced by inbreeding.

Q

Q arm - The long arm of chromosome.

R

random mating - Selection of a mate without regard to genotype; panmixia.

reading frame - The reading of an RNA sequence as a series of codons of three nucleotides each.

recessive gene - A gene that is expressed only if the individual is homozygous for it .

reciplocal cross - A genetic cross that can be symbolized as A male X B female and B male X a female, where A and B represent different genotypes.

recombinant DNA - Artificially constructed DNA in which a DNA sequent from one organism is inserted into the genome of another organism.

recombination - The formation of a new combination of alleles following meiosis.

reduction division - The first half of the meiotic process, in which the paired homologues segregate to different nuclei, thus reducing the chromosome number by half.

regulator gene - A gene whose product controls the activity of distant genes.

RAD - A unit of energy like the roentgen, but based on absorbed energy and applicable to both ionizing and nonionizing radiation, usually equivalent to a roentgen.

REM - The quantity of ionizing radiation that is equivalent in biological damage to one rad.

repressor - The protein product of a regulator gene that, when bound to the operation, prevents transcription of the operon.

restriction enzyme - An enzyme that cleaves DNA at specific nucleotide sequences.

RNA ribonucleic acid - A polymer of ribonucleotides.

ribosomes - A cytoplasmic structure composed of RNA and protein, on which polypeptide synthesis from mRNA occurs.

roentgen - A unit of ionizing radiation that produces 2×10^9 ion-pairs percubic centimeter of air or 1.6 ion-pairs per cubic centimeter of water.

S

segregation - The separation of the members of a homologous pair of chromosomes into different gametes, through the process of meiosis.

structural gene - The gene that codes for the amino acid sequence of a polypeptide chain.

structural proteins - The product of the structural gene that does not have enzymatic function.

suppressor - A matuation that reverses the effects of a mutation at a distant locus.

syntanic - Pertaining to loci on the same chemosomes; linkage.

T

telomere - The tip of the chromosome arm.

terminating codon - A codon that terminates the elongation of a polypeptide chain.

testcross - A cross involving a heterozygote crossed back to a homozygous recessive individual.

transcription - The synthesis of RNA from a DNA termplate; or, in certain viruses, the synthesis of DNA from an RNA template.

tRNA - Transfer RNA; a small molecule that binds to mRNA, the ribosome, and an amino acid.

transformation - The genetic alternation of a cell by the transfer of the DNA in the medium across the membrane and into the cell, where it combines with the host cell's DNA.

transition mutation - A base-pair substitution in which the orientation of the purine and pyrimidine bases on each DNA strand remain the same. (i.e., AT to GC).

translation - The process of converting the information contained in a sequence of RNA bases into a sequence of amino acids.

transversion mutation - A base-pair substitution in which the purine-pyrimidine orientation on each DNA STRAND IS REVERSED (i.e., A–T to T–A).

triploid - Having three sets of the basic haploid chromosome complement.

trisomy - The diploid condition plus one extra chromosome.

W

wild type - The penotype most frequently observed in nature and the one arbitrarily designated as normal.

Z

zygotene - The meiostic stage in which homologus chromosomes pair.

PLANT BREEDING

A

Adaptation – It is the process by which organisms become more suited to survive and function in a given environment. It also refers to the result of this process.

Addition Line – An addition line has one pair of chromosomes from another variety or species in addition to the normal somatic chromosome complement of the species.

Alien-Addition Line – It has one pair of chromosomes from a related wild species in addition to the normal somatic chromosome complement ($2n$) of the species.

Allele – Alleles are alternative forms of the same gene, and are located at the same point (locus) in homologous chromosomes.

Allogamy – In allogamy, pollen grains from flowers of one plant pollinate the flowers of other plants (syn. cross-pollination).

Allopolyploid – It has two copies of each of the two or more different genomes present. Thus an amphidiploid has the somatic chromosome complement of two diploid species.

Aneuploid – An aneuploid organism has a chromosome number that is not an exact multiple of the basic chromosome number (x).

Anther Culture – Culture of anthers (or pollen grains) on a suitable medium for production of callus and/or haploid plants.

Anthesis – The first opening of a flower.

Apogamy – Development of embryo from synergids or antipodal cells without fertilization; a form of apomixis.

Apomixis – Development of embryo (and seed) without fertilization.

Apospory -A form of apomixis in which the embryo sac develops from a vegetative cell of the ovule.

Artificial Selection – Selection by man.

Asexual Reproduction – Asexual reproduction does not involve fusion of male and female gametes.

Autogamy – Pollen grains of a flower pollinate the same flower (Syn. self-pollination).

Autopolyploid – A polyploid that has more than two copies of the same genome.

B

B_1 and B_2 – Backcrosses of F_1 to P_1 (first parent) P_2 (second parent), respectively, of hybrid.

Backcross – A cross between a hybrid and one of its parents.

Backcross Breeding – Breeding method based on repeated backcrossing of the F_1 and the subsequent generations to the recurrent parent.

BC_1 BC_2 BC_3 etc – Progeny from F_1 × recurrent parent, BC_1 × recurrent parent, BC_2 recurrent parent etc., respectively.

Basic Number – The haploid chromosome number of the ancestral diploid species of polyploids. Represented by x. Also haploid chromosome number of diploid species. x chromosomes constitute a genome.

Biometry – It consists of application of statistical procedures to biology.

Biotype – Strains of a species of pathogen, particularly an insect pest, differing in their ability to attack different varieties of the same host species (Syn. physiological races).

Bivalent – An association of two homologous chromosomes at meiosis.

Bridging Species – A species used in gene transfer from one species to another sexually incompatible species; bridging species is compatible with both the donor and recipient species; it may be a natural or synthetic species.

Bud Mutation – Mutation in somatic tissues, usually affecting an axillary bud.

Bud Selection – A form of clonal selection in which mutant buds are selected.

Bulk Method – In this method, F_2 and the subsequent generations are grown in bulk, usually without artificial selection. In the end, pureline varieties are developed through individual plant selection.

C

Certified Seeds – Seed produced from foundation, registered or certified seed. Its purity is certified by a seed certification agency, and is usually used for commercial crop production.

Centres of Diversity – Areas where cultivated plant species and /or their wild relatives show much greater variation than anywhere in the rest of the world.

Centres of Origin – Areas where cultivated plant species are supposed to have originated based on centres of diversity.

Centromere – The localised region of chromosome where spindle is attached at metaphase stage.

Character – Morphological, anatomical or physiological feature of an organism; usually a product of genotype and environment interaction.

Chiasma – The point of contact between nonsister chromatids of homologous chromosomes during and after diplotene.

Chimera – An individual that is composed of cells of two or more genotypes.

Chromatid – Substructures of chromosomes produced by replication of preexisting chromoatids. Each chro-mosome has two chromatids which are united at the centromere.

Chromosome – Thread like structures present in nucleus which show distinct change in their morphology during cell division. They are deeply stained with some dyes, particularly with the Feulgen's reagent. Genes are located in chromosomes in a linear order.

Chromosome Manipulation Techniques – Techniques used for promoting gene transfer from chromosomes of a related species to those of a cultivated species.

Chromosome Substitution – The pro-cedure or the process of replacing one pair of chromosomes of a variety with those of antoher variety to a related species. The latter is *Alien substitution*.

Cleistogamy – Flowers do not open at all, i.e., there is no anthesis.

Clone – Individuals obtained from a single plant through asexual reproduction.

Clonal Selection – A method of selection based on clones. It does not involve sexual reproduction.

Combination Breeding – It involves transfer of one or a few characters, usually governed by oligogenes, from one variety to another.

Combining Ability – Ability of a strain to produce superior progeny upon hybridization with other strains.

Complex Cross – See convergent Cross.

Composite Varieties – Varieties produced by open-pollination among a number of outstanding strains usually not tested for combining ability with each other.

Convergent Cross – A cross involving more than two parents (Syn. complex cross).

Correlated Response to Selection – Change in one or more quantitative characters due to selection for another character.

Coupling Phase – Linkage between dominant alleles of two or more genes.

Cross – Mating of two or more individuals or strains having different genotypes. Also the product of such a mating.

Cross-Polination – See allogamy.

Crossing Over – Exchange of homologous segments between nonsister chromatids of homologous chromosomes; takes place during diploteno.

Cytogenetics – Study of chromosomes in relation to genetics.

Cytoplasmic Inheritance – Transmission of characters through cytoplasm. It is due to DNA present in cytoplasmic organelles.

Cytoplasmic Male Sterility – Male sterility showing cytoplasmic inheritance.

Cytoplasmic Genetic Male Sterility – Cytoplasmic male sterility for which a restorer gene is available or known.

D

Deficiency – Loss of a part of chromosme.

Detasselling – Removal of the tassel (the male inflorescence) in maize. An easy method of emasculation.

Diallel Cross – Crossing of a number of genotypes in all possible combinations.

Dichogamy – Maturation of male and female reproductive organs of a hermaphrodite flower at different times.

Dihaploid – Haploid derived from an amphidiploid i.e., monoploid for two distinct genomes.

Dihybrid – An individual heterozygous for two genes.

Dioecious – Plant species in which unisexual (male and female) flowers occur on different plants. Such species have male and female plants.

Diploid – An organism having two genomes, i.e., with chromosome number of 2x.

Diplospory – A form of apomixis in which the embryo sac develops from the megaspore.

Diplotene – Stage of meiosis following pachytene in which homologous chromosomes forming a bivalent (or multivalent) begin to move away from each other. Chiasma is seen during this stage.

Disease – An abnormal condition produced by an organism.

Disjunction – Separation of chromosomes at anaphase.

Distant Cross – See Wide Cross.

Distant Hybridization – Hybridization between individuals belonging to two different species of the same genus or of different genera.

Disomic – An individual with two homologues for each chromosome of the genome.

Dockage – Per cent impurity in a seed sample.

Dominance – Ability of an allele to express itself in the heterozygous state.

Domestication – The process of bringing a wild species under human management.

Donor Parent – In backcross breeding; the parent from which one or few genes are transferred to the recurrent parent (Syn.nonrecurrent parent).

Double Cross – Cross between two single crosses (i.e., between two 'F₁'s from two single crosses)

Drift – Random change in gene and genotype frequencies in small populations.

Duplication – Occurrence of a chromosome segment more than twice in a diploid chromosome complement. The duplicated segment may occur in the same or in a different chromosome.

E

Emasculation – Removal of anthers (or androecium) from a hermaphrodite flower.

Embryo Culture – Removal of developing embryo from seed and cultivation in vitro.

Embryo Sac–Cell derived from megaspore. It contains egg cell, synergids, antipodals and polar nuclei (Syn. megagametophyte).

Environment – Sum total of external conditions which influence an organism.

Epidemic – Development of a disease from a low infection level to a high intensity of infection. *Artificial epidemics* are created by man by providing inoculum and suitable environment for disease development and used in disease resistance tests.

Epiphytotic – See Epidemic.

Epistasis – Interaction between two different genes, that is, expression of one gene is suppressed by another gene.

Error Variance – Variance due to factors beyond the control of the experimenter. (It is used as the denominator in F-test).

Escape – More commonly disease escape, phenomenon of susceptible plants avoiding disease attack. Also plants showing escape.

Euploid – Individual whose chromosome number is an exact multiple of the basic number.

Excitation – Movement of an electron to an outer orbit of a higher energy level.

Exotic – A variety or species introduced from a foreign country.

Exploration – A trip for collection of germplasm of cultivated and related wild species (Syn. Expedition).

Expressivity – Ability of a gene to express itself uniformly in the individuals that carry it in the appropriate genotype.

F

F_1 – The progeny obtained by crossing two different genotypes. The first generation from a cross.

F_2 – Progeny obtained by self-fertilization or intermating of F_1 individuals.

F_3, F_4 etc – Selfed progeny of F_2, F_3 etc., respectively.

Factor – See Gene.

Family – A group of individuals sharing a common parent/ancestor.

Fertility – Ability to produce viable progeny.

Fertilization – Union of male and female gametes in sexual reporduction.

Foundation Seed – Seed produced from breeder seed. It is the source of registered and certified seeds.

Free Radical – Electrically neutral molecules with an unpaired electron in the outer orbit. Free radicals are highly reactive.

G

Gamete – A specialised cell produced by gametogenesis and participating in fertilization. Gametes are produced through meiosis.

Gametic Chromosome Number – Chromosome number present in gametes of a species.

Gametogenesis – Production of gametes from spores.

Gene – Functionally, gene is unit of inheritance; one or more genes control expression of a character; and one gene codes for a polypeptide. *Structurally*, gene is a segment of DNA which codes for one polypeptide, ribosomal or transfer RNA.

Gene Bank – Large collection of germplasm representing materials from various parts of the world (*Syn.* World Collection).

Gene Interaction – Modification of

expression of a gene by one or more nonallelic oligogenes.

Gene Pool – Sum total of genes present in a Medelian/panmictic population.

Gene Sanctuary – An area within the centre of diversity protected from human interference.

Genetic Advance – Improvement in the performance of selected lines over the original population.

Genetic Emasculation – Use of genetic factors to make the male gamete nonfunctional in self-and or cross-pollination.

Genetic Equilibrium – In a random mating population; the stage in which genotype frequencies do not change frome one generation to another. It has reference to one or more genes.

Genetic Erosion – Gradual disappearance of various form of a cultivated species and of its wild relatives.

Genetic Load – The sum total of deleterious (harmful) alleles present in a random mating or Mendelian population.

Genetic Purity of Seed – Freedom of seed from seeds of weeds, other crops and other varieties of the same crop.

Genetic Variance – Variance due to genotype of different plants or strains. It has additive, dominance and epistatic components.

Genetics – A study of the mechanism of transmission of characters from parents to offspring, origin of variation and gene action.

General Combining Ability – Average performance of a strain in a series of cross combinations. Estimated from the performance of 'F₁'s from the crosses.

Genome – Genome consists of all the chromosomes of a diploid species that are distinct from each other with respect to their gene content, and often morphology. Members of a genome do not pair.

Genotype – Genetic constitution of an organism.

Germplasm – The sum total of hereditary material or genes present in a species.

Germplasm Complex – Open-pollinated progeny from a mixture of a number of strains of diverse origin.

H

Haploid – An Individual with the gametic chromosome number.

Heritability – The proportion of total phenotypic variance that is due to genotype.

Hermaphrodite – A flower having both male and female reproductive organs.

Heterogeneous – A mixture of different types, usually different genotypes.

Heterocaryosis – In fungi, a condition in which a hypha has nuclei of two different genotypes.

Heteroploid – An individual with a chromosome number other than the normal diploid (2x) number.

Heterosis – Superiority of F_1 over the parents (or even inferiority to both the parents, e.g., earliness), (Syn. Hybrid Vigour).

Heterostyly – Occurrenc of styles and stamens of different lengths in flowers from different plants of a single species.

Heterozygous – An individual having dissimilar alleles or a gene.

Homoeologous Chromosomes – Partially homologous chromosomes; generally show reduced or lack of pairing.

Homologous Chromosomes – Chromosomes identical with each other in gene content and morphology; two homologous chromosomes pair to form one bivalent during meiosis.

Homogeneous – Consisting of individuals of the same genotype or phenotype.

Homozygous – An individual having

two or more identical alleles of the same gene.

Horizontal Resistance – Resistance governed by polygenes and is pathotype nonspecific.

Host – The organism attacked by a pathogen.

Hybrid – Progeny from hybridization between two or more strains.

Hybrid Substance – New isozyme present in the hybrid which is different from those present in the parents.

Hybrid Varieties – Hybrid varieties are F_1 generation from a cross between two different strains.

Hybrid Vigour – See Heterosis.

Hybridization – Mating of two different strains.

I

I₁, I₂, I₃ etc – Generations after one, two, three etc. respectively, generations of inbreeding.

Immune Reaction – Complete absence of symptoms of a disease.

Improved Seed – Seed of an improved variety having a very high genetic and physical purity and germination.

In vitro Techniques – The technique of culturing plant cells and organs on artificial media in vitro.

Inbred Line – A nearly homozygous line developed by continued inbreeding, usually selfing, accompained by selection.

Inbreeding – Mating of individuals more closely related by ancestry than would be expected under random mating.

Inbreeding Depression – Loss in vigour due to inbredding.

Inbred-Variety Cross – Cross between an inbred and an open-pollinated variety.

Inert Matter – In a seed sample; all non-living matter, disease and insect infested seeds and broken (damaged) seeds.

Interspecific Hybridization – Hybridization between plants belonging to two species of the same genus.

Introduction, Plant – Taking a variety or a species into an area where it was not grown before ; even within a country.

Introgression – Transfer of a few genes from one species into the full diploid chromosome complement of another species.

Inversion – A segment of a chromosome is rotated by 180° so that the gene order on the segment is reversed.

Ion – An atom that has lost or gained an electron in comparison to the number of electrons normally present.

Ionization – Loss or grain of an electron by an atom.

Irradiation – Exposure of biological materials to radiations, more particularly to mutagenic radiations.

Isogenic Lines – Lines identical in genotype except for one gene.

Isolation – Separation of two or more plants, strains or populations to prevent mating among them. Usually achieved by distance and /or border rows.

K

Kinetochore – See Centromere.

L

LD₅₀, The dose of a mutagen (or any other agent) which will kill 50 per cent of the treated individuals.

Lethal Gene – Usually recessive; lethal genes kill each. and every individual that carries them in homozygous state.

Line – A group of individuals having common parents or ancestors.

Linkage – Tendency of two or more genes to be inherited together.

Linkage Map – A linear (or circular) map of chromosomes showing the position of various genes present in them as determined by their linkage relationship.

Locus – The position at which a gene is located in a chromosome.

M

M₁, M₂, M₃, etc – After treatment with a mutagen; first, second, third etc, generations, respectively, derived by selfing or asexual reproduction.

Maintainer Line – Line used for maintaininng a cytoplasmic male sterile line. It has the same nuclear genotype as the male sterile line.

Male Sterility – Absence of functional male gametes (pollen grains in plants).

Mass-Pedigrec Method – In this method, the population is maintained as a bulk till such time when condittions suitable for selection occur; at this stage pedigree method of breeding is applied.

Mass Selection – In mass selection, several plants are selected on the basis of their phenotype and their seed is composited to raise the next generation.

Mating System – Scheme according to which individualos or lines are mated to produce sexual progency.

Mean – It is the arithmetic average of a set of observations.

Megagametogenesis – Production of the female gemetophyte (embryo sac) from megaspore (through mitosis).

Meiosis – Two cell divisions, occuring one after another, after only one DNA replication; shows chromosome pairing and reduction in the first division and chromatid separation in the second. Leads to production of haploid cells (spores/gametes).

Meristem Culture – Cultivation of apical meristems, particularly shoot apical meristem, for production of shoots and plantlets.

Mericloning – Vegetative multiplication through meristem culture.

Metaphase – A stage during cell division where the chromosomes lie at the metaphase plate (equatorial plate).

Metaxenia – Effect of pollen grains on maternal tissues of fruits.

Microcentre – Areas within centres of origin that show greater diversity than the remaining centre of origin.

Microgametogeneis – Production of the male gametophyte from microspore (through mitosis).

Microsporogenesis – Production of pollen grains (microspores) from microspore mother cell (through meiosis).

Migration – Movement of individuals from one population into another.

Mitosis – Cell division in which the two chromatids of each chromosome move to the opposite poles producing two 2n daughter nuclei. Every mitotic division follows DNA replication (compare with meiosis).

Modifying Genes – Genes with no phenotypic effect of their own, but hanve the expression of some oligogene. They have small, cumulative effect producing a continuous range of phenotypes.

Monoecious – A plant having both male and female flowers.

Monohybrid – An individual heterozygous for one gene.

Monoploid – An individual with the basic chromosome number (x), i.e., with one genome.

Monosomic – An individual with one chromosome less than the somatic chromosome number (2n-1).

Multiline Varieries – Mixtures of several similiar purelines having different genes for disease resistance.

Multiple Alleles – More than two alternative forms of a single gene.

Multivalent – The structure formed by association of more than two homologous chromosomes as a consequence of synapsis during meiosis.

Mutagen – A chemical or physical agent that induces mutation.

Mutagenesis – Induction of mutations with the aid of mutagens.

Mutations – A sudden heritable

change in characteristics of an organism. *Macromutation* produces a sufficiently large change to be easily identifiable, while *Micromutations* produce small changes (usually in quantitative traits) and are studied in terms of mean and variance.

Mutation Breeding – Breeding method utilizing variation created by mutagenesis.

N

n – Gametic chrmosome number of a species.

2n – Somatic chromosome number . The chromosome number present in somatic cells of a species.

Natural Selection – Selection due to natural forces, i.e., enivronment.

Nobilisation of Sugarcane – Introgression of genes and chromosomes from noble canes and wild *Saccharum* species into Indian canes (*S. barberi*).

Nonrecurrent Parent – In backcros method; the parent from which one or few genes are transferred to the recurrent parent (*Syn.* donor parent).

Nullisomic – An individual having a pair of homologous chromosomes less than the somatic chromosome number of the species. (2n-2).

O

Oligogenes – Genes having large individual effects producing distinct phenotypes.

Open-Pollination – In cross-pollinated species, pollination occurring naturally without restriction.

Outcross – Natural cross between two different genotypes.

P

P₁, P₂, P₃, etc – First, second, third etc. parents, respectively, used for producing a hybrid or a series of hybrids. Also the first, second, third etc, generations of parents obtained by selfing.

Pachytene – Stage in meiosis; when pairing between homologous chromosome is complete giving a single-stranded appearance to the bivalent.

Panmixis – Random mating without restriction.

Parameter – A numerical quantity which describes some characteristics of a population.

Parthenogenesis – Egg cell of an embryo sac derived from megaspore develops into an embryo without fertilization.

Pathogen – An organism producing a disease.

Pathotype – Strain of a pathogen virulent toward a specific resistance gene.

Pedigree – A record of ancestry.

Pedigree method – A method of breeding in which individual plants are selected in F_2 and subsequent generations and a pedigree record is maintained. The pedigree record can be used as a basis of selection in later generations.

Penetrance – Ability of a gene to express itself in the individuals carrying that gene. Penetrance may be compete or incomplete.

Phenotype – External appearance of an individual with reference to a single character or a number of characters.

Physical Purity – Freedom of seed from inert matter, including broken seeds.

Physiological Races – Strains of a pathogen species that differ in their ability to attack different varieties of the host species. Unlike pathotype, it has no reference to the resistance genes.

Plant Breeding – The branch of biology concerned with changing the genotype of plants so that they become more useful.

Pleiotropy – Phenomenon of a single gene affecting two or more different characters.

Pollen Culture – Cultivation of pollen grains in vitro for producing haploid plants (generally referred to as anther culture).

Pollination – It consists of pollen grains reaching stigma of a flower.

Polycross – Open-pollination (assumed to be random mating) in isolation among a number of selected genotypes.

Polygenes – Genes with small and cumulative effect; their expression is greatly affected by environment. They govern quantitative traits.

Polygenic Inheritance – Inheritance of quantitative characters due to action of polygenes.

Polyploid – An genetics, all the individuals mating at random. In *statistics,* infinitely large number of observations of the same type.

Population Cross – A cross between two open-pollinated varieties (*Syn.* varietal cross).

Population Improvement – Improvement of random mating populations through a scheme of selection with or without progeny test. It is essential to keep inbreeding to a low level.

Prepotency – Ability of an individual to produce progeny which are similar to each other and to itself.

Probability – The proportion of times an event may be expected to occur if infinitely large number of trials are made.

Progeny Test – Evaluation of the genotypic value of an individual on the basis of the performance of its progeny.

Propagule – Plant part used for propagation.

Protandry – Dehiscence of anthers before the stigma of that flower has become receptive.

Protogyny – Maturation of stigma (becoming receptive) before the dehiscence of anthers of that flower.

Pureline – Progeny of a single homozygous self-pollinated plant.

Pureline Selection – Isolation of purelines from a mixture of purelines (*Syn.* individual plant selection).

Q

Quadrivalent – Structure produced by pairing among four homologous chromosomes during meiosis.

Qualitative Character – Character showing distinct classes and little or no effect of environment.

Quantitative Character – Character showing continuous variation and considerable effect of environment.

Quarantine – Isolation of an organism for observation on weeds, diseases and pests and for preventing their spread.

R

Random – An event whose occurrence is determined solely by chance and there is no discrimination.

Random Drift – See Drift.

Random Mating – A system of mating is which an individual has equal chance of mating with every other individual of the same population.

Real Value of Seed – The percentage (by weight) of the seed sample which would produce seedlings of the variety under certification.

Recessive – An allele unable to express itself in heterozygous state.

Reciprocal Cross – Cross in which the line previously used as male is used as a female,while that previously used as female is used as male.

Recipient Parent – In backcross breeding, the parent to which one of few genes from the donor parent are transferred (*Syn.* recurrent parent).

Recombination – Production of new combinations of genes as a result of independent segregation of non-alleles or crossing over between linked genes; usually the latter.

Recurrent Parent – See Recipient Parent.

Recurrent Selection – In cross-pollinated population; schemes of selection (on the basis of phenotype or progeny test), followed by intermating (in all combinations) of the selected plants or their selfed progeny to produce the population for the next cycle of selection. More than one cycle of selection practised.

Registered seed – Progency of foundation seed.

Reproduction – Production of a new generation of individuals (progency) by sexual or asexual means.

Repulsion Phase – Linkage between dominant allele of one gene with the recesive allele of another gene.

Roguing – Removal of off-type plants of the same species.

Resistance – Ability of a host strain to restrict or even prevent production of disease symptoms by a pathogen; due to a gene for resistance.

Restorer Gene – A gene, usually dominant, that effectively overcomes the effect of male sterile cytoplasm on male fertility, i.e., produces functional male gametes even in the presence of the male sterile cytoplasm.

S

S_1, S_2, S_3, etc – First, second, third etc. generations derived by selfing of a plant (SÉ) from a random mating population.

Sample – A set of random observations taken from a population.

Sampling Error – Deviation of a sample value from that of the population due to small size of the sample.

Seeds – part of a plant used for raising a seed crop or a commercial crop.

Seed Test – A series of tests on purity, moisture content and germination of a seed sample to determine its quality.

Seed Certification – Seed certification consists of field inspection and seed tests to ensure genetic and physical purity and good germination of seed lots, and issuance of a certificate to that effect.

Segregation – Separation of alleles at the time of gamete formation so that each gamete receives only one of the two alleles of a gene.

Selection – Differential reproduction rates of different genotypes.

Self-fertility – Ability to set seeds on self-pollination.

Self-Fertilization – Union of male and female gametes from the same individual.

Self-Incompatibility – Lack of seed set on self-pollination.

Self-Pollination – See autogamy.

Short- Term Inbred. A line derived by one or a few generations of inbreeding. Such a line is not homozygous or even nearly homozygous.

Sibs – Individuals having both the parents common, but derived from different gametes, i.e., they are not identical twins. *Half-sibs* are individuals having one common parent.

Sib Mating – Mating between sibs.

Single Cross – A cross between two lines, usually inbred lines.

Species – A unit of taxonomic classification. Members of a species are more like each other than those from different species, and do not show barriers in sexual reproduction (except, of course, in cases of self-incompatibility and male sterility etc).

Specific Combining Ability – Deviation in performance of a cross combination from that predicted on the basis of the general combining abilities of the parents involved in the cross combination from that predicted on the basis of the general combining abilities of the parents involved in the cross.

Sporogenesis – Production of mega- and micro-spores from mega-and micro-spore mother cells, respectively (through meiosis).

Standared Deviation – Square root of variance (see variance). It is a measure of spread (variation) of a sample or population.

Standard Error – Standard deviation of a sample of means estimated from a number of random samples drawn from a single population. It is estimated as standard deviation in a sample.

Strain – A group of individuals similar in phenotype and often in genotype. A strain is known as variety when released for commercial cultivation by a variety release committee.

Substitution Line – A line in which a pair of chromosomes has been replaced by a pair from another variety of the same species. Used in aneuploid analysis.

Syn₀ – The parental lines of a synthetic variety.

Syn₁ – The population derived by intermating of Syn° lines in all possible combinations. Also, by open-pollination among the Syn° lines (parental lines).

Syn₂, Syn₃, etc – Generations derived through random mating from Syn₁, Syn₂ etc, generations, respectively.

Synapsis – Pairing between homologous chromosomes during meiosis, more particularly during zygotene stage of Prophase I.

Synthetic Allopolyploid – Allopolyploid produced experimentally by man.

Syntheic Variety– In cross-pollinated species, a variety obtained by mating in all possible combinations a number of lines that combine well with each other; Syn₁ or, more generally, Syn₂ and later generations.

T

Telophase – The last stage in cell division after the chromosomes/chromatids reach the opposite poles

and before the nuclear mem-brane reappears.

Test Cross – In genetics, cross between a hybrid and the homozygous recessive strain. *In plant breeding,* cross between a plant or line and a tester (the tester may be an inbred, hybrid, synthetic or open-pollinated variety).

Tetraploid – An individual having four identical or distinct genomes.

Teterasomic – An individual having one pair of chromosomes in addition to the normal somatic chromosome complement (2n+2).

Three-Way Cross – A cross between a single cross and an inbred.

Tissue Culture – Cultivation of plant cells and tissues in vitro on artificial media.

Tolerance – Ability of a host to avoid loss in productivity although it has been infected by a pathogen.

Top Cross– See Inbred-Variety Cross.

Trait – See Character.

Transgressive Breeding – Breeding methods designed to utilize transgressive segregants.

Transgressive Segregation – Appearance of individuals in the progeny from a hybrid which exceed either of the two parents of the hybrid with respect to one or more characters.

Translocation – Incorporation of a chromosome segment in a different chromosome.

Triploid – An individual having three identical or distinct genomes.

Trisomic – An individual having one chromosome in andition to the normal diploid chromosome complement (2n+1).

Trivalent – Structure formed by pairing among three homologous chromosomes during meiosis.

U

Univalent – During meiosis; an unpaired chromosome.

V

Variance – Average of the squares of deviations of the observations from the mean of a sample drawn from a population.

Variation – Differences among individuals belonging to a single species or to different species. Variation may be due to environment, or due to both genotype and environment.

Varietal Cross – See Population Cross.

Variety – In plant breeding, a strain released for commercial cultivation by a variety release committee. *In botany*, a sub-division of species based on form, function or both.

Vegetative Reproduction – Production of new individuals from vegetative parts.

Virulence – Ability of a pathogen strain (pathotype or physiological race) to attack a strain of the host.

Vertical Rssistance – Resistance, usually immunity, towards a virrulent race/jpathotype; the specific virulent race/pathotype produces susceptibe rsponse. Generally, cotrolled by oligogenes.

W

Wild Cross – Cross between two species of the same genus or of World Collection. *See Gene Bank*.

X

x – See Basic Number.

X_1, X_2, X_3 etc – First second, third etc. generation, respectively, after irradiation (with X-rays or some other mutagenic radiation obtained through selfing or clonal multiplication.

Xenia – Effect of the genotype of pollen grain on the phenotype of seed tissues (embryo and endosperm).

Z

Zygote – The immediate product of fusion of male and female gametes.

Zygotene – A stage during Prophase I of meiosis during which homologous chromosomes pair with each other.

PLANT PHYSIOLOGY

A

Abscissin – a kind of hormone with accelerating effect on leaf fall.

Abscission layer – layer formed at the leaf base in woody dicotyledonous plants in which parenchyma cells become separated from one another due to disorganisation of middle lamella. The leaf fall occurs due to it.

Accerescent – the plants which continue to grow after flowering.

Acellular – Organism not dividing into separate cells.

Amino acid – An organic acid with an amino group, i.e., possessing both acidic and basic linkages.

Anthranilic acid – O amino benjoic acid.
aminosuccinic acid – aminosuccinic acid.
cysteic acid – α-aminc-β sulphopropionic acid
nicotinic acid – niacin
pantothenic acid – D (d) – N – (α γ dihydr-B, β-dimethylbutyryl O-β alanine. (Vit. B. soluble B-complex)

Aconitase – aconitate hydratase.

Adenosine – a nucleoside containing a heterocyclic nitrogen base (adenine) and a pentose sugar.

Adenosine diphosphate (ADP) – adenine nucleoside with two phosphate molecules of which terminal one is energy rich.

Adenosine monophosphate (AMP) – adeninue nucleoside with one phosphate molecule.

Adenosine tri-phosphate (ATP) – adenine nucleoside with three phosphate molecules, the last two molecules are energy rich.

Aerobic respiration – respiration which occurs in presence of oxygen.

Aleurone – a kind of protein found in plants, particularly in the peripheral region of seeds.

Aleurone grains – protein granules stored in plants, commonly in the seeds.

Aleuroplast – colourless plastid (leucoplast) which stores protein as reserve food material, found very commonly in seeds.

Amidase – an enzyme which catalyses the hydrolysis of carboxylic acid amides to carboxylic acid and ammonia (e.g. formandiase, phosphamidase).

Aminase – an enzyme catalysing reactions that produce urea or ammonia e.g. glutaminase, glycocyminase, transaminase etc.

deaminase – also called aminohydrolase; an enzyme that catalyses the liberation of ammonia by hydrolysis.

Aminidase – an enzyme catalysing the hydrolysis or terminal links in various mucopolysaccharide.

Amyle – meas starch.

Amylaceous – starchy.

Amylase – an enzyme that hydrolyses 1, 4-glucan links, in starch, glycogen, and related polysaccharide, e.g. – glucoamylase -enzyme that removes glucose unit fromthe non-reducing ends of polysaccharide chains.

Amyloplast – colourless plastid (leucoplast) which stores starch, common in cotyledons, potato tuber and endosperm.

Anaerobic – living without oxygen.

Anthesis – the opening of flower.

Anthocyanins – blue and red glycoside pigments (flavonoid) found dissolved in cell sap.

Assimilation – uptake and formation of simple food stuffs.

ATP – see adenosine triposphate.

ATPase – also called myosin, an enzyme catalysing TTP, CTP, GTP. UTP and ATP to yield orthophosphate and the corresponding diphosphates.

Autolysis – due to action of own enzymes self-dissolution of tissues which undergo after death of their cells.

Autonomic – movement which occurs as a result of internal stimuli, e.g., nutation.

Auxin – a group of hormones of plants which are synthetized by growing tips of stems and roots and regulate many aspects of plant growth, e.g., IAA.

B

Biotin – Vitamin H.

Bio-assay – the quantituative estimation of biologically active substances in standardised conditions by the amount of growth of micro-organisms.

Bio-chemistry – study of chemical processes of living things.

Bioluminescence–production of light by living organisms.

C

Callose – a kind of carbohydrate deposited on the sieve plate of sieve tubes.

Callus – Superficial tissue which develops in woody plants in response to wounding.

Carbohydrate – compound containing carbon, hydrogen and oxygen with general formula $C_x (H_2 O)_y$, e.g. sugar; starch, cellulose.

Carboxylase – a kind of ligase enzyme which catalyses the attachment of carbon dioxide.

decarboxylase – an enzyme catalysing removal of carbon dioxide.

Carotenoids – Yellow, orange or red pigments, soluble in oil. These pigments are found in the chloroplast and in plastids, sometimes found in the region where chlorophyll is absent such as carrot roots, floral petals etc.

Catabolism – process in which complex organic substances are broken into simple sustances, with the liberation of energy.

Catalase – an enzyme consisting of haemoprotein, sometimes referred to a group of haemoprotein enzyme; catalysing the production of oxygen from hydrogen peroxide or of water and aldehyde from hydrogen peroxide and alcohol.

CDP – cytidine diphosphate.

Cell – cell is the structural and furctional unit of plant. It is a discrete mass of protoplasm bounded by a plasma membrane.

Cell membrane – outer bounding membrane of the protoplasm, also called plasma membrane. Each membrane consists of two unit membranes of lipoproteinaceous nature.

Cellulose – a constituent of cell wall consisting of long chain of polysaccharides made up of glucose units.

Cell wall – in plant cells the outer layer surrounding the plasma membrane. It consists of cellulose, hemicellulose and pectic substances.

Chemoautotrophic – (also called chemosynthetic). Organisms which obtain energy from simple inorganic reaction, e.g., *Thiobacillus*.

Chemotaxis – movement (tactic) in response io chemical concentration.

Chemotropism – movement (tropic) in response to gradient of chemical concentration.

Chlorophyll – Green pigment found in autotrophic plants. It's a magnesium containing porphyrin compound and by this plants become capable of manufacturing own food.

Chloroplast – Chlorophyll containing plastid bodies and the site of photosynthesis.

Chlorosis – disease of green plants in which green colour is lost and characteristic yellow colour appears. The condition may occurs due to failure of chlorophyll synthesis in absence of light.

Chondriome – the sum total of granules found in the cell.

Chondrion – an intracellular inclusion.

Chondriosome–also called mitochondria.

Chrome – colouring material, a pigment.

Chromoplast – (or chromatophore) pigmented plastid of plant cell other than green. May be red, orange or yellow.

Cytochylema – the total content of plant cell that includes protoplasm and vacuoles.

CTD – Cytidine triphosphate.

Cycle – a repetitive metabolic process.

—**biochemical cycle** – in the biosphere, a circulation through organisms of chemical elements.

—**Carbon cycle** – a cycle involving the proces of conversion of Co_2 into carboohydrate and carbohydrate into CO_2.

—**Citric acid cycle** – also called tricarboxylic acid cycle; the cycle which starts with the production of citric acid with pyruvic acid through acetyl CoA and ending in the production of citric acid through oxaloacetic acid.

—**energy cycle** – a term used generally for the successive metabolic processes.

—**glyoxylate cycle** – a cycle which consists of carbohydrate mechanism involving the entry of glyoxylate into citric acid cycle.

—**Krebs cycle** – also called tricarboxylic acid cycle (see citric acid cycle).

—**nitrogen cycle** – a cycle, which begins with the fixation of atmospheric nitrogen, involves the synthesis of organic compounds and

the final return of gas (N_2) to the atmosphere as the end product of decomposition.

Cuticle– superficial non-cellular layer covering epidermis and protection against mechanical injury and prevents excessive water loss.

Cyclosis – circulation of protoplasm in cell.

Cytochrome – a hemoprotein enzyme (protein with an iron containing prosthetic group).

Cyt – a-a cytochrome in which the Fe is in a formylporphyrin linkage.

Cyt – b-a cytochrome in which Fe is in a protoporphyrin linkage.

Cyt – c-a cytochrome in which Fe is in a mesoporphyrin linkage.

Cyt – d-a cytochrome in which Fe is in a dehydroporphrin linkage.

Cytokinins – a group of hormones with stimulatory effect on divisions of plant cells, e.g., Kinetin and Zeatin.

Cytoplasm – All the protoplasm of a cell excluding nucleus.

D

Death – the end of life from an organic complex of otherwise known structure.

Denitrifying bacteria – the soil bacteria which break nitrate and nitrite with the evolution of free nitrogen in an oxygenless atmosphere.

—**interstitial growth** – growth which occurs within the substance of tissue.

—**intrusive growth** – the growth which involves the intrusion of wall of growing cell between the walls of other; inter-positional growth.

G

GTP – guanosine triphosphate.

Guanine – 2-aminohypoxanthine; a purine base and component of DNA molecule.

Guttation – loss of water in form of water drops through hydathode.

H

Heliotropism – phototropism.

Heredity – transmission of character from one generation to next generation (parent of offspring).

Hormone – hormones are organic substances which are produced in exceedingly minute quantities in one part of plant and transported to other parts where they exert a profound effect on growth.

Hydathode – water excreting end occurring on the edges or tips of leaves.

Hydrogenase – a group of enzymes which utilises a molecular hydrogen.

Hydrolase – a group of enzymes catalysing a hydrolytic reaction.

I

larovization – vernalisation.

IAA – indole acetic acid.

Inhibitor – a substance that inhibits the process.

Inositol – (hexahydroxy cyclohexane) a kind of water soluble nutrient; commonly classified with vitamins and active in the control of glucose metabolism.

Isoenzyme(isozyme) – An enzyme that occurs in different structural forms (this has same function).

Isomer – two compounds identical but different in structure are called isomers.

Isomerase – A group of enzymes catalysing reactions leading to the internal reorganisation of molecules, such as triose phosphate isomerase which catalyses the production of dihydroxyacetone phosphate from glyceraldehyde O-phosphate.

K

Kinase – a group of enzymes which catalyses pnosphorylation with energy derived from the conversion of ATP to ADP, e.g., Phosphoglucokinase which catalyses the production of Glucose 1, 6 diphosphate from Glucose, 1 phosphate.

Krebs cycle – also called citric acid cycle.

L

Lecithin – a fatty substance which contains glycerol, fatty acid, choline and phosphoric acid and present in all plant cells.

Ligase – a group of enzymes which catalyses the combination of two molecules using energy derived from ATP molecules.

Lignin – a complex aromatic compound which is deposited in the cell walls of sclerenchyma, xylem vessels and tracheids.

Lipase – a group of enzymes which catalyses hydrolysis to triglycerides into diglycerides and fatty acids.

Lipoid – it is used for three different senses.
(i) a substance which resembles fats in solubility, but not containing fatty acid, e.g. carotene, terpene etc.
(ii) a substance which resembles fat in solubility, but may contain fatty acid, e.g. phospholipid (excluding .... fats).
(iii) Occasionally fat.

Long day plant – plants which flower during long days.

Lysosome – a submicroscopic one unit membrane bounded by particles containing hydrolytic enzymes.

Lysozyme – a group of enzymes which destroys or weakens the cell wall leading to rupture and death of protoplast.

M

Maltose – a disaccharide sugar formed during the breakdown of starch.

Mesosome – invagination of plasma membranne in procaryotic cells such as bacteria.

Messenger RNA – a type of RNA formed from the DNA by transcription and bring the message of it.

Metabolism – the chemical processes

occuring whithin the organism.

Metabolite – substance that takes part in process of metabolism.

Microsome – Particle found in the cytoplasm. they consist mostly of vesicles with attached ribosmes and formed from the broken up endoplasmic reticulum.

Mutase – a term used for enzymes which catalyse intramolecular transfers.

N

NAD – nicotinamide-adenine dinuclotide (a frequent receptor in oxidoreductase system).

NADH – reduced NAD.

NADPH – rduced NADP.

Nitrogen fixation – a process of conversion of atmospheric nitrogen into organic compound by soil inhabiting bacteria and certain blue green algae.

NMN – nicotinamide mono nucleotide.

Nucleic acid – macro-molecule consisting of large number of nucleotides. Two types of nucleic acids, in general, are found, e.g., DNA and RNA.

Nucleotide – a unit of nuclei. Each consists of one N- base, one pentose sugar and one phosphoric acid molecules.

Nutation – a phenomenon related with the twisting of the growing parts of a plant organ.

O

Organelle – a descrete portion of a cell which has a specific function.

Osmosis – a phenomenon which refers to the passage of solvent from a region of low solute concentration to one of high solute concentration through a selectively permeable membrane.

—**endosomosis** – in plants diffusion of solvent inside the cell.

—**exosmosis** – reverse of endosmosis.

Oxidase – a group of oxidoreductase enzymes in which oxygen is used as a hydrogen acceptor and the byproduct is hydrogen peroxide or water, e.g., malate oxidase which catalyses the production of oxaloacetic acid from malic acid.

Oxidative phosphoryl – on – formation of energy rich phosphate bonds (ATP) utilising energy released during respiration.

Oxygenases – a group of enzymes catalysing the direct transfer of molecular oxygen to a molecule.

P

Paraplasm – protoplasm in a sol condition.

Parasites – a group of plants which for nutrition depends on other plants.

Pectic compounds – polysaccharide carbohydrate found present in cellwalls of unlignified tissues.

Peroxidase – a group of enzymes which catalyses oxidations. Here peroxides act as hydrogen donor and water is,therfore, by product, e.g., NAD peroxidase which oxidises NADH and hydrogen peroxide to NADT.

Phosphatase – group of enzymes which catalyses hydrolysis or phosphate ester linkages, e.g., diphosphoglycerate phosphatase which catalyses the hydrolysis of 2, 3 diphosphoglyceric acid to yield 3 phosphoglyceric acid and orthophosphate.

Phosphorylase – a group of transferase enzymes which catalyses reactions involving orthophosphate, e.g., maltose phosphorylase which catalyses the production of Glucose I, phosphate and glucose from maltose and orthophosphate.

Photonasty – response to light in general (non directional).

Photoperiodism – response of plant to relative day and night length.

Photosynthesis– synthesis or organic compounds from CO_2 and H_2O using sun light energy by chlorophyll. Photosynthesis is the basis virtually for all other kinds of life.

Phototropism – tropic movement in response to light.

Phycobilins – a group of pigments found in certain groups of algae such as Rhodophyceae & Myxophyceae.

Phytochrome – proteinaceous, photoreversible pigment energised by red light and responsible for germination, flowering, etc.

Plasmolysis – shrinkage of protoplasm away from its cell wall. The condition is observed when a cell is placed in hypertonic solution.

Plastids – pigments or reserve food material containing cytoplasmic bodies.

Polarity – polar behaviour.

Polioplasm – part of protoplasm of a plant cell which flows.

Procaryotic – a cell with primitive type of nucleus (no true nucleus). e.g., Bacteria.

Protoplasm – the sum total of the contents of a living cell.

Prosthetic group – a non-protein-aceous part of an enzyme may be some organic substance or metal.

Protoplast – a smaller bit of protoplasm separated by cell wall.

Q

Q_{10} – temperature coefficient the response of rising $10°C$ temperature on the rate of metabolic process.

R

Reductase – a group of enzymes catalysing reduction reaction, e.g., glyoxylate reductase which reduces glyoxylate to glycollate using reduced NAD as the hydrogen donor.

Respiration – the sum total of method or methods by which an organism utilises an oxidation process in which complex compounds are broken down in simpler compounds while liberating the energy.

Respiratory quotient – it is the rartio of volume of CO_2 released and O_2 consumed during the same time.

Riboflavin – Vitamin B_2 (belonging to Vit. B group).

Ribosomes – granules consisting of nucleic acid (RNA) and protein. They participate in protein/enzyme synthesis.

S

Seismonasty – response to a non-directional shock stimulus, e.g., as in *Mimosa pudica*.

Skotophite – dark loving.

Starch – the principle reserve food materials of the green plants, frequently found in colourless plastids.

Stoma – pore in the epidermis of plants, frequently found in the green region of the plants. It's helpful in gaseous exchange.

Symbiosis – association of two dissimilar organisms to their mutual adavntage, e.g., root nodules leguminous plants (Bacteria are present in root nodules).

T

Tissue-culture – technique for maintanirig tissue part or separated cells alive being removed from the organism.

Tonoplasm – also called vacuolar membrane.

Trasferase – a group of enzymes catalysing the transfer of atoms between molecules.

Transipartion – loss of water in the form of vapour by the aerial parts of terrestrial plants.

Tropism – response to external stimulus (light, gravity) by growth curvature.

U

Unit membrane – unit of cyto-membrane consisting of two layers of lipids and two layers of proteins.

V

Vernalisation – treatment of seeds with low temperature to shorten the maturation period.

Vitamin – an organic substance necessaary in minute amounts for normal growth of the plants or animals.

Z

Zymase – an enzyme system which breaks down hexose sugars into alcohol and carbon dioxide, found in yeast. Zymase is the first cell free enzyme preparation.

ECOLOGY

A

Abiotic – the non-living components in an ecosystem.

Abundance – the number of individuals per unit area calculated on the basis of number of quadrats of occurrence.

Abyssal – deep sea region ranging between 2000-6000 m. depth.

Acclimatization – Adaptation to a changed or new environment.

Accretion – Gradual addition in reference to land, inorganic bodies or minerals in soil.

Acre – land area 43,560 sq. ft., Acre feet =1 ft. thick column of water over 1 acre area.

Adaptation – settling or adhesion of a substance on the surface of another substance, usually at very small particles or ionic levels.

Aeolian – soil transported by wind action (also eolian).

Aerosol – suspended fine particle, usually liquids in the atmosphere.

Allelopathy – inhibitory activity of one plant species commonly through exudation of chemicals from their roots on the seed germination or growth of other associated species. Also known as allelochemic interaction and antibiasis.

Alluvium – soil carried and deposited by water current.

Alpine – very cold climate or referring to high mountains such as Alps or Himalayas.

Ambient – the outside air.

Amphiphytes – plants that grow in two contrasting conditions such as on land and in water.

Aquifer – underground natural water storages, sometimes flowing out through artesam wells due to sufficient pressure on the storage water.

Associates – species occuring together.

Atmosphere – the gaseous envelope around the earth, the air. It has many gases but nitrogen, oxygen and carbon dioxide and water vapour are most important. It exerts a pressure of 14.7 pounds per sq. inch or in terms of column of mercury equivalent to 760 mm.

Autecology – study of a species or its population with an emphasis on ecological life-history.

Autotrophic – related to producing its own food such as the green plants.

B

Barrier – hurdles in the migration of flora on geographical scale such as mountains and oceans.

Bathyal – ocean zone between 200-2000 m. depth.

Beaufort scale – wind velocity scale named after the inventor -Sir F. Beaufort of the British Navy.

Bench mark – A reference point of permanent structure to measure elevation.

Benthic – of bottom of a water body; benthos for organisms living on the bottom of lakes, river or sea.

Bioassay – use of some organisms to know the biological effect of any substance or factor.

Biological control – use of some organisms to control the growth of some other pests or of weeds, etc.

Biological magnification – successive increase of some chemicals through different food level organisms such as of DDT from water to algae to herbivor fish to carnivore fish to fish eating birds and to man.

Biological spectrum – the range and ratio of different life forms in an area such as of phanerophyte, therophyte, etc. or the range of different levels of biotic organization from sub-cellular to cellular, to organ, individual, population and community.

Biome – a large unit of climax communities usually representing well defined climatic zone.

Biosphere – all parts of earth where organisms live.

Biosphere reserve – major vegetational zones or characteristic habitats set aside and protected atgainst human disturbance to act as a reference area of natural habitat.

C

Canopy – the foliage cover at the top layer of the plant.

Carnivore – organisms feeding on other living organisms or flesh.

Climax – the community that perpetuates itself on a habitat indefinitely, i.e., in balance with the environment.

Climograph – rainfall-temperature relationships diagram.

Coliform bacteria – those infesting the intestine of man or other animals: fecal coliform bacteria occur in fecal material: *coliform index* is for expressing the relative purity of water based on these bacterial counts.

Coluvium – soil deposits at the base of hills, transported by gravity.

Commensalism – association of two kinds of organisms in which one or both are benefitted.

Compensation point – At certain depth in a water column where the reduced light intensity is just sufficient to allow only as much photosynthetic production as in

completely used in the respiratory process.

Contour – lines on a map connecting points of equal value of depth, ups and down on a land surface or the bed of an aquatic body.

D

DBH – diameter (of a tree) at breast height, i.e., at 1.3 m. or 4.5 ft.

dB – decibell, a unit of sound intensity on a logarithmic scale.

Delta – alluvial sand deposits at the moutn of a river, usually triangular in shape and hence named after the Greek letter delta (d) .

Density – number of individual per unit area.

Desalination – removing salt from sea water.

Desertification – the process of desert formation, permanent loss of plant life in a region, loss of capacity to sustain good plant growth.

Diversity – number of different species growing together.

Dredging – removal of river of lake bottom mud to deepen the same.

DDT – one of the commonest chlorinated haloginated hydrocarbon, Dichlorodi-phenyltrichloroethane which is used as an insecticide, particularly for mos-quito control. It reaches non target organisms such as man through food chain and accumulates in the fatty tissues.

E

Eacd – or ecophene is a form variation within a species associated with some environmental condition but the character is not genetically fixed.

Ecological bomerang or back lash – the unforeseen cancelling of projected gain or unforeseen adverse or catastrophic reaction from a natural system in response to human interference, management or developmental activities.

Eco-development – development taking prior consideration of the holistic impacts on the concerned system.

Ecological pyramid – diagrammatic representation of numbers, biomass, energy or productivity of different trophic level organism with autotrophs at the base superimposed by those of herbivores, carnivores of the first order, second order and so on.

Ecosystem – the structural and functional entity of biotic communities and their environment. It may be of any size.

Ecotone – the transition zone between two different adjacent communitees showing characteristic edge effect,.

Ecotype – genetically distinct serological races within the frame work of species and variation in character is associated with variation in some environmental condition.

Effluent – liquid discharges or flowing out materials that usually pollute the environment.

Electrostatic precipitator – device to collect dust from factory chimney smoke by imparting electrical charge to dust particles to reduce air pollution.

Emmission standard – the maximum permissible or safe limit of discharge of pollutants.

Endemism – restricted area of distribution of a species to an isolated region.

Energy conserving efficiency (ECE) – percentage of half of the incident solar radiation that is fixed through photosynthesis by plants in unit area and time.

Environment – the totality of all kinds of influencing forces that surround an organism, both non-living and living.

Eollan – same as aeolian; wind transported soil.

Erosion – the physical removal of top soil by wind or water; wearing under the forces of wind, rain beating, water current, tramplings, etc.

Eutrophic – lakes rich in nutrients and productivity, i.e., choked with rich growth of vegetation.

F

Fall out – the descending radioactive fission products from atmosphere to earth's surface.

Fault – deep cracks in earths surface, fractures in rock layers, etc. due to geological events.

Fecundity – the rate of offspring production at individual level.

Feed back – the reverse direction flow (of information or any other thing) from the products back to the reactants.

Fission – splitting of atom into lighter fragments and release of energy and neutrons.

Flow metre – the instrument used to measure the quantity of water or effluents flowing out of a system over unit time.

Flu gas – the mixture of gases emitted from the chimney due to combustion.

Fly ash – the dust and other small fragments of burnt up material dispersed by gas mixture; usually emitted through chimneys.

Food chain – the sequence of different trophic level organisms starting from primary producers to top carnivores.

Food wed – the network of food chains in an ecosystem.

Fossil fuel – the fuel produced out of fossil plants and animals such as coal, petroleum and petroleum gas.

Fresh water – drinking water; less than 0.2% salt content.

Fusion – combining of nuclear or other things; combination of atoms such as of hydrogen forming helium and associated with release of energy (as in sun).

G

Green house effect – the heating effect through readiation. In glass house the sunlight enters, but on reradiation the heat is trapped inside the glass house thus increasing the inside temperature. Similar effect is caused by atmosphere, dust, water-vapour, ozone, carbon dioxide, etc. which allow the sun light to reach ground surface but trap the reradiating heat waves.

Gross production -the total increase in organic matter, energy fixation due to photosynthesis over a unit area (or volume) and time. If the loss due to respiration is deducted it is *net producting,*.

H

Half life – the time taken by a radioactive isotopes (or by pesticides) to come down to its half strenghth or decay to its half level from the initial.

Halophytes – plants adapted to grow in saline soils or sea coast marshes.

Heliophytes – plants adapted to grow in open sun.

Holistic – taking all the aspects and components of ecosystem into consideration.

Homeostasis – the capacity of ecosystem to resist changes due to disturbances or to return to balanced state.

Humic acid – the different kinds of acids formed due to partial decay of litter in the soil. These are decay resistant dark coloured acids.

Humus – dark coloured fine powder of organic matter formed after the decomposition of litter.

Hydrology – the study of water storages, distribution and movement on earth's surface.

Hygrograph – instrument to measure relative humidity.

I

Incineration – complete combustion of materials in a controlled condition.

Infiltration rate – the maximum rate of water entering the soil system through soil pores when it is abundantly available at its surface.

Infra-red radiation – wavelengths just larger than the red wave between 0.7 to 1.0 mm. It has a heating effect.

Ionsphere – the upper most layer of atmosphere above the stratosphere, where iozization takes place. In day time its lower limit is 56 km and in night 96 km above the earth's surface. It reflects radio signals.

Isotherm – Lines on a map connecting places with equal temperature.

Isotope – forms of an element with same number of protons but different number of neutrons. They differ in radioactive behavior and atomic weights but have the same atomic number and chemical properties.

L

LC 50 – indicates the concentration of chemicals that would kill 50% of the target organisms.

Leaching – study of fresh waters for its various physico-chemical and biological characteristics.

Littoral – shallow zone of a water body; also on sea coast; the zone receiving low to high tides.

Lotic – running water such as streams and rivers.

Lux – a measure of light intensity obtained at 1 m. distance from a standard candle, 10.76 Lux is equal to 1 Foot candle.

M

Meandering – Loop-like changing course of rivers from time to time.

Merological – study in parts as against hological, i.e., holistic study.

Microcosm – a miniature ecosystem often controlled to study the cause-effect relationship of environmental factors on test organisms or populations. Often used in Aquatic ecology.

Migration – movement of plant diaspores or of animals such as birds to distant places along some set routes.

Mimicry – adaptation of structure and colour to merge into its habitat in order to avoid detection by predators.

Mist – fine dropets larger than fog (i.e., from 40 mm to 500 mm) of liquids in the air.

Monsoon – the rain laden seasonal winds originating in Indian ocean and Bay of Bengal.

Mulch – the organic debris left on soil surface to retain moisture, enrich soil and avoid weeds.

Mycorrhizae – fungi assoicated with roots in soil.

N

Necrosis – death of plant parts and its consequent discolouration.

Neritic – the zone of sea bed from margin to 200 m depth (including the overlying column of water.

Net Primary Production – accumulated biomass over a unit area and time (GPP-respiration).

Niche – the location and function of an organism in the context of its ecosystem.

Nitrogen fixation – conversion of free atmospheric nitrogen into combined form.

O

Oligotrophic – nutrient wise a poor lake with clear water.

Overburden – the deposits of surface soil or other material over-lying usefully mined material like coal or other ores by open cast mining.

Oxbow – lakes formed by the left over loop of a river which has changed its earlier course and moved away.

P

PAN – peroxy acetyl nitrate, a secondary pollutant in smog formed by the action of light on hydrocabons

and nitrogen oxides in air.

Pedology – the study of soils.

Pelagic – the marine communities free from bottom or shore effects.

Periphyton – aquatic organisms attached to stem and leaves of macrophyte.

Permafrost–permanently frozen zone of the ground.

Pesticides – chemicals used to kill pests such as herbicides for killing weeds, *fungicides* for fungi, inesecticides for insects, rodenticides for killing mice, rats, etc. Permanent pesticides continue their killing property even after one year of its spray.

Photoperiodism – control of phenological events by the day lenth time, such as leaf fall, flowering, migration of birds, etc.

Phytosociology – the study of community structure and composition of a vegetation.

Plume – the visible smoke coming out of a chimney.

Pollution – undesirable effects of for drinking and cooking.

Ppb and Ppm – parts per billion or 1 mg/1 and part per million or 1mg/1.

Primary Production – gain in biomass or energy by photosynthetic organisms (green plants).

Psammophyte – plants adapted to grow on sand.

R

Radio carbon dating – determining the age of material by measuring the proportion of C^{14} in its total carbon content, useful upto 30,000 years old specimen.

Recalcitrant – not easily degraded by the microorganism.

Remote sensing – obtaining information about any aspect of ecosystems from far away recording device such as by aerial photography, lasers, ultraviolet and infrared detectors, radars, etc.

Resource – materials useful to man

and to other organisms.

Rhizosphere – zone of soil around the roots and its influenced by the root exudates.

Riparian – related to rivers and their banks.

S

Savanna (h) – grasslands interspersed by isolated and scattered growth of shrubs and/or trees.

Scrubber – a liquid spray device to control air pollution.

SCUBA – abbreviation for Self Contained Under Water Breathing Apparatus.

Secchi disc – a painted disc to measure turbidity of water.

Seral – a successional stage of a community.

Sewarage – the system of sewer collection, treatment and disposal.

Soil profile – a section of soil column from surface to lower horizons.

Sonic boom – the booming sound and shock waves (emitted by jet planes) that reach the ground surface.

Surfactant – the chemicals producing lather in detergents; these are

phosphate rich and cause eutrophication.

Synergistic effect – enhanced effects of two combining substances than the total effect of both in isolation.

T

Thermocline – the sharp zone separating the warm epilimnion and cold hypolimnion in water body.

Trophic – related to food and feeding habit.

Troposphere – the zone of atmosphere containing clouds and moisture.

W

Watershed – the area drained by a stream or river.

Water table – the upper layer of ground water.

Z

ZPG – Zero Population Growth, i.e., each couple producing only two children so as to stabilize the population at the existing level.

ECONOMIC BOTANY

A

Abaca – Manila hemp; inner fibres of *Musa textilis*.

Addiction – Unable to resist indulgence in some habits especially strong dependence on a drug.

Adulteration – The addition (intentional or unintentional) of impurities or inferior products to a drug, food , etc, or the removal or valuable portions from such commodities.

Aflatoxin – Specific mycotoxin produced by *Aspergillus flavus;* a potent liver toxin that can induce liver tumours.

Agar – Dry mucilaginous extract of certain marine algae especially of *Gelidium cartilagineum.*

Annual Ring (of Wood) – Alternative title for "Growth Ring".

Algin – Ectracted from kelps, chiefly *Laminaria* spp; is sodium salt of alginic acid.

Alkaloid – Basic organic nitrogenous compound of plant origin that is pharmacologically active and bitter to taste.

Analgesic – Pain reliever that does not induce loss of consciousness.

Anise – Fruit (seed) of *Pimpinella anisum;* source of anise oil.

Antacid – A drug which neutralises the acidity of the gastric juice.

Apotracheal Parenchyma (of Wood) – Soft tissue of a type that does not touch the pores or pore groups as in Ceylon Ebony wood. Thin lines of soft tissue, or terminal parenchyma lines, are often apotracheal.

Aliform Parenchyma (of Wood) – A type of soft tissue arrangement. When aliform, the tissue forms a sheath around the pore, but the sheath extends tangentially on two sides so that it tends to assume a diamond shape. Aliform parenchyma is to be seen in European Ash and Oak, and Afara, among other timbers.

Aggregate Rays (of Wood) – Very large rays, which when viewed under a lens, appear to be made up of several rays fused into one. Aggregate rays are not of common occurrence. They are to be found in Alder, Hornbeam and a few other timbers, and form an important diagnostic feature.

Anatto – A colouring matter consisting of dry pulpy aril surrounding the seeds of *Bixa orellana*

Antiperiodic – A drug that cures periodic attacks.

Antiphlogistic – A drug which counteracts inflammation.

Antipyretic – Agent relieving or reducing feaver.

Anthelmintic – A drug that kills intestinal worms.

Antispasmodic – A drug which counteracts spasmodic disorders.

Anthocyanins – Glycosides of plants constituting the blue, red and violet colouring matters, responsible for the reds and purples of autumn leaves.

Anthralinic Acid – O-aminobenzoic acid; used as tracer or marker in squalene to determine if used to adulterate olive oil.

Aphrodisiac – Stimulator of sexual desire.

Aromatic – A drug which is fragrant, spicy and mildly stimulant.

Arthritis – Inflammation of joints.

Astringent – A drug which asserts secretion or bleeding.

Aquiculture – Hydroponics; soilless gardening.

Arrack – Alcoholic liquor distilled from fermented rice, dates, palm juice, *mahua* flowers, etc.

Artificial Oil of Wintergreen – Methyl salicylate.

Aspirin – Acetyle salicycilic acid.

Atta – Wheat flour.

Attar – Rose oil from *Rosa* spp.

B

Badam – Almond.

Bael – *Aegle marmelos.*

Bagasse – Sugarcane crushed stalk after expression of its juice.

Bagh – Garden.

Bajra – Spiked millet.

Banjwain – *Sesli* indicum.

Banyan – *Ficus religiosa.*

Biseriate Rays (or Wood) – Rays that, when viewed in end-section are two cells in width.

Bast – Phloem.

Bende – *Hibiscus esculentus.*

Bengal Beans – *Stizolobium* spp.

Bengal Quince – Bael.

Benzaline Seed – Behenut, from *Moringa oleifera.*

Benzoin – A balsam (gum) obtained from *Styrax benzoin,* & S. tonkinensis.

Berry – Fruit characterized by having thin epicarp, thick fleshy pulpy mesocarp and wtih small hard seeds loosely embedded in the latter, ex: grape, capsicum.

Berseem – *Trifolium* spp.

Betel (Leaf) – *From Piper betle.*

Betel (Nut) – *Areca catechu.*

Bhahji – Spinach; sometimes applied to any vegetable.

Bhang – *Cannabis*; dried leaves and young stems in combination with sugar, milk and water.

Bhindi – *Okra.*

Bidi – A cheap rough tobacco cigarette made from leaf of *Diospyros mela-*

noxylon.

Brandy – A distilled liquor made from wine by steam distillation.

Bread Fruit – *Artocarpus altilis.*

Breast Tea – Mixed herbs and used as tea for coughs, colds, bronchitis and other chest diseases.

Brewer's Yeast– Yeast formed during the process of brewing in which nutriment is composed of cereal grain extract with hops.

Brinjal – Egg plant.

British Gum – dextrin.

Bromelin – Proteolylic enzyme of pineapple fruit.

Bronchitis Plant – *Artemisia heterophylla.*

Brume Corn – *Sorghum halepense variety.*

Brucine – Alkaloid from nux vomica seeds.

C

Camphor – Dextro-camphor, obtained by distilling all parts of tree of *Cinnamomum camphora*

Camphor Basil – The tree *Abies balsamea* or its oleoresinous product.

Caprification – The process of artificially pollinating figs.

Capsaicin – Active pungent crystalline principle of *Capsicum.*

Capsule – Dry fruit of more than one carpel, typically opening at maturity along one or more lines of dehiscence; e.g. poppy.

Caramel – Burnt sugar; used as colouring or flavouring of foods, & liquors.

Carbooja – Melon.

Carminative – Substance that relives excessive amount of gas in the stomach or intestine.

Carrageenan – Purified extract of marine alga, *Chondrus* (or Irish moss).

Caryology – Science of nuts or Kernels.

Caryopsis–Dry indehiscent one-celled fruit of cereal grasses; e.g. oat.

Centres of origin – N.I. Vavilov's &

centres of origin or cultivated plants (See appendix for list of plants).

Cereals – The fruits of grasses; also the grasses which produce them.

Chaff – The glumes, paleas, lodicules, rachis, rachilla, bristles or a grass spikelet collectively.

Chapati – Unleaved wheaten bread in thin flat griddle cakes.

Charas – *Cannabis.*

Chicle Gum – Dried latex from: 1. *Acharas zapota* (sapodilla). 2. *Mimusops balata.*

Chocolate – Solid mass made by grinding seeds of *Theobroma cacao*, without removal of any aprecialbe quantity of fatty oil.

Chutney (Chatni) – Pickle preserve made with capsicum and green mangoes or vegetables.

Clover – *Trifolium* spp.

Coir – Elastic but stiff fibers of outer husk of coconut.

Cold Cream – Rose water ointment.

Cologne Spirits– Deodorized alcohol.

Cologne Water – A toilet water used to impart fragrance to skin.

Copra – Dry meat of the coconut.

Confluent Parenchyma (of Wood) – Soft tissue that consists of a series of connected aliform shapes, that sometimes make well defined bands. Short confluent lines or chains are to be found in the European Ash.

Clustered Arrangement (of Wood) – A type of pore arrangement in which the proes fall into well-defined clusters which may be either large or small.

Cover Crop – Crop planted on land to prevent erosion and weed growth and to improve soil conditions.

Cremocarp – A type of fruit found in and typical of Apiaceae; it consists of two achenes or mericarps borne on a forked stalk, the carpophore; ex. aniseed, coriander, caraway.

Crocus – *Crocus sativus*, true saffron; dried stigmas contain crocin, picrocrocin and volatile oil.

Catch Crop – One planted to protect the soil in winter, to fix nitrogen in soil etc. (ex. clover).

Crude Drugs – Medicinal materials found in the raw form.

Cultigen – Species of organism known only in cultivation.

Curing – Process of proper storage so as to permit the enzymes present to act upon the constituents and transform them and thus improve the quality of the material; ex. coca, tobacco, vanilla beans.

Currants – Fruits of *Ribes rubrum, R. nigrum* ex. 1. popular curry oriental food, ex. essentially spiced and garnished rice. 2. condiments used in such foods.

Curry Leaf Tree – *Murraya Koenigil.*

D

Dal (Dhal) – Kind of pulse. chiefly applied to *Cajanus indicus*, used in food

Dextrin – Carbohydrate intermediate product obtained by partaial hydrolysis of starch.

Dhurra – Indian corn or millet. diastase (*of malt*) a mixture of amylolotic (or starch splitting) enzymes obtained from malt.

Diffuse Porous Structure (of Wood) – One of the two most common types of general pore arrangement, the second type being Ring porous Structure (q.v.). With a diffuse porous arrangement the pores, viewed in end-section in any growth ring, do not show any marked variation in size throughout that ring.

Density (of Wood) – Weight per cubic foot.

Diffuse Parenchyma (of Wood) – A type of soft tissue in which the parenchyma strands are seattered haphazardly throughout the fibres. Examples are to be seen in Chestnut and Hornbeam.

Diagonal Grain (of Wood) – Term applied to a timber in which the wood elements tend to run slightly diagonally whebn compared with the longitudianl edges of the board.

Degrade (of Wood) – A lowering of the quality of a timber due to seasoning defects. Such degrade is usually avoidable.

Diaphoretic – That which promotes perspiration, especially when it is profuse.

Digitonin – Saponin-like glycoside of *Digitalis*.

Dill – *Anethum graveolens.*

Dilling Herb – Dill.

Distillation – Process of heating a substance (in retort). then cooling the vapours in a condenser, and collecting such distillate product.

Diuretic – A drug which increases the secretion and distilling without adding water to product in retort so as to destroy original substance in large part the vapours of newly formed substances being condemned and collected.

Drug – A crude medicinal substance.

Drupe – Type of l-celled fruit in which mesocarp is more of less succulent but endocarp is stony or leathery, and it encloses l seed; ex. peach. Mango.

Durra – Jowar or bajra.

E

East India Gum – Gum obtained by *Feronia elephantum.*

Ergot – Dried sclerotium of the fungus *Claviceps purpurea,* contains ergonovine, ergatoxine and ergotamine.

Essence – Volatile oil.

Etherial Oil – Volatile oil.

Emetic – A drung which induces vomiting.

Early – **Wood** – An alternative term to describe "spring-wood" (see Growth Rings).

Emmenagogue – A drug which promotes menstruation or regulates the menstrual periods.

Ergotism – Chronic poisomning marked by spasms, cramps, dry gangrene, and various cerebrospinal symptoms due to excessive use of medicinal ergot or of eating ergot -contaminated grains. The contamination is due to the fungus *Claviceps purpurea.*

Expectorant – A drug that promotes the removal of catarrhal matter and phlegm from the brochial tubes.

F

Fenugreck – *Trigonella foenum graecum.*

Fibre – A hardened scleren-chymatous elongated cell, generally with tapered ends.

Fibres, Hard – Those extending lengthwise in leaves of stems monocot plants of very hard and wiry in nature, ex, abaca, sisal, henequeen.

Fibres, Soft – Relatively flexible and elastic fibres found in stems and inner barks of dicots. ex. hemp, jute, ramie.

Flecking (of Wood) – A type of ray figure to be seen on the quarter sawn face of timbers (such as Beech) that save fine rays. The ray flecking is often very inconspicuous.

Fine Lines of Parenchyma (of Wood) – Parenchyma that is in narrow lines of the terminal type, though not delimiting growth boundaries.

Figure (or Wood) – The pattern on the longitudinal faces of the wood.

Flat-Sawn Face (of Wood) – That face of a piece of timber that has been sawn at right-angles to the rays, producing a growth-ring figure with inconspicuous ray markings. A flat-sawn face is alos descriged as a "tangential surface."

Flamboyant Tree – *Delonix regia.*

Flax – Linsed, *Linum,* especially L. *usitatissimum.*

Floss – Fluffy substance; ex. silk cotton, vegetable silk.

Flour – Finely ground and bolted cereal grain.

Fodder – Herbaceous livestock food after being cured and dried, ex, hay, silage.

Foxglove – *Digitalis.*

Food of the Gods – *Asafoetida.*

Forage – Herbage food for livestock as grazed in the field; pasture green feed.

Fractional Distillation – Process in which the lower and higher boiling point portions of a substance being (re) distilled are vaporized at different termperatures and collected separately; commonly used to separate components of volatile oil etc.

G

Gambir – Pale catechu, from *Uncaria gambir.*

Gambog (Gum) – Gum-resin from *Garcinia spp*

Ganjah – Addictive drug, *Cannabis.*

Geraniol – Colurless fragrant liquid, chief constituent of rose, geranium and palmarosa oils.

Ghur – Crude sugar in lumps.

Gingelly – Seasme, *Sesamum indicum.*

Ginger Nut – *Hyphaene crinita.*

Gitalin – Glycoside of *Digitalis;* considered best drug for congestive heart failure.

Gluten – Mixture of proteins obtained from wheat and other cereal grains.

Glycoside – A compound which on hydrolysis breaks down to produce simple sugars and a non-sugar component, the glycone.

Gomuti Sugar Palm – *Arenga pinnata.*

Gossypol – Toxic component of cotton seed *Gossypol.*

Gunny – Strong coarse jute sacking.

Grain (of Wood) – The general direction of the wood elements in relation to the longitudinal axis of the tree, long or plank. Often incorrectly used in the sense of "figure".

Gum Cannals (of Wood) – Small dark cavities in the rays, filled with gum.

Gutta parcha – latix from *Palaquium gutta* (f. sapotaceal) contains toxic saponins.

Growth Rings (or Wood) – An alternative name for "annual rings." The concentric layers of wood which are usually clearly visible on the end-section of a long, each layer of which represents a growth season.

H

Heartwood – The central core of wood in a tree. In this wood the pores have ceased to conduct plant food and have : undergone certain chemical changes. The hearrtwood is usually heavier, darker in colour, and more resistant to insect and fungal attack than than the sapwood, but in other instances the colour difference between heart had sap is only slight.

Homogeneous Rays (of Wood) – Rays that are all of aproximately the same width in any given specimen.

Hallucinogen – Agent inducing fabre perceptions that occur without true sensory stimuli.

Haemorrhage. – Bleeding especially profuse, from any part of the body.

Heavy Oil – Higher boiling point fraction from distillation of coal tar.

Hectare – (H.A.) 1 sq. Hectometer; 10,000 sq. meter; c. 2.47 acres; the common metric unit of area used.

Hegari – *Sorghum vulgare* v. *coffrorum.*

Hemp, Indian – *Cannabis sativa.*

Hemp, Sunn – *Crotalaria juncea.*

Henbane – *Hyoscyamus niger.*

Heterogeneous Rays (of Wood) – Any timber in which the rays can be readily classified into two or more distinct widths (e.g. very fine and coarse) is said to contain heterogeneous rays. Beech, Alder and Oak are examples of common European timbers that possess heterogeneous rays.

Herb – Two above-ground parts of a plant which do not persist through the winter.

Heroian – Diacetylmorphine.

Hesperidin – Glycoside from *Citrus spp.*

Hessian – Coarse sacking made of hemp or jute.

Hop – *Humulus lupulus.*

Hornmeal – Nitrogenous manure manufactured from hoofs, horns and clacos of animals.

Hydroponics – Soil-less gardening.

I

Indigenous – Native to an area, country, continent etc.

Ink Root – *Limonium carolinianum.*

Ink Berry – *Ilex glabra.*

Intelligence Plant – *Hydrocotyle asiatica.*

Invertase – Enzyme from yeast, etc. Which splits sucrose into glucose and fructose.

IVY – *Hedera helix.*

Interlocked Grained (of Wood) – Descriptive of a timber in which successive layers of fibres are inclined in different directions.

J

Jaggary – Gur. raw-brown sugar in lumps.

Jatamansi, – *Nardostachys* spp.

Jawar – Jowar.

Jowar – Great millet, *Sorghum vulgare.*

K

KaramKala – Cabbage.

Karanja – *Pongamia glabra.*

Karaya Gum – Gum from *Sterculia* spp.

Kelps – Seaweeds of genera *Laminaria, Nereoystis, Porphyra.*

Kusum – *Carthamus tinctorius.*

Kusum Tree – *Schleichera trijuga,* source of resin.

Kutera-Gum – 1. Karaya. 2. Tragacanth

L

Lanolin – Hydrous wool fat.

Larkspur – *Delphinium ajacis and D. consolida.*

Latex – Milky juice of plants carried in special ducts ca' 'd laticiferous tubules; ex. gutta , cha, rubber, opium.

Lentil – *Lens culinaris.*

Leprosy Oil – chaulmoogra oil.

Lichenin – Polysaccharide found in *Cetraia* and a number of other lichen genera.

Licorice–Root of *Glycyrrhiza glabra.*

Latewood – An alternative term to describe 'summerwood' (see Growth Rings).

Lipid – Fatty oil like substance found in plants and animals.

Liquorice – See licorice.

Litmus – Colouring matter obtained from lichen *Rocella tinctoria* etc.

Loquat – Fruit of *Eriobotrya japonica.*

Lards and Ladies – *Arum moculatum.*

Lupulin – Glandular trichomes of hopes, *Humulus lupulus.*

Lustre (of Wood) – Small canals (rather similar to pores) that secrete latex.

M

Mace – Aril of fruit of nutmeg, *Myristica fragrans.*

Madar – *Calotropis gigantea.*

Madras Gum – A very red form of Indian Gum *Acacia.*

Malt – Fermented grain of *Hordeum vulgare.*

Malt Extract – Preparation made by extracting of malt with hot water.

Metatracheal Parenchyma (of Wood) – Soft tissue in layers or bands that can very from narrow to fairlh broad.

Manna – Saccharine exudation from plants which contains mannitol, ex. *Fraxinus ornus, Setaria italica* etc..

Manure – Ordure of animals, mixed with straw etc. valuable fertilizer of plants.

Marking Nut – Fruit pericarp of *Semicarpus anacardium*

Mash – *Phaseolus mungo.*

Masur – *Lens esculenta.*

Meal – A coarsely ground grain such acorn.

Menthol – Peppermint camphor ($C_{10}H_{20}O$) obtained from the mint plant.

Mesquite – *Prosopis* spp.

Methi – Fenugreek.

Millet – *Panicum miliaceum.*

Millet, Indian – *Pennisetum spicatum, P. glaucum; Sorghum vulgure.*

Millet, Bulrush – *Pennisetum spicatum.*

Millet Etyptian – *Sorghum halepense,*

Millet, Finger – *Eleusine coracana.*

Millet Great – *Sorghum vulgare.*

Millet, Italian – *Setaria italica.*

Millet, Pearl – 1. *Panicum miliaceum,* 2. *Echinochloa crusa-galli.*

Millet, Pearl – *Pennisetum americanum*

Momira – *Thymus* sp. used as flavour and spice in curries.

Moong – Mung

Mulch – Any substance such as straw, leaves etc., spread on the ground to protect plant roots from cold etc.

Myrrh – Gum-resin from *Commiphora myrrha.*

Multiseriate Rays (of Wood) – Rays that, when viewed in end-section have a width greater than two cells.

N

Net-Like Pattern (of Wood) – A rays-cum-parenchyma pattern in which the rays and fine lines of parenchyma are approximately the same distance apart and form of net-like arrangement on the end-section of the wood.

Nectar – Sweet liquid saccharine secretion of plant nectaries, chief source of honey.

Neem – *Melia azadirachta.*

Neroli – Orange flowers; source of neroli oil.

Nut – A fruit characterised by being dry indehiscent with one seed (the kernel, portion eaten if nut is edible) and with a hard stony pericarp (or shell).

Nutmeg – Ripe seed of *Myristica fragrans.*

O

Olibanum – Indian frankincense; ole-resin from *Boswellia serrata.*

P

Pahari Pudina – Spear mint.

Paan – Popular masticatory composed of betelnut, lime spices as cardamom, cloves etc. gold leaf, etc. wrapped in a betle leaf.

Papain – Digestive enzyme mixture from latex of *Carica papaya.*

Paprika – *Capsicum frutescens* and other *Capsicum* species bearing large fruits.

Parched – Roasted, as peanuts.

Parslen – *Petroselinum crispum.*

Patent Medicine – Package drug produced which has been patented in patent office.

Phaseoliy – A glosulin protein found in seeds of *Phaseolas* spp.

Phenology – Study of relation of climate to periodic phenomena in plants (as flowering, fruiting etc.).

Paratracheal Parenchyma (of Wood) – Soft tissue (of the vasicentric, aliform, etc., types) that actually surrounds the individual pores or pore groups.

Pith Flecks (of Wood) – Faint brown lines running the way of the grain on the longitudinal faces of timber species.

Parenchyma (of Wood) – A soft tissue responsible for food storage in the growing tree.

Permeable (of Wood) – Readily treated with preservatives, the fluid penetrating the fibres as well as the pore cavities.

Pipal – *Ficus religiosa.*

Podina – *Mentha arvensis* or *M.*

spicata.

Pore (of Wood) – A hollow, tubular wood cell used by the growing tree for the conduction of sap.

Q

Quarter-Sawn Face (of Wood) – That face of a plank that has been sawn parallel to the rays, so as to display the latter to the best advantage.

R

Radial Chains of Pores (of Wood) – A type of pore arrangement in which the pores are connected in chains, the length of which run parallel to the rays (e.g. Hornbeam).

Rays (of Wood) – An alternative title for "Medullary Rays". Strands of horizontal storage tissue which, on the end-section of the log, form lines radiating from the centre of the log to the bark.

Radial Oblique Arrangement (of Wood) – A type of pore arrangement in which the pores are connected in chains of varying length, the chains passing obliquely across the rays. This type of pore arrangement is to be seen in Chestnut.

Ragi – *Eleusine coracana* small millet.

Rasin – Dried grape fruit *Vitis vinifera.*

Rajmahal Hemp – 1. Fibres of *Marsdenia tenacissima.* W&A. 2. Cortical fibers of *Calotropis gigantea.*

Ramad – A fain quality of karaya gum.

Ratoon – Stalk or shoot of a perennial plant especially from second year's growth or later (ex. sugarcane, ginger).

Rawa – Raw material for macaroni, vermicelli etc.

Retting – A process of rotting way of the soft tissues of stems leaving the phloem fibres behind, the process is accomplished by laying the fresh material in water, where bacteria (such as *Clostridium felsinium, Plectridum pectinorum* in case of hemp and flax) feed upon the tissues.

Ripple Marks (of Wood) – A type of ray arrangement to be seen on the tangential surface of certain timbers (e.g. Horse Chestnut and Central American Mahogany).

Refractory (of Wood) – Difficult to treat. Usually applied (when necessary) to the seasoning properties of a wood that teads to develop shakes, splits or pronounced warp during the drying processes.

Radial Groups of Pores (of Wood) – Small groups or clusters of pores, the main axis of the group being parallel to the rays.

Radial (of Wood) – Parallel to the rays. Thus the radial surface of a timber is that which is known in the trade as te quarter-sawn face.

Ring Porous Structures (of Wood) – One of the two main types of overall pore within a growth ring, and one which is characteristic of such well-known timber species as Ash Chestnut, Elm, Oak and Teak. With this type of structure, the spring wood zone of the growth ring carries a well-defined band of large pores, the ring being two, three or more pores wide.

Rotenone – Complex organic compounds representing active insecticidal constituent of many plants including *Derris, Pachyrrhizus, Tephrosia* etc.

Rutin – Vitamin P analog; flavanol glycoside found commonly in Rutaceae, Polygonaceae and other families; mostly prepared from buck wheat; used to decrease capillary fragility, in pulmonary hemorrhage, vascular purpurea, rheumatic epistaxis etc.

Rubifacient – A mid counter-irritant.

Rye – *Secale cereale.*

Rye, Italian – *Lolium multiflorum.*

Rye, Wild – *Elymus* spp.

S

Solitary Pores (of Wood) – A type of pore arrangement in which the ovewhelming majority of pores are separate entities, and are not connected one to the other, subdivided, etc. When this arrangement is present the pores are described as solitary, irrespective of their size or the number that may be present in a typical area of the end-surface. Norway Maple and Nox are examples of timbers in which the pores are solitary.

Silver-Grain (of Wood) – Name commonly applied to the distinctive figure to be seen on the quarter sawn face of timbers (such as Oak) that possess broad or very broad rays.

Semi Ring Porous Structures (of Wood) – A type of pore structure in which, though not truly diffuse porous, the broad distinctive ring or ring porous structure is also lacking.

Safflower – *Carthamus tinctorius.*

Saffron – *Crocus sativus.*

Saag – Any pot herb.

Sage – Leaves of *Salvia officinalis.*

Sago – Starch from the stem of sago palm, *Metroxylon laevae.*

Saki – Rice wine.

Sal – *Shorea robusta.*

Sarsaparrilla–Root of several *Smilax* spp.

Semolina – The purified middlings of high quality *durum* wheats; used in preparation of spaghetti, macaroni, vermicelli ect.

Shakar – Crude granulated sugar.

Sherbet – Flavoured   sweetened beverage.

Silage – Fodder (either green or ripened) changed into succulent winter feed by fermentative process, using a cold chamber, the silo.

Scalariform   Perforations   (of Wood) – Pores are normally open at both ends, but in a few cases the ends of the pores are connected by small bands of tissue.

Spiral Grained (of Wood) – Timbers in which the fibres are in clockwise or anti-clockwise spirals. Such timbers may develop faults in

seasoning, or show structural weaknesses in service.

Sedative – a drug which reduces excitement, irritation and pain.

Simple Perforation (of Wood) – The normal type of pore opening. (see Scalariform perforations).

Sisal – Fibre from *Agave sisalana.*

Sissoo – *Delbergia* spp.

pruce – *Picea* spp.

Squash – Length of cotton fibre; short staple under 25 mm., medium staple 25-30 mm., long staple 30-40 mm.

Straw Process– Hay process; method of producing plant principles by growing for a relatively short period, then cutting down and extracting the plants with volatile solvents; applied to Cinchona, Opium, Camphor etc.

Sunnhemp – *Crotolaria juncea.*

Soft Tissue (of Wood) – Alternative description for "parenchyma".

Sapwood – The wood coming from the outer sheath of the 'growing tree.

Straight Grained (of Wood) – Timbers in which the fibres run parallel to the longitudinal axis of the plank.

T

Tabak – Tobacco.

Tamarind – *Tamarindus indica.*

Tan – Tanning agent; substance applied to skins and hides to preserve and make resistant to decay, infestation etc.

Tanbark – Any bark useful for tanning purposes.

Tannin – Class name for complex organic compounds many of which are glycosides.

Taramira – *Eruca sativa.*

Til – *Sesamum indicum.*

Tangential (of Wood) – At right-angles to the rays. Thus the tangential surface of a timber is that which is known in the trade as the "flat-sawn" or "plain-sawn" face.

Tangential Arrangement of Pores (of Wood) – A type of pore arangement in which the pores are connected in lines or chains that cross the rays approximately at right-angles. This arrangement can be seen in Elm.

Terminal Parenchyma (of Wood) – A thin line of soft tissue (usually only one cell wide) developed at the end of a growing season and therefore marking the boundary of the growth ring. Examples of terminal parenchyma are to be found in Alder, Birch, Horse Chestnut and Norway Maple, among many other timbers.

Texture (of Wood) – The relative fineness of a timber, determined by the size of its pore cavities.

Tor – Kind of pulse.

Toria – Oil seed.

Tulsi – *Ocimum sanctum.*

Turmeric – *Curcume longa.*

Tyloses (of Wood) – An ingrowth into the cavity of a pore caused by pressure on the pitted wall is described as a tylosis (in the plural, tyloses).

U

Uniseriate Rays (of Wood) – Rays of one cell width when viewed in end-section (i.e., very fine).

V

Vinegar – An acetous liquid made by fermentation of alcoholic liquid.

Volative Oil – Essential oil.

Vessel Deposites (of Wood) – Gum, resin, silica, etc., to be found in many of the vessels of certain species, and usually very distinct when seen under a hand lens.

Vessel (of Wood) – An alternative term to "pore", but whereas the word pore is used of the wood element when viewed in end-section; the description vessel is used when the element is viewed longitudinally.

Vessel Lines (of Wood) – In some cases the vessels can be seen on the longitudinal surfaces, and somewhat resemble grooves or scratches of varying degree of coarseness.

Vermifuge – A drug which expels intestinal worms.

Vasicentric Parenchyma (of Wood) – A type of tissue arrangement in which the parenchyma makes a sheath or halo around the pore. The sheath is complete, but may be variable in width. Vasicentric parenchyma occurs in such species as Ash and Chestnut.

W

Wine – Fermented grape juice, with alcohol content up to 15%.

Wing-Like Parenchym (of Wood) – Alternative description for "Aliform parenchyma" (q.u.).

Wavy Grained (of Wood) – Timbers in which the fibres tend to form undulating lines.

Y

Yeast – *Saccharomyces* spp.

Yeast, Top – Variety of brewer's yeast which grows at or near the surface of the vat and does best at summer temperature, causing much CO_2 formation and the froth which is characteristic of many brews.

List of Cultivated Plants Classified under Primary Centres of Origin according to N.I. Vavilov (1951)

I. GROUP

Aconitum wilsonil Hort - Aconite.

Aleurites montana Wilson - Wood oil tree.

Allium fistulosum L., - Spanish Onion.

Andropogon Sorghum Brot - Kaoliang.

Avena muda L., - Naked Oats.

Basella cardifolia Lam - Chinese Spinach.

Bohemeria nivea Hook and Arn - Ramie.

Brassica chinensis L., - Pak choc.

B. juncea. Czern - Leaf mustard. (Secondary centre of origin).

Brassica rapa L., - Turnips.

Camellia sinensis L., - Tea.

Chrysanthemum coronarium L., - Chrysanthemum.

Cinnamomum cassia - Chinese Cinnamon.

C. nobilis Lour. - Sweet Orange.

Citrus sinensis Osb. - Orange (secondary centre)

Cucumis sativus L. - Cucumber.

Cucurbita moschata var *Toonasa* Makino, - watery squash (secondary centre).

Cycas revoluta Thunb - Cycas revoluta Thunb.

Dioscorea batatas Decne - Chinese Yam.

Diospyros kaki L., - Cucumber.

Fagopyrum esculentum Moench - Buckwheat.

Glycine hispida Maxim - Soya Bean.

A group of endemic hull-less awnless barley varieties (*Hordeum hexastichum, L.*).

Juglans sinensis Dode - Walnut.

Lactuca sp. - Stem Lettuce.

Litchi chinensis Sonu - Litchi.

Malus asiatica Nakai. - Chinese Apple.

Melia azadirach L., - China Berry.

Metroxylon sago Rottb - Sago palm.

Nelumbo nucifera Gaertn - Lotus.

Panicum frumentaceum - Fr. and Sar. Japanese barnyard millet.

P. italicum L., - Italian Millet.

P. miliaceum L., - Broom Corn Millet.

Papaver somniferum L., - Opium poppy.

Phaseolus angularis Wight - Adzuki Bean.

Phaseolus vulgaris L., - Bean (secondary centre).

Phyllostachys tuberula - Munro and other species.

P. armeniaca L., - Apricot.

Prunus persica L., - Peach.

Pyrus serotina Rehd - Chinese Pear.

Raphanus sativus L., - Radish.

Saccharum sinense Roxb - Endemic group of Sugareane varieties.

Sagittaria sagittifolia L., - Arrowbead.

Sesamum indicum L., - Sesame (secondary centre).

Solamum melongena L., - Eggplant.

Trachycarpus excelsus Makino - Fibre Palm.

Trapa bispinosa Rox - Water Chesnut.

Vigna sinensis Endl - Subsp. Sesquipedalis.-Cowpea (secondary centre).

Groups of Waxy Maize varieties - *Zea mays* - (secondary centre).

Zizyphus sativa Gaertn. - Chinese jujube.

II. GROUP

Acacia arabica Wild - Gum Arabic.

Aegle marmelos - Correa.

A. gangeticus L., - Amaranth.

Amaranthus speciosus Sims - Amaranth.

Aamaranthus speciosus Sims - Amaranth.

Amorphophallus campanulatus Blume - Elephant Yam.

Andropogon sorghum Brot - Sorghum.

Areca catechu L., - Areccanut.

Srenga saccharifera Labill - Sugar Palm.

Artocarpus integra Merr - Jack Fruit.

Averrhoa bilimbi L., - Bilimbi.

Bombax malabaricum DC.

Brassica juncea Czern - (Possibly, secondary centre of origin).

B. nigra Czern - Black mustar.

Cajanus indicus Spreng - pigeonpea.

Canavalia gladiata D.C. - Swordbean.

Cannabis indica L., - Hemp.

Carissa carandas L.

Carthamus tinctorius L., - Safflower.

Cassia angustifolia Vahl - Senna.

Cicer arietinum L., - Chickpea.

Cinnamomum zeylanlcum Breyn - Cinnanion.

Citrus aurantifolia L. - Sour Lime.

C. aurantium L., - Sour Orange.

C. limonia Osb. - Lemon.

C. medica L., - Citron.

C. nobilis Lour. - Sweet Orange.

Citrus sinensis Osb. - Orange.

Cocos nucifera L., - Coconut Palm.

Colocasia antiquorum Schott.

Corchorus capsularis L. - Jute.

C. olitorius L. - Jute.

Cucumis sativus L., - Cucumber.

Cuminum cymimm L., - Cumin.

Crotalaria juncea L., - Sunnhemp.

Cyamopsis psoralioides D.C. - Guar.

Cymbopogon nardus Rendle - Citronella Grass.

Dioscorea alata L., - Yam.

Dolichos biflorus L. - Horsegram.

D. lablab L., - Hyacinthbean.

Elettaria cardamomum Maton and White - Cardamom.

Eleusine coracana Gaerte - African Millet.

Eugenia jambolana Lam - Jambo.

Feronia elephantum Correa - Woood apple.

Gossypium arboreum L., - Cotton.

G. obtusifolium Roxb.

Hibiscus cannabinus L., - Bimilipatam Jute.

H. sabdariffa Roselle.

Lactuca indica L. - Indiago.

Lagenaria vulgaris Ser - Bottle Gourd.

Lawsonia albo Lam - Henna.

Luffa acutangula Roxb - Ribbed Gourd.

Mangifera indica L., - Mango.

Mimusops elengi L.

Momordica charantia L., - Bitter Gourd.

Murraya Keonigii Ser - Curry Leaf.

Paspalum scrobiculatum L., - Kodo Millet.

Oryza sativa L., - Rice.

Phaseolus aconitifollus Jack - Mothbean.

P. aureus (Roxb) - Mungbean.

P. mungo L., - Blackgram.

Phyllanthus emblica L., - Myrobalan.

Phoenix sylvestris Roxb - Wild Date.

Piper betle L., - Betlenut, *p. longum.* L., -Long peppper.

Piper nigrum L., - Black peper.

Rubia tinctorum Madder.

Saccharum officinarum L., - Sugarcane.

Santalum album L., - Sandal-wood.

Sesamum indicum L., - Seasame.

Sesbania aculeata L., - Daincha.

Tamarindus indica L., - Tamarind.

Terminalia catappa L., - Indian Almond.

Trichosanthes anguina L., - Snake Gourd.

Trigonella foenum-graecum L., - Sanke Gourd.

Vigna sinensis Endle-Cowpea.

II-A. GROUP

Aleurites moluccana Wild - Candlenut.

Artocarpus communis Frost - Breadfruit.

Areca catechu L., - Arecanut.

Citrus gradis Osb - Pumelo.

Citrus microcarpa Bge.

Coleus tuberosus Benth.

Cocos mucifera L., - Cocoanut.

Curcuma longa L., - Turmeric.

Dendrocalamus asper Backer - Giant Bamboo.

Dioscorea alta L.

Elettaria cardamommum Maton and White - Cardamom.

Garcinia mangostana L., - Mangosteen.

Musa paradisaca L.

M. sapientum L., - Banana.

Musa textiles Nee - Manila hemp.

Myristica fragrans Houtt - Nutmeg.

Piper nigrum L., - Black Pepper.

Saccharum officinarum L., - Sugarcane.

Vetiveria zizanioides stapf - Vetiver.

Zingiber officinale Rose - Ginger.

III. GROUP

Allium cepa L., - Onion.

Allium sativum L., - Garlie.

Amygdalus communis L., - Almond.

Brassica campestris. subsp. *oleifera* Metzg - Rape. (Secondary Centre).

Brassica campestris L., - subvar. *rapifera.* - Turnip.

B. juncea Czern - Mustard.

Cannabis indica L., - Hemp.

Carthamus tinctorius L., - Safflower.

Carum copticum Benth and Hoom.

Cicer arietinum L., - Chickpea.

Coriandrum sativum L., - Coriander (one of the centres of origin).

Cucumis melo L., - (Secondary centre).

Daucus carota L., - Carrot.

Gossypium herbaceum L., - Hemp.

Lagenaria vulgaris Ser - Bottle Gourd. (Secondary Centre).

Lathyrus sativus L.,

Lens Lens esculenta Moench - Lentil.

Linum usitatissimum L., - Flax (One of the centres of origin).

Phaseolus aureus Roxb - Mungbean.

P. mungo L., - Blackgram.

Pistacia vera L., - Pistachio.

Raphanus sativus L., - Radish (One of the centres of origin).

Prunus armeniaca L., - Apricot.

Pyrus communis L., - Pear.

Raphanus sativus L., - Radish (one of the centres of origin).

Secale creale L., - Rye (Secondary centre).

Sesamum indicum L., - Sesame (One of the centres of origin).

Triticum compactum Host - Club Wheat.

T. sphaerococcum Pere - Short Wheat.

Triticum vulgare Vill - Common wheat.

Vicia faba L., Broad bean.

IV. GROUP

Avena byzantina C. Koch. - Mediterranean oats.

A. sativa L., - Common oats. (Endemic varieties as weeds).

Beta vulgaris L., - Garden Beet.

Brassica campestris L., - subsp. oleifera. - (Secondary Centre).

Cicer arietinum subsp. *pisiforme* G. pop - (Secondary Centre).

Coriandrum sativum L., - Coriander. (One of the Centres).

Crocus sativus L., - Saffron.

Cucurbita pepp L., - Pumpkin.

(Greatest diversity in Asia Minor).

Ficus carica L., - Fig.

Juglans regia L., - Walnut.

Linum usitatssimum L., - Flax.

Pimpinella anisum L., - Falx.

Pisum sativum L., - Pea (Secondary Centre).

Punica   granatum   L., Pomegranate.

Pyrus comumunis L., - Pear.

Trifolium resupinatum L,. - Persian Clover.

T. durum subsp. *Aav-Durm* Wheat (2*n* = 28).

Triticum monococcum L., - Einkorn Wheat (2*n* = 14).

T. persicum Vall - Persian Wheat (2*n* = 28).

T. timophevii Zhuk - (2*n* = 28).

T. turgidum L., - Poulard Wheat (2*n* = 28).

T. vulgare Vill., - (2n=42).

Vicia sativa L., - Crop Vetch.

V. GROUP

Allium cepa L., - Large Onion (Secondary Centre).

A. porruc L., - Leek.

A. sativum L., - Garlic (Secondary Centre).

Beta vulgaris L., - Garden Beet.

B. campestris L., subv. *rapifera* Meizg - Turnip. (Basic Centre).

Brassica oleracea L., - Cabbage.

Cicer arietinum L., - chickpea.

Cichorium intybus L., - Chicory.

Cumimum cyminum L., - Cumin.

Hordeum sativum Jess - Coarse B-Arley. (Secondary Centre).

Linum usitatissimum L. - subsp. mediterranium. Vav.-Flax.

Lupinus albus L., - Lupines.

Mentha piperita L., - Peppermint.

Pisum sativum L., - Pea.

Sinapis alba L., - White Mustard.

Trifolium alexandrinum L., - Egyptian Clover.

Triticum dicoccum Schrank - (One of the Centres).

Triticum durum Desf - Durum Wheat.

T. polonicum L., - Polish Wheat. (One of the Centres).

T. spelta L., - Spelt Wheat.

VI. GROUP

Andropogon sorghum Link.

Carthamus   tinctorius   L., - Sanflower.

Cicer arietinum L., - Chichkpea.

Coffea arabica L., - Coffea.

Dolichos lablab L., - Lablab bean.

Eleusine coracana Gaertn - Finger millet.

Hordeum sativum - Barley.

Hibiscus esculentus L., - Okra.

Lens esculentus Moench - Lentil.

Limum usitatissimum L., - Flax.

Musa ensete J.F. Gmel - Abyssinian Banana.

Pennisetum   americanum - Pearl Millet.

Sesamum indicum L., - Sesame. (Basic Centre)

Triticum dicoccum subsp. *Abyssinicum* Stol - Emmer Wheat.

Triticum durum subsp. *Abyssinicum* var - Abyssinian Hard What.

Hordeum sativum - Barley.

VII. GROUP

Achras sapota Miller - Sapota.

Amaranthus paniculatus L.

Anona squamosa L., *A. reticulata* L., *a. muricata.* L. - Custard apple; Bullock's heart.

Anacardium   occidentale   L., - Cashew.

Bixa orellana L., - Annatto.

Canavalia ensiformis D.C. - Jackbean.

Capsicum annum L., - Chilli.

Carica papaya L., - Papaya.

Cucurbita   moschata   Duch - Pumpkin.

Gossypium hirsutum L., - Upland Cotton.

Ipomoea batatas Poiret - Sweet Potato.

Maranta arundinacea L., - Arrowroot.
Nicotiana rustica L., - Tobacco.
Opuntia sp. - Prickly pear.
Phaseolus lunatus L., - Limabean.
P. vulgaris L., - Bean.
Sechium edule Swartz - Chowchow.
Theobroma cacao L., - Cacao.
Zea mays L., - Corn.

VIII. GROUP

Capsicum frutescents L.,- Greenpepper.
Cinchona succirubra Pav., - Quinine tree.
Cucurbita maxima Duch - Pumpkin.
Cyphomandra betacea Sendtu - Tree Tomato.
Gossypium barbadense L., -Egyptian Cotton.
Lycopersion esculentum Mill - Tomato.
L. peruvianum Mill - Tomato.
Nicotiana tabacum L., -Tobacco.
Phaseolus lunatus L., - Limabean. (Secondary Centre)
Solanum andigenum - Juz et Buk. Other endemic cultivated potatoes.
Solanum cuencanum Juz and Buck - (2n=24).
S. Kesselbrenneri Juz and Buck - (2n=24).
S. ajanhurri Juz and Buk - (2n=24).

S. prucifiorum Juz and Buk - (2n=24).
S. stenotomum Juz and Buk - (2n=24).
S. goniocalyz Juz and Buk - (2n=24).
S. rybinii Juz and Buk - (2n=243).
S. buyacense Juz and Buk - (2n=24).
S. juzepexukii Juz and Buk - (2n=36).
S. tenuifilamentum Juz and Buk - (2n=36).
S. mamilliferum Juz and buk - (2n=36).
S. choclo Juz and Buk - (2n=36).
S. riobacbense Juz and Buk - (2n=36).
S. curtilobum Juz and Buk - (2n=36).
Zea mays L., - Starchy Mize. (Secondary Centre).

VIII-A. GROUP

Solamum tuberosum L., - Common potato.

VIII-B. GROUP

Anacardium occidentale L., - Cashew.
Arachis hypogea L., - Peanut.
Hevea brasiliensis Mull - Rubber tree.
Manihot utilissima Pohl - Tapioca.
Theobroma cacao L., - Secondary Centre.

MEDICINAL BOTANY

A

Abortifacient – An agent that produces abortion.

Alterative – A drug which corrects disordered process of nutrition and restores the normal fuction of an organ or of the system.

Amenorrhoea – Abnormal suppresion of menses.

Anaemia – A deficiency of blood or of red blood-cells.

Angina pectoris – A disease of the heart marked by severe constricting pains in the chest.

Anodyne – A drung that relieves pain.

Anthelmintic – A drug that kills intestinal worms.

Antithdrotic – A drug which checks sweating.

Antilithic – A drug which counteracts the development of stone.

Antiperiodic – A drung that cures periodic attacks.

Antipyretic – A drug which reduces fever.

Antiscorbutie – A drug which cures scurvy.

Antspasmodic – A drug which counteracts spasmodic disorders.

Aphrodisiac– A drug which promotes sexual desire.

Aromatic – A drug which is fragrant, spicy and mildly stimulant.

Asthma – A chronic disorder of the bronchial tubes.

Astringent – A drug which checks secretion or bleeding.

B

Beriberi– A deficiency disease caused by lack of vitamins especially B_1.

C

Bronchitis – An inflammation of the air passages.

Calculus – A hard and solid concretion formed in the body, especially in the urinary organs; it may be sand, gravel or stone.

Cancer – Any maliganant growth.

Caries – Decay of teeth.

Carminative – A drug which relieves flatulence.

Cathartic – A drug which induces active movement of the bowels.

Cholagogne – A drug which promotes flow of bile.

Coli – Pain due to spasmodic contraction of the abdomen.

Congestion – An abnormal collection of blood in the blood vessels of any organ oir part of the body.

Conjunctivitis – Inflammation of the conjuctiva, the mucous membrane covering the cyeball and lining the eyelids.

D

Dandruff – An inflamed condition of the scalp characterized by the presence of white scales in the hair due to the exfoliation of the horny cells of the scalp.

Demulcent – An agent having a soothing effect on the skin and mucous membranes.

Decobstruent – A drug that removes an obstruction to secretion or excretion by opening the natural passages or pores of the body.

Diabetes – A wasting diesease of metabolism; abundant sugar is present continuously in the urine.

Dropsy – A disease marked by an excessive collection of a watery fluid in the tissues or the cavities of the body.

Dysentery – Usually painful and difficult menstruation.

Dyspepsia – Indigestion.

E

Eczema – A skin disease acompanied by swelling, redness and exudation of lymph.

Elephantasis – A disease of the skin caused by a tiny worm and attended with hypertrophy of the affected parts.

Emetic – A drug which induces vomiting.

Emmenagogue – A drug which promotes menstruation or regulates the menstrual periods.

Emollient – A drug which allays irritation of the skin and alleviates swelling and pain.

Enteritis – Inflammation of the intestines.

Epilepsy – A chronic nervous disorder marked by attacks of unconsciousness or convulsions.

Expectorant – A drug that promotes the removal of catarrhal matter and phlegm from the bronchial tubes.

F

Febrifuge – An agent used for reducing fever.

Flatulence – A disorder in which there is an excessive collection of the gas in the stomach.

G

Galactagogue – An agent that promotes secretion and flow of milk.

Gleet – A chronic discharge from the urethra.

Goitre – A chronic enlargement of the thyroid gland.

Gonorrhoea – An infectious venereal disease marked by an inflammatory discharge from the genital organs.

Haemoptysis – Spitting of blood from the lungs or bronchial tubes.

Haemorrhage – Bleeding, especially profuse, from any part of the body.

Heartburn – A burning feeling in the regions of the chest and stomach, generally due to indigestion.

Hepatitis – Inflammation of the liver.

Hysteria – A disease in which a physically healthy patient has lost control over acts and feeling and suffers from imaginary ailments.

I

Intermittent fever – Fever which is marked by intervals of normal temperature between periods of rise of temperature.

J

Jaundice – A diseased condition in which there is a yellowish staining of the tissues and excretions with bile.

Lactagogue – Galactagogue.

Laryngitis – Inflammation of the larynx.

Leprosy – A chronic wasting disease caused by germ; the disease generally results in mutilations and deformities.

Leucoderma – A condition of the skin in which there is loss of pigment wholly or partially.

Lithontriptic – A drug used for removing calculi of stones formed in the urinary system.

M

Malaria – A recurrent disease marked by bouts of shivering, sudden rise of temperature and general aching of the body.

N

Narcotic – A drug which induces deep sleep.

Nausea – A feeling that vomiting is about to take place.

Nephritis – Inflammation of the Kidney.

Neuralgia – Pain felt above a nerve.

O

Ophthalmia – Conjuctivitis.

Orchitis – Inflammation of the testicles.

P

Paralysis – A disease in which there is loss of power of voluntary movement in any part of the body.

Pectoral – A drug to cure disorders of the chest.

Pharyngitis – Inflammation of the pharynx.

Phythisis – consumption; tuberculosis of the lungs.

Prophylactic – An agent that prevents disease.

R

Refrigerant – A drug which relieves feverishness or produces a feeling of coolness.

Rheumatism – An indefinite term used for pains in the muscles, joints and certain tissues.

Rubefacient – A mild counter-irritant.

S

Scabies – An itching skin disease caused by a mite.

Seorbutic – Suffering from scurvy.

Seurvy – A deficiency disease due to lack of vitamin C.

Sedative – A drug which reduces excitement, irritation and pain.

Sialagogue – A drug which promotes salivation.

Soporific – A drug that induces sleep.

Stomachic – A drug that strengthens the stomach and promotes its action.

Styptic – An agent which checks bleeding.

Syphilis – A chronic venereal disease.

T

Tetanus – An infectious disease, marked by painful contraction in the muscles.

Tonsillitis – Inflammation of the tonsils.

U

Ulcer – An open sore on the skin.

V

Vermifuge – A drug which expels intestinal worms.

W

Whooping Cough – An acute infectious disease of coughing.

SOME OTHER IMPORTANT MEDICAL TERMS

Acne – A pimple-like eruption of the sebaceous glands of the skin, with accumulation of yellow secretion and black overgrowth of the horny layer of the skin;

After-Pains. Painful contraction of the womb after child-birth;

Alopecia – A disease of the scalp resulting in complete or partain baldness;

Anasarca – Dropsy;

Antacid – A drug which neutraqlizes the acidity of the gastric juice;

Antiphlogistic – A drug which counteracts inflammation;

Aperient – A mild purgative;

Aphthae – Minute white ulcers on the tongue and in the mouth;

Apoplexy – Sudden los of consciousness with some paralysis;

Ardor – round worms;

Arine – A burining sensation on urinating;

Ascaris – Intestinal parasitic round worms;

Atony – Lack jof muscular power;

Bechic – A remedy for cough;

Bedsores – Ulceration on any part of the body exposed to pressure of bed-ridden patient;

Bronchorrhoea – Excessive discharge from the bronchial mucous membrane;

Colic – Pain due to spasmodic contraction of the abdomen;

Contusion – An injury to the soft parts without breaking the skin;

Cystitis – Inflammation of the bladder;

Depilatory – An agent that removes or destroys hair

Diphtheria – An infectious disease of the throat and the air passage;

Discutient – A drugh which disperses or absorbs a tumour or any coagulated fluid in the body;

Dysuyria – Painful and difficult urination;

Fistula – An abnormal channel which connects one cavity of the body, with another, or which opens out from a cavity to the surface of the body,

Freckles – Coloured spots on the exposed parts of the skin:

Glycosuria – A diseased condition of the urine in which sugar is excreted;

Hemiplegia – Paralysis of one side of the body;

Hepatic – Pertaining to the liver.

Hernia – Rupture; Protrusion through its covering of any organ of the body;

Hypnotic – A drug which induces sleep;

Hepochondriasis – A mental disorder in which the patient is tormented by meloncholy views, particularly about its health;

Itch – An infectious skin disease, caused by a mite, without specific lesions and marked by excessive itching;

Lithontriptic – A drug for removing stonges stones fromed in the urinary systems;

Menorrhagia – Abnormally excessive menstruation:

Metrorrhagia – Bleeding from the womb:

Migraine – Periodic attack of headeache affecting one side of the head;

Night-blindness – A disease in which the patient is incapable of seeing in the dark;

Otitis – Inflammation of the ear;

Psoriasis – A common chronic inflammatiou of the skin, marked by rounded reddened patches which are covered with dry silvery scales;

Pyrrhoea – A disease marked by purulent discharge from the gums;

Ringworm – A parasitic skin disease usually marked by red, scaly, circular patches;

Sciatica – An inflammation of the sciatic nerve at the back of the thigh;

Stomatitis – Inflammation of the mouth;

Typhoid fever – An acute infectionus disease characterized by ulceration of the intestines, eruption of rose-coloured spots, and a typical course of temperature;

Urethritis – Inflammation of the urethra;

Vulnerary – A drug which promotes healing of wounds.

MICROBIOLOGY

A

Abscess – A circumscribed pus-filled lesion charcteristic of staphylococcal skin disease.

Acyclovir (Zovirax) – A drug used as a topical ointment for herpes simplex and injected for herpes encephalitis.

Adjuvant – A substance such as aluminium sulphate that increases the efficiency of a vaccine.

Agar – A drivative of marine seaweed used as a solidifying agent in many microbiological media.

Agglutination – A type of antigen-antibody reaction that results in visible clumps of organisms or other material.

Agglutinins – Antibodies that participate in agglutination reactions.

Agranulocytes – The destruction of neutrophils (granulocytes) resulting from the reaction of antibodies with antigens on the neutrophil surface, a form of type II hypersensitivity.

Alginate – A carbohydrate thickening agent used in icecream. soups and other foods, industrially produced by microograisms.

Allergen – An antigenic substance that stimulates an alleergic reaction in the body.

Alloantigens – Antigens that exist in certain but not all members of a given species, examples are the A.B and Rh factors in humans.

Allograft – A tissue graft between two members of the same species, such as between two humans.

Anaphylaxis – A life-threatening allergic reaction in which series of mediators cause contractions of smooth muscle throughout the body.

Antibody – A highly specific protein molecule produced by plasma cells in the immune system. antibodies function in humoral immunity.

Antigen – Any chemical substance that elicits a response by the body's immune system.

Antiglobulin antibody – Any antibody that reacts with human antibodies.

Antitoxins – Antibodies that circulate in the blood stream and provide protection against toxins by neutralizing them.

Autoantigens – A person's own proteins and other organic compounds that clicit a specific response in the body.

Autograft – Tissue taken from one part of the body and grafted to another.

B

Bacteriocins – A group of bacterial proteins toxic to other bacteria.

Blanching – A process in which food is subjected to stream for five minutes in order to destroy cellular enzymes and enhance presservation.

C

Cancer – A condition characterized by the readiating spread of cells that reproduce at an uncontrolled rate.

Carcinogens – Cancer-casuing substances.

Cellular immunity – Immunity arising from the activity of T-lymphocytes on or nber the body cells, also called tissue immunity and cell-mediated immunity.

Chancre – A circular purplish hard ulcer with a raised margin that occurs during primary syphilis.

Chimera – A plasmid enginecred to contain a fragment of foreign DNA.

Clone – A collection, or colony, of identical cells arising from a single cell.

Colostrum – The first milk secreted from the mammary gland of animals or humans.

Complement – A group of proteins that functions in a cascading series of reactions during the response by the body to certain antigens, the complement cascade is stimulated by antigen-antibody activity.

D

Disease – Any change from the general state of good health.

Disinfectant – A chemical used to kill pathogenic microganisns or a lifeless object wuch as a table top.

Droplets– Airbome particles of mucus and sputum from the respiratiory tract that contain disease organisns.

E

Eaton agent – An alternative name for *Mycoplasma pneumoniae*.

Edema – A swelling of the tissues brough about by an accumulation of fluid.

Endotoxin – A metabolic poison produced chiefly by Gram-negatlve becteria, endotoxins are part of the bacterial cell wall and, consequently, are released on cell disintegration, they are composed of lipid-polysaccharide-peptid complexes.

Enierotoxin – A toxin that is active in the gastrointestinal tract of the host.

Enterovirus – A virus that infects intestinal cells.

Episome – A plasmid attached to the chromosome of a bacterium.

Erythema – A zone of redness in the skin due to accumulation of blood.

Exotoxin – A metabobic posion produced chiefly by Gram-positive bacteria, exotoxins are released to the environment on production, they are composed of protein and affect various organs and systems of the body.

F

Fab fragment – The portion of the antibody molecule that combines with the determinat sites of the antigen.

Fe fragment – The portion of the antibody molecule that combines with phagocytes, viral receptor sites and complement.

Fermentation – Anaerobic respiration in which intermediaries in the process are used as electron acceptors, also refers to the industrial use by bacteria for attachment, sometimes used as a an alternative expression for pili.

Formites – Inanimate objects such as clothing or utensils that cárry disease organisms.

Fungemia – Dissemination of fungi through the circulatory system.

G

Genome – The nucleic acid core of the virus.

Gonococcus – A colloquial expression for *expression of Neisseria gonorrhoeae*.

H

Hansen's disease – An altermative name for leprosy.

Hemagglutination – The agglutination of red blood cells.

Hemagglutinin – An enzymne on the surface spikes of certain influenza viruses that allows the virus to bind to red blood stream.

Hybridoma – A mass of cells produced by the fusion of myeloma cells with antigen-stimulated plasma cells. produces monoclonal antibodies.

I

Icosahedron – A symmetrical figure composed of 20 triangular faces and 12 points, one of the major shapes taken by the virus.

Imidazoles – A groups of antifungal drugs that interfere with sterol synthesis in fungal cell membranes, includes miconazoles and ketoconazoles.

Immunoglobulin – An alternative term for antibody.

Inducer – A substance that may activate the operon of the cell by combining with and negating the repressor protein.

Inflammation – A non-specific defensive respones to injury usually characterized by red color from blood accumulation warmth from the heat of blood. Swelling from fluid accumulation and pain from injury to local nerves.

Interferon – An antiviral protein produced by body cells on exposure to viruses, interferon triggers production of a second protein that binds to mRNA coded by the virus and thereby inhibits viral replication.

Interleukins – Lymphokines produced by white blood cells that act on other white blood cell important in cellular immunity.

Iodophores – Complexes of iodine and detergents that release iodine.

Isograft – Tissue taken from an identical twin and grafted to the other wtin.

J

Job's syndrome – An immune disorder characteriazed by defective chemotaxis between phagocyte and microorganism. That attacks and destroys cells altered by the presence of antigens, important in the destruction of cancer cells, also called a killer cell.

K

Kelbs-Loffer bacillus – A common term for *Corynebacterium diphtheriae.*

Koch-Weeks bacillus – An alterative name for *Haempohilvs aegypticus.*

Leukemia – A cancer the white blood cells.

Leukocidin–An enzyme that destroys phagocytes thereby preventing phagocytosis of the parasite.

Lymphoblast – The young cell to which the T-lymphocyte reverts. lymphoblasts secrete lymphokines.

Lymophocyte – A type of leukocyte that functions in the immune system.

Lymphokines – Proteins that increase the efficiency of phagocytosis at the antigen sites in cellular immunity.

Lymphopoietic cells– Primitive cells that arise from stem cells and are modified to form B-lymphocytes or T-lymphocytes.

Lysogeny – The phenomenon in which a virus remains in the cell cytoplasm as fragment of DNA or attaches to the chromosome, but fails to replicate in or destory the cell.

M

Macrophages – Large cells derived from monocytes and found within the tisues, macrophages actively phagcytize foreign bodies and comprise the reticuloendothelial system (RES).

Magnetosome – A cytoplasmic body in certain bacteria that assists orientation to the environment by aligin with the magnetic field.

Malt – Digested barley grain used in beer fermentations.

Mast cells – Connective tissue cells to which IgE fixes in type I hypersensitvity reactions, the cells degranulate and release histamine during allergic attacks.

Memory cells – Cells derived from B-lymphocytes or T-lymphocytes that react rapidly upon the future recurrence of antigens in the tissues.

Meningitis – A general term for infection of the meninges due to any of several bacteria, fungi, viruses, or protoza.

Mesophiles – Organisms that grow at the temperature range of 20-40°C.

Microorganism – A microscopic form of life including bacteria, viruses, fungi, protozoa and some multicellular parasites.

Mole – The quantity of a substance whose weight in grams in numerically equivalent to the molecular weight of that substance.

Monocyte – A lekocyte with a large bean-shaped nucleus, functions in phagocytosis.

Mycoplasmas – A microscopic form of life including bacteria, viruses, fugi protozoa and some multicellular parasites.

Myeloma – A mass of cancerous cells.

N

Nanometer – A unit of measurement equivalent to one billionth of a meter, the unit is designated as nm and is often used in measuring viruses and the wavelenth or energy.

Neoplasm – An uncontrolled growth of cells, often called a tumour.

Neurotoxin – A toxin that is active in the nervous system of the host.

Neutralization – A type of antigen-antibody reaction in which the activity taking place between reactants is not visible.

Night soil – Human feces sometimes used as an agricultural fertilizer.

O

Okazaki fragments – Segments of DNA that combine with one another to form a DNA molecule during chromosomal duplication.

Oncogene – A region of DNA in human cells thought to induce uncontrolled growth of the cell if per-

mitted to function.

Oncology – The study of tumors and cancers.

Operon – The unit of gene activity that expresses a particular trait, also the unit that controls proteir synthesis.

Opsonins – Antibodies or complement components that encourage phagocytosis.

Opsonization – Enhanced phagocytosis due to the activity of antibodies or complement.

P

Papilloma – A tumour of the skin tissue.

Pasteurization – A heating process that destroys pathogenic bacteria in a fluid such as milk and lowers the overall number of bacteria in the fluid.

pH – An abbreviation for the negative logarithm of amount of hydrogen ion concentration in 1 litre of solution, the pH scale extends from 1 to 14 and indicates the degree of acidity or alkalinity of a solution.

Phagocyte – A cell that practices phagocytosis.

Phagocytosis – A process in which solid particles are taken into the cell, important in nutritional processes and in defenc against disease.

Phagosome – A vesicle that contains particles of phagocytized material.

Picornavirus – A small virus containing RNA in its genome.

Pili – Short, hairlike appendages of bacteria that anchor the cell to a suface, pili are also involved in conjugations between bacteria.

Pinocytosis – A type of phagocytosis in which materials dissolved in fluid are taken into the cell.

Plaque – A clear area on a lawn of bacteria where viruses have destroyed that bacteria, also the gummy layer of gelatinous material consisting of bacteria and organic matter on the teeth.

Plasmid – A small, closed loop m,olecule of DNA apart from the chromosome, plasmids carry genes for drug resistance and pilus formation and are used in genetic engineering experiments.

Potable water – Water fit to drink.

Precipitins – Antibodies that participate in precipitation reactions.

Prions – Infectious particles of protein, possibly involved in human diseases of brain.

Properdin – A protein that functions in the alternative pathway of complemtn activation.

Prophase – The DNA ssegment of a temperate phase.

R

R factors – Plasmids that occur frequently in Gram-negative bacteria and carry genes for drug resistance.

Reagin – An alternatigve name of the IgE that stimulates anaphy-laxis in the body.

Repressor protein – A protein that inhibits the activity of certain genes, lysogeny is established when repressor protein is produced under direction of a virus.

Resolving power – the numerical value of a lens system that indicates the size of the smallest object that can be seen clearly when using that system.

Retrovirus– An RNA virus that uses reverse transcriptase to synthesis DNA from RNA.

Reverse transcriptase – An enzyme that sythexizes a DNA molecule from the code supplied by an RNA molecule.

Ropy bread – Bread that has become soft and stringy due to capsular material deposited by a bacterium such as *Bacillus subtilis*.

Rum – Distilled spirits produced by fementation of molasses.

S

Sabin vaccine – A type of polio vaccine prepared with attenuated viruses, the vaccine is taken orally.

Sake – A type of rice beer produced primarily in the Orient.

Sarcoma – A tumour of the connective tissues.

Septicemia – A generalized bacterial infection of the blood stream due to any of several organisms including streptococci and staphylococci, once known as blood poisoning.

Serology – The branch of immunology that studies serological reactions.

Serotype – A rank of classification below the species level based on a organism's reaction with antibodies in serum, used for several bacteria, especially *Salmonella*.

Sexduction – A process of recombination in which chromosomal genes pass from a donor cell to a recipient cell while attached to F factor.

Silage – A type of animal feed produced by fermenging grains and other plants in silos, the huge cylindrical structures that often stand next to barns.

Starter culture – A quantity of cbacteria added to milk in the industrial production of dairy products.

Stem cell – A primordial cell of the bone marrow from which hemapoietic and lymphopoietic cells develop.

Sterilization -the removal of all life forms, especially bacterial spores.

Stoirmy fermentation – Femnnentation and curding of milk accompanied by gas accumulation that forces the curd apart.

Sppressor T-lymphocyte – A T-lymphocyte that interfers with the activity of B-lymphcyte.

Synthetic vaccine – A vaccine that contains chemically synthexized parts of microorganisms. such as proteins normally found in viral capsids.

T

Target cells – Cells to which IgG fixes during type II hypersensitivity.

Thermal death point – The temperature required to kill an organism in given length of time.

Thermophiles– Organisms that grow at a high temperature ranges of 40°C to 90°C.

Tissue typing – An immunological procedure used to locate compatible tissue types for transplantation.

Titer – The most dilute concentration of antibody that will yield a positive reaction with the system of cellular immunity also called a T-cell.

Toxin – A poisonous substance produced by a species of microorganism, bacterial toxins are classified. as exotoxins or endotoxins.

Toxiod – An immunizing agent produced from an exotoxin that elicites antitoxin production by the body.

Transduction – A type of bacterial recombination in which a virus transports fragments of DNA from a donor cell to a recipient cell.

Transposon – A segment of DNA that moves from one site on a DNA molecule to antoher site, transposons carry information for protein syntheis also known as jumping genes.

U

Ultrapasteurization – A pasteurization process in which milk is heated at 82°C for 3 seconds.

V

Vaccinia – The alternative name for cow-pox.

Vaginitis – A general term for disease of the vagina.

Varicella – An alternative name for chicken pox. menans "little vessel", a reference to small chickenpox lesions.

Varicella-Zoster immune globulin (VZIG) – A preparation of purified antibodies from blood donors that give some protection to chickenpox.

Vector – A living organism that transmits the agents of disease.

Viroion – A completely assembled virus outside its host cell.

Viroids – Tiny fragments of nucleic acid associated with certain plant diseases, possibly associated with animal disease.

V-Z virus – the name given to the virus that causes varicellia (chickenpox) and herpes zoster (shingles).

W

Wandering cells – Cells of the reticulo-endotheilial system that move about actively within the tissues.

Whiskey – distilled spirits produced by fermentation of malted cereal grains.

Wort – The fluid portion of mashed barely grain usd in beer production.

X

Xenograft – A tissue graft between members of different species such as between an animal and a human.

Y

Yeast – A type of fungus that is unicellular and resembles bacteria in culture.

Z

Zoonosis–an animal disease that may be transmitted to humans.

Appendix

Selected microbiologists who won the Nobel Prize in Physiology or Medicine

Nobel laureate	Country*	Year of award	Accomplishment
Emil A. von Behring	Germany	1901	Development of serum therapy for diphtheria
Sir Ronald Ross	England	1902	Studies on the cause and transmission of malaria
Robert Koch	Germany	1905	Cultivation of the tubercle bacillus
Paul Ehrlich	Germany	1908	Theories on the development of immunity
Elie Metchnikoff	France (USSR)		Description of phagocytosis
Charles R. Richet	France	1913	Investigations on anaphylaxis
Charles J.H. Nicolle	France	1928	Studies on the cause and transmission of epidemic type
Gerhard Domagk	Germany	1939	Discovery of chemotherapeutic effects of prontosil
Sir Alexander Fleming	England	1945	Discovery and development of penicillin
Ernst Boris Chain	England (Germany)		
Sir Howard W. Florey	England (Australia)	1951	Development of vaccine for yellow fever
Max Theiler	U.S.A. (South Africa)		
Selman A. Waksman	U.S.A.	1952	Discovery and development of streptomycin
Fritz A. Lipman	U.S.A (Germany)	1954	Cultivation of polio viruses in cell culture
John F. Enders	U.S.A.		
Thomas H. Weller	U.S.A		
Frederick C Robbins	U.S.A.		Investigations on antihistamines
Daniel Bovet	Italy (Switzerland)	1957	
Joshua Lederberg	U.S.A.	1958	Studies on the biochemistry of microbial genetics
George W. Beadle	U.S.A.	1959	Discoveries of the mechanism of synthesis of DNA and RNA
Severo Ochoa	U.S.A. (Spain)		
Arthur Komberg	U.S.A.		Studies on immunologic tolerance
Sir F. Macfarlane Burnet	Australia	1960	

Nobel laureate	Country*	Year of award	Accomplishment
Peter Brian Medawar	England (Brazil)	1962	Determination of the structure of DNA
James D. Watson	U.S.A.		
Francis H.C. Crick	England	1965	Studies on the regulation of gene activity in the cell
M.H.J. Wilkins	England		
Francois Jacob	France		
Jacques Monad	France	1968	Discovery of the genetic codes for amino acids
Robert Holley	U.S.A.		
Har Gobind Khorana	U.S.A. (India)		
Marshall W. Nirenberg	U.S.A.	1969	Studies on the mechanism of viral infection of cells
Max Delbruck	U.S.A. (Germany)		
Alfred D.Hershey	U.S.A.	1972	Elucidation of the nature and structure of antibody molecules
Salvadore E.Luria	U.S.A. (Italy)		
Gerald M. Edelman	U.S.A.		
Rodney R. Porter	U.K.	1975	Studies on the transformation of cells by tumor viruses
David Baltimore	U.S.A.		
Howard Temin	U.S.A.		Discovery of the Australia antigen
Renato Dulbecco	U.S.A. (Italy)		
Baruch Blumberg	U.S.A.		
D. Carleton Gajdusek	U.S.A.	1976	Description of slow virus diseases
Roslayn Yalow	U.S.A.	1977	Development of the radio-immunoassay procedure
Roger C.L. Guilemin	U.S.A. (France)		Synthesis of peoptide hormones
Andrew V.Schally	U.S.A. (Polland)	1978	Studies on restriction enzymes and their use in genetic engineering
Daniel Nathans	U.S.A.		
Hamilton Smith	U.S.A.	1980	Discovery of the histo-compatibility antigens used in tissue typing
Werner Arber	Switzerland		
Baruj Benacerraf	U.S.A.		
George, D.Snell	U.S.A.		
Jean Dausset	France		
Barbara Mc-Clintock	U.S.A.	1983	Studies on transposable genetic elements
Georges J.F.Kohler	Switzerland (Germany)		Research in immunology
Niels K.Jeme	Switzerland (Denmark)		
Susumu-Tonegawa	Japan	1987	Genetic principle for generation of antibody diversity.

*Country in which the work was done: country of birth in parentheses.